T0192395

Surfactants in Tribology

Surfactants in Tribology

Volume 6

Edited by
Girma Biresaw
K.L. Mittal

CRC Press
Taylor & Francis Group
Boca Raton London New York

CRC Press is an imprint of the
Taylor & Francis Group, an **informa** business

CRC Press
Taylor & Francis Group
6000 Broken Sound Parkway NW, Suite 300
Boca Raton, FL 33487-2742

First issued in paperback 2022

© 2020 by Taylor & Francis Group, LLC
CRC Press is an imprint of Taylor & Francis Group, an Informa business

No claim to original U.S. Government works

ISBN: 978-1-03-240133-1 (pbk)
ISBN: 978-1-138-58457-0 (hbk)
ISBN: 978-0-429-28770-1 (ebk)

DOI: 10.1201/9780429287701

Contents

SECTION I Ionic Liquids, Hydrogels, and Biosurfactants: Applications in Tribological Phenomenon

SECTION II Adhesion, Wetting, and Their Relevance to Tribology

SECTION III   Green, Nano-, and Biotribology

Preface

Surfactants perform a wide range of functions in tribology that includes control of basic lubrication functions (such as friction and wear) as well as numerous critical lubricant properties, such as emulsification/demulsification, bioresistance, oxidation resistance, rust/corrosion prevention, and so on. Surfactants also spontaneously form a wide range of organized assemblies in polar and non-polar solvents. These include monolayers, normal/reverse micelles, o/w and w/o microemulsions, hexagonal and lamellar lyotropic liquid crystals, and uni- and multi-lamellar vesicles. However, the tribological properties of these organized surfactant structures are not yet fully investigated or understood. Recently, there has been a great deal of interest in certain organized assemblies such as self-assembled monolayers (SAMs). These structures are expected to play critical roles in the lubrication of a wide range of products, including microelectromechanical systems (MEMS) and nanoelectromechanical systems (NEMS).

Whereas there is a great deal of literature on topics of both surfactants and tribology separately, there is very little information on the subject of surfactants and tribology together. This is despite the fact that surfactants play many critical roles in tribology and lubrication. In order to bridge this gap in the literature linking surfactants and tribology, we decided to organize a Surfactants in Tribology symposium as a part of the 16th International Symposium on Surfactants in Solution (SIS) in Seoul, Korea, June 4–9, 2006 (SIS-2006).

The SIS series of biennial symposia began in 1976 and have since been held in many corners of the globe and are attended by "who is who" in the surfactant community. These meetings are recognized by the international community as the premier forum for discussing the latest research findings on surfactants in solution. In keeping with the SIS tradition, leading researchers from around the world engaged in unraveling the importance and relevance of surfactants in tribological phenomena were invited to present their latest findings at the first symposium on this topic held in 2006. We also invited leading scientists working in this area, who may or may not have participated in the symposium, to submit written accounts (chapters) of their recent research findings in this field, which culminated in the publication of the first volume of *Surfactants in Tribology* in 2008.

Since the first symposium, interest in the relevance of surfactants in tribology has continued to grow among scientists and engineers working in the areas of both surfactants and tribology. So we decided to organize follow-up symposia on this topic at subsequent SIS meetings. Concomitantly, Surfactants in Tribology symposia were held during SIS-2008 (Berlin, Germany); SIS-2010 (Melbourne, Australia); SIS-2012 (Edmonton, Canada); SIS-2014 (Coimbra, Portugal); and SIS-2016 (Jinan, China). Each of these symposia has been followed by publication of the next volume in *Surfactants in Tribology*, Volumes 2 through 6, respectively.

Volume 6 (the current volume), based on the symposium held in Jinan, China, in 2016, comprises a total of 10 chapters dealing with various aspects of surfactants and tribology, some of which had not been covered at all in previous volumes in this series. These 10 chapters have been logically grouped into three theme areas as follows: Section 1 consists of four chapters dealing with the relevance of ionic liquids, hydrogels, and biosurfactants in tribological phenomena. Topics covered in Section 1 include tribological performance of cyano-based ionic liquids as functional fluids; reduction and control of friction in hydrogels; properties and applications of biosurfactants; and an overview of potential tribological applications of microbial lipids. Section 2 consists of three chapters dealing with adhesion and wetting and their relevance to tribology. Topics discussed in Section 2 include problems relevant to tribology resulting from moving contact lines of droplets on structured surfaces; a quantitative approach to measure work of adhesion; and adhesion of bacteria to solid surfaces. Section 3 comprises three chapters dealing with green, nano-, and biotribology. Topics discussed in Section 3 include tribological behavior of nanoadditives in lubricants; application of tribology test for quality assessment of fabric softeners based on cationic surfactants; green approach to tribology.

Surface science and tribology play many critical roles in various industries. Manufacture and use of consumer and industrial products rely on the application of advanced surface and tribological knowledge. Examples of major industrial sectors that rely on these two disciplines include mining, agriculture, manufacturing (metals, plastics, wood, automotive, computers, MEMS, NEMS, appliances, planes, rails, etc.), construction (homes, roads, bridges, etc.), transportation (cars, boats, rails, airplanes), and medical arena (instruments and diagnostic devices; transplants for knee, hips, and other body parts). The chapters in *Surfactants in Tribology*, Volume 6 discuss some of the underlying tribological and surface science issues relevant to many situations in diverse industries. We believe that the information compiled in this book will be a valuable resource to scientists and technologists working or entering in the fields of tribology and surface science.

This volume and its predecessors (Volumes 1 through 5) contain bountiful information and reflect the latest developments highlighting the relevance of surfactants in various tribological phenomena pertaining to many different situations. As we learn more about the connection between surfactants and tribology, new and improved ways to control lubrication, friction, and wear utilizing surfactants will emerge.

Now it is our pleasant task to thank all those who helped in materializing this book. First and foremost, we are very thankful to the contributors for their interest, enthusiasm, and cooperation, as well as for sharing their findings, without which this book would not have seen the light of day. Also we would like to extend our appreciation to Barbara Knott (Taylor & Francis, publisher) for her steadfast interest in and unwavering support for this book project, and to Danielle Zarfati and other members of the staff at Taylor & Francis for giving this book a body form.

Girma Biresaw
Bio-Oils Research Unit
NCAUR-MWA-ARS-USDA
Peoria, IL

K.L. Mittal
Hopewell Jct., NY

Editors

Girma Biresaw received a PhD in physical-organic chemistry from the University of California, Davis. As a postdoctoral research fellow at the University of California, Santa Barbara, he investigated reaction kinetics and products in surfactant-based organized assemblies for four years. As a scientist at the Aluminum Company of America (Alcoa), he conducted research in tribology, surface/colloid science, and adhesion for 12 years. Girma is currently a research chemist/lead scientist at the Agricultural Research Service (ARS) of the US Department of Agriculture in Peoria, Illinois, where he is conducting research in tribology, adhesion, and surface/colloid science in support of programs aimed at developing biobased products from farm-based raw materials. Girma has authored/coauthored more than 315 scientific publications, including more than 93 peer-reviewed manuscripts, 7 patents, and 8 edited books. Girma is a fellow of the STLE and an editorial board member for the *Journal of Biobased Materials and Bioenergy.*

K.L. Mittal received his PhD from the University of Southern California in 1970 and was associated with the IBM Corp. from 1972 to 1994. He is currently teaching and consulting worldwide in the areas of adhesion and surface cleaning. He is the editor of 130 published books, as well as others that are in the process of publication, within the realms of surface and colloid science and of adhesion. He has received many awards and honors and is listed in many biographical reference works. Dr. Mittal was a founding editor of the *Journal of Adhesion Science and Technology* and was its editor-in-chief until April 2012. He has served on the editorial boards of a number of scientific and technical journals. He was recognized for his contributions and accomplishments by the international adhesion community that organized the First International Congress on Adhesion Science and Technology in Amsterdam in 1995 on the occasion of his 50th birthday (235 papers from 38 countries were presented). In 2002, he was honored by the global surfactant community, which instituted the Kash Mittal Award in the surfactant field in his honor. In 2003, he was honored by the Maria Curie-Sklodowska University, Lublin, Poland, which awarded him the title of doctor *honoris causa*. In 2010, he was honored by both the adhesion and surfactant communities on the occasion of publication of his 100th edited book. In 2012, he initiated a new journal titled *Reviews of Adhesion and Adhesives.* In 2014, two books were published in his honor: *Recent Advances in Adhesion Science and Technology*, and *Surfactant Science and Technology: Retrospects and Prospects.*

Contributors

Richard D. Ashby
Eastern Regional Research Center
Agricultural Research Service
U.S. Department of Agriculture
Wyndmoor, Pennsylvania

Girma Biresaw
National Center for Agricultural
 Utilization Research
Agricultural Research Service
U.S. Department of Agriculture
Peoria, Illinois

Ratul Das
Dan F. Smith Department of Chemical
 Engineering
Lamar University
Beaumont, Texas

Semih Gulec
Dan F. Smith Department of Chemical
 Engineering
Lamar University
Beaumont, Texas

David R.K. Harding
Institute of Fundamental Sciences
Massey University
Palmerston North, New Zealand

Yuki Hirata
Department of Mechanical Engineering
Tokyo University of Science
Tokyo, Japan

Dina A. Ismail
Petrochemicals Department
Egyptian Petroleum Research Institute
 (EPRI)
Nasr City, Egypt

Maram T.H. Abou Kana
National Institute of LASER Enhanced
 Science
Cairo University
Giza, Egypt

Nadia G. Kandile
Department of Chemistry
Ain Shams University
Cairo, Egypt

Shouhei Kawada
Department of Mechanical Engineering
Tokyo University of Science
Tokyo, Japan

Jie Liu
Dan F. Smith Department of Chemical
 Engineering
Lamar University
Beaumont, Texas

Sahar A. Moustafa
Petrochemicals Department
Egyptian Petroleum Research Institute
 (EPRI)
Nasr City, Egypt

Nabel A. Negm
Petrochemicals Department
Egyptian Petroleum Research Institute
 (EPRI)
Nasr City, Egypt

Marta Ogorzalek
Department of Chemistry
Faculty of Materials Science and
 Design
University of Technology and
 Humanities in Radom
Radom, Poland

Shinya Sasaki
Department of Mechanical Engineering
Tokyo University of Science
Tokyo, Japan

Daniel K.Y. Solaiman
Eastern Regional Research Center
Agricultural Research Service
U.S. Department of Agriculture
Wyndmoor, Pennsylvania

Rafael Tadmor
Dan F. Smith Department of Chemical
    Engineering
Lamar University
Beaumont, Texas

Chiharu Tadokoro
Department of Mechanical Engineering
Saitama University
Saitama, Japan

Sirui Tang
Dan F. Smith Department of Chemical
    Engineering
Lamar University
Beaumont, Texas

Ryo Tsuboi
Department of Mechanical Engineering
Daido University
Nagoya-shi, Japan

Zhanlong Wang
State Key Laboratory of Nonlinear
    Mechanics
Institute of Mechanics
Chinese Academy of Sciences
and
School of Engineering Science
University of Chinese Academy of
    Sciences
Beijing, China

Tomasz Wasilewski
Department of Chemistry
Faculty of Materials Science and
    Design
University of Technology and
    Humanities in Radom
Radom, Poland

Seiya Watanabe
Department of Mechanical Engineering
Tokyo University of Science
Tokyo, Japan

Sakshi B. Yadav
Dan F. Smith Department of Chemical
    Engineering
Lamar University
Beaumont, Texas

Tetsuo Yamaguchi
Department of Mechanical Engineering
    and I^{2}CNER
Kyushu University
Fukuoka, Japan

Guangbin Yang
Engineering Research Center for
    Nanomaterials
Henan University
Kaifeng, China

Mona A. Youssif
Petrochemicals Department
Egyptian Petroleum Research Institute
    (EPRI)
Nasr City, Egypt

Pingyu Zhang
Engineering Research Center for
    Nanomaterials
Henan University
Kaifeng, China

Shengmao Zhang
Engineering Research Center for
    Nanomaterials
Henan University
Kaifeng, China

Zhijun Zhang
Engineering Research Center for
    Nanomaterials
Henan University
Kaifeng, China

Ya-Pu Zhao
State Key Laboratory of Nonlinear
    Mechanics
Institute of Mechanics
Chinese Academy of Sciences
and
School of Engineering Science
University of Chinese Academy of
    Sciences
Beijing, China

Section I

Ionic Liquids, Hydrogels, and Biosurfactants: Applications in Tribological Phenomenon

1 Tribological Performance of Cyano-Based Ionic Liquids as Functional Fluid

Shouhei Kawada, Seiya Watanabe, Yuki Hirata, Chiharu Tadokoro, Ryo Tsuboi, and Shinya Sasaki

CONTENTS

1.1 INTRODUCTION

Currently, environmental issues such as global warming are well recognized throughout the world because they are reported extensively by the mass media [1]. In its *Fifth Assessment Report: Climate Change 2013*, the Intergovernmental Panel on Climate Change (IPCC, Stockholm, Sweden) reported that the global average temperature rose by 0.85°C between 1880 and 2012 [1]. This report also warned of a possible temperature increase of 4.8°C by 2100. To help address this problem by minimizing inefficient production processes, it is essential to improve the energy efficiency of machine systems. One strategy to accomplish this goal is to reduce friction loss to its utmost limit in sliding applications. However, technological improvements to existing lubrication systems have struggled to meet this stringent requirement, which is why novel lubrication systems are urgently needed. To this end, ionic liquids are expected to function as novel functional fluids.

A characteristic feature of ionic liquids (ILs) is that they are molten salts at ordinary temperatures and pressures. Compared to conventional oils, ILs have many superior properties, such as high thermal stability, low vapor pressure, and low melting point [2–11]. In addition, combining different types of ions can be easily controlled, giving rise to a series of ILs with low friction coefficient under various applications. In the first investigation of ILs as lubricants, Kondo and colleagues applied them to magnetic film media in 1989 [12–14], and in 2001 Ye et al. reported the first instance of ILs as liquid lubricants against aluminum alloys for engines, plain bearings, compressors, and refrigerators [7]. Subsequently, the tribological properties of ILs have been reported regularly. Most of the investigations of IL properties concerned halogen-based ILs such as tetrafluoroborate [BF_4] and hexafluorophosphate [PF_6] [2–10,15,16]. Relative to existing lubricants, these halogen-based ILs exhibited good tribological properties such as low friction and low wear rate [7,15]. It was believed that the halogen atoms reacted with the metal sliding surfaces to form a boundary layer. However, it was found that the halogen species also caused corrosion of the metal materials [17]. In particular, [BF_4] and [PF_6] were formed to hydrolyze and generate hydrogen fluoride [18–20]. Thus, it is desirable to develop "halogen-free" ionic liquids.

Since 2010, investigations of sulfur- and phosphate-based ILs as additives have increased considerably [21–24]. These ILs are related to traditional engine oil additives such as zinc dialkyl dithiophosphate (ZDDP) and molybdenum dithiocarbamate (MoDTC) [25–27], and exhibit similar tribological performance as halogen-based ILs [22,23]. However, in order to reduce the environmental impact caused by lubricant usage, the use of halogen- as well as sulfur- and phosphate-based ILs should be minimized. To this end, the focus of this study concerns the tribological performance of cyano-based ILs, which are composed of light elements (e.g., hydrogen, boron, carbon, and nitrogen) and may therefore reduce environmental burden relative to other types of ILs and lubricants [17,28,29].

1.2 TRIBOLOGICAL PERFORMANCE OF CYANO- VERSUS HALOGEN-BASED IONIC LIQUIDS

Table 1.1 lists the molecular structures of the three types of ILs investigated in this study: 1-butyl-3-methylimidazolium iodide ([BMIM][I]), 1-ethyl-3-ethylimidazolium dicyanoamide ([EMIM][DCN]), and 1-butyl-3-methylimidazolium tricyanomethane ([BMIM][TCC]). Note that [BMIM][I] is a halogen-based IL, whereas the other two ILs are cyano-based. All of these ILs were purchased from Merck Chemicals, Germany. The principal properties of these ILs are listed in Table 1.2.

The tribological performance of each IL was evaluated using an SRV oscillating sliding tester (Optimol, Germany). The sliding conditions are listed in Table 1.3. The specifications of the test specimens were as follows: disk diameter × thickness = 25 × 7.9 mm; ball diameter = 10 mm; the disk and ball were made of bearing steel (AISI52100). The wear scar diameter of the ball specimen was measured by optical microscopy (VHX-100, Keyence, Japan).

The friction behavior of each IL at 50°C is shown in Figure 1.1. As shown in Figure 1.1, [BMIM][I] exhibited the lowest friction coefficient. In contrast, the cyano-based ILs exhibited higher friction coefficients, with [BMIM][TCC] exhibiting the highest value. From these results, it is clear that changing the anion of the IL affects its tribological performance. Even though the same cation was used in [BMIM][I] and [BMIM][TCC], very different friction coefficients were obtained.

TABLE 1.1
Structures of Ionic Liquids

1-butyl-3-methylimidazolium iodide [BMIM][I]

1-ethyl-3-methylimidazolium dicyanoamide [EMIM][DCN]

1-butyl-3-methylimidazolium tricyanomethane [BMIM][TCC]

TABLE 1.2
Physical Properties of Ionic Liquids Used

Ionic Liquids	Melting Point (°C)	Viscosity (mPa s)		Decomposition Temperature[a] (°C)
		40°C	100°C	
[BMIM][I]	−58	303	30.2	>200
[EMIM][DCN]	<−50	9.51	3.35	>200
[BMIM][TCC]	<−50	15.3	3.87	>260

[a] TGA results.

TABLE 1.3
Sliding Test Conditions

Frequency (Hz)	50
Amplitude (mm)	1
Lubricants (µL)	30
Temperature (°C)	50, 100, 150, 200
Load (N)	50
Sliding time (s)	3600

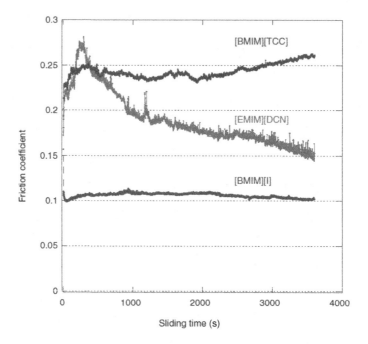

FIGURE 1.1   Friction behavior of each ionic liquid on steel/steel contacts at 50°C.

Figure 1.2 compares the friction behaviors of [BMIM][I] and [BMIM][TCC] at 50°C, 100°C, 150°C, and 200°C. The results reveal the effect of the anion on the tribological performance at high temperature since, as noted, these ILs have the same cation (i.e., [BMIM]). For [BMIM][I], the friction coefficient was stable at temperatures below 150°C and became unstable at 200°C. For [BMIM][TCC], the friction coefficient was higher than 0.2 at all temperatures.

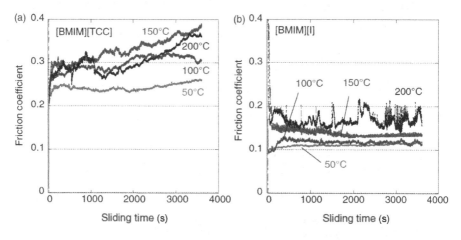

FIGURE 1.2   The effect of ionic liquid anion chemical structure on the friction behavior of (a) [BMIM][TCC] and (b) [BMIM][I] at 50°C, 100°C, 150°C, and 200°C.

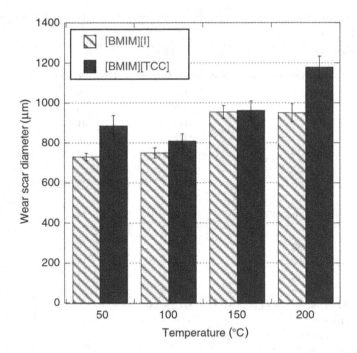

FIGURE 1.3 Wear scar diameters of ball specimens at different temperatures obtained from steel/steel contact.

Figure 1.3 shows the wear scar diameters of the ball specimen as a function of temperature. It shows that the wear scar diameter increased with increasing temperature. IL also showed that the wear scar diameters for [BMIM][I] were smaller than those for [BMIM][TCC] at all temperatures. On the basis of these results, the cyano-based ILs demonstrate poor tribological performance relative to the halogen-based IL. Clearly, the tribological performance of the cyano-based ILs must be improved before they can be put to practical use.

1.3 TRIBOLOGICAL PERFORMANCE OF CYANO-BASED IONIC LIQUIDS ON AISI 52100 WITH HARD COATINGS

To improve the tribological performance of the cyano-based ILs, several sliding specimens with different hard coating materials were used. In this regard, hard coatings are often used in various applications throughout the manufacturing industry. Thus, an improvement in the tribological performance of cyano-based ILs with these coated materials was expected.

The same cyano-based ILs presented in Section 1.2 (i.e., [EMIM][DCN] and [BMIM][TCC]) were used to investigate four types of nitride coatings, three types of diamond-like carbon (DLC) coatings, and three types of sintered ceramics as sliding materials. The physical properties of these materials are listed in Table 1.4.

TABLE 1.4

Physical Properties of Coating Materials Used

Coating[a]	Deposition Method	Roughness, R_a (μm)	Hardness (GPa)	Remark
CrN	AIP[b]	0.028	26	
TiN	AIP	0.019	30	
TiCN	AIP	0.025	41	
TiAlN	AIP	0.008	16	
H-free DLC	AIP	0.012	73	H = 0 at. % Interlayer: Cr
PVD-DLC	UBMS[c]	0.032	28	H = 20 at. % Interlayer: Cr
CVD-DLC	RF plasma CVD[d]	0.006	24	H = 30 at. % Inter layer: Si
Al_2O_3	Sintering	0.253	18	
Si_3N_4	Sintering	0.109	14	
SiC	Sintering	0.020	22	

[a] Thickness of all coatings was 1 μm on AISI 52100 steel disk.

[b] AIP: arc ion plating

[c] UBMS: unbalanced magnetron sputtering

[d] CVD: chemical vapor deposition

For these experiments, the sliding conditions were identical to those reported in Section 1.2. The ball and disk specimens were made of bearing steel (AISI 52100), and the disk specimens were coated with the aforementioned types of hard materials.

Figure 1.4 shows the results of friction behavior against the steel/ceramics contacts. The friction behaviors of both cyano-based ILs, except SiC lubricated with [EMIM][DCN], showed high stability and a low friction coefficient compared to the bearing steel/steel contacts (Section 1.2). In particular, the tribological performance of Al_2O_3 and Si_3N_4 coated disks lubricated with [EMIM][DCN] were similar to that bearing steel/steel contacts lubricated with [BMIM][I] (Section 1.2).

Figure 1.5 shows the average friction coefficient for the last 5 minutes of the tests and the wear scar diameters of the ball specimens. As can be seen, the wear scar diameter was much less than the values for steel/steel contacts shown in Figure 1.3.

Figure 1.6 shows the average friction coefficients and wear scar diameters on tests with bearing steel ball against nitride-coated disk. The performance of both cyano-based ILs, especially the wear property, became worse with the nitride coating than that with the ceramic materials. [EMIM][DCN] exhibited a higher friction coefficient than that with the uncoated bearing steel. Clearly, the hard materials consistently demonstrated good tribological performance with the ILs as lubricants.

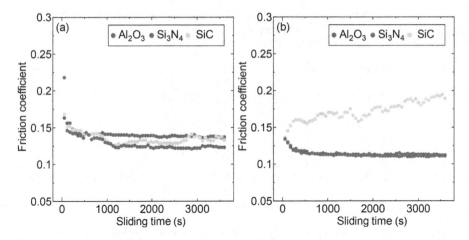

FIGURE 1.4 Friction coefficient for tests on AISI 52100 steel ball against ceramic disks (a) [BMIM][TCC], (b) [EMIM][DCN].

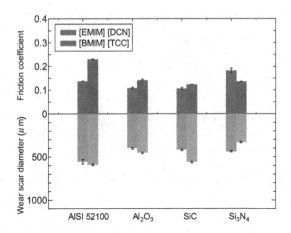

FIGURE 1.5 Average friction coefficient and wear scar diameter of ball specimens against AISI 52100/ceramics contacts.

Figure 1.7 shows the average friction coefficient and wear scar diameter for tests with the uncoated ball against the DLC coated disk. All DLC-coated specimens lubricated with [EMIM][DCN] showed low friction coefficients. Notably, the friction coefficient of the H-free DLC was less than 0.1. However, [BMIM][TCC] showed high friction coefficients against all DLC specimens.

From these results, it is clear that the tribological performance of cyano-based ILs can be improved by changing the surface chemistry of the sliding materials. H-free DLC showed the best tribological performance with [EMIM][DCN] ionic liquid.

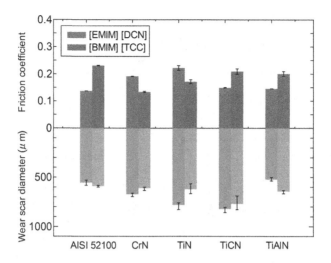

FIGURE 1.6 Average friction coefficient and wear scar diameter of ball specimens against AISI 52100/nitride coatings contacts.

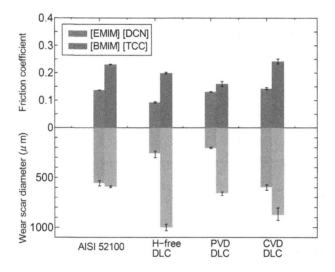

FIGURE 1.7 Average friction coefficient and wear scar diameter of ball specimens against AISI 52100/DLC contacts.

1.4 ULTRALOW FRICTION OF CYANO-BASED IONIC LIQUIDS

DLC lubricated with cyano-based ILs showed excellent friction and wear properties, suggesting that this system has great potential as a novel lubricant. Thus, the tribological performance of the cyano-based ILs on the steel/steel and DLC/DLC contacts was further investigated.

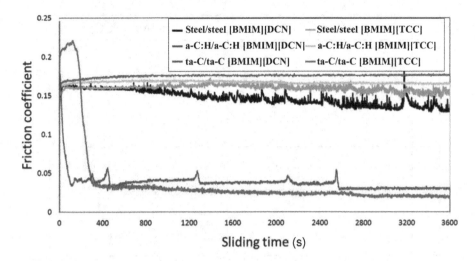

FIGURE 1.8 Friction behaviors of steel/steel and DLC/DLC contacts lubricated with cyano-based ionic liquids.

To investigate the effects of the IL anions on tribological performance, two types of cyano-based ILs with the same cation were used as the lubricants: [BMIM][DCN] and [BMIM][TCC]. Their chemical structures are shown in Table 1.1.

The tribological performance of each IL was evaluated using an SRV oscillating sliding tester (Optimol, Germany). The specifications of the test specimens were as follows: disk diameter × thickness = 24 × 7.9 mm; cylinder diameter × length = 6 × 8 mm; the disk and ball were made of bearing steel (AISI52100). Tetrahedral amorphous carbon (ta-C = H-free) and hydrogenated amorphous carbon (a-C:H) films were coated on the bearing steel.

Figure 1.8 shows the friction behavior of all tribopairs lubricated with the cyano-based ILs. The ta-C/ta-C contacts exhibited the lowest friction coefficient; [BMIM][DCN] and [BMIM][TCC] showed values of approximately 0.03 and 0.018, respectively. However, the steel/steel and a-C:H/a-C:H contacts showed high friction coefficients of more than 0.15.

Figure 1.9 shows the laser microscopy images of the wear tracks on the disk specimens. For the bearing steel surface, a film resulting from chemical reaction of the IL was observed on the edge of the wear tracks. This film was not observed on any of the DLC-coated surfaces.

Figure 1.10 shows the topography, lateral force mapping (LFM), and surface roughness R_a of disk specimens before and after the sliding tests. The steel surfaces lubricated with both ILs confirmed the tribochemical reaction. Both disk surfaces lubricated with the ILs became smoother than they were before the sliding tests.

Figure 1.11 shows the results of the force curve measurements. The pull-off force was hardly detected on the steel and DLC surfaces before the sliding tests. After the sliding tests, however, the pull-off force was indeed detected. Additionally, viscous products were present on the sliding surface. On the whole, the combination of a smooth surface and viscous products achieved ultralow friction.

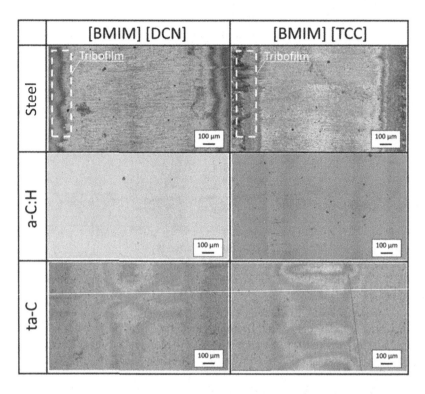

FIGURE 1.9 Laser microscopy images of uncoated and DLC-coated steel disk specimens.

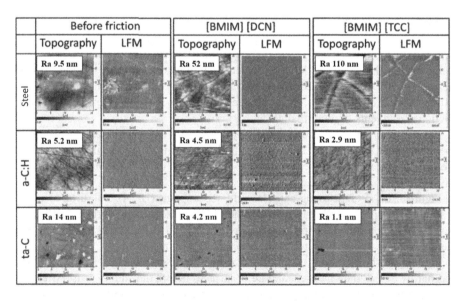

FIGURE 1.10 AFM topography, lateral force mapping (LFM), and surface roughness R_a of disk surfaces before and after steel/steel and DLC/DLC contacts tests.

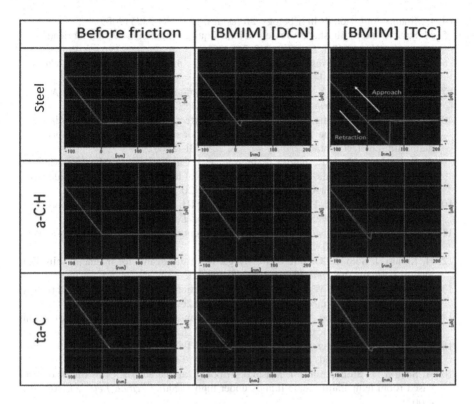

FIGURE 1.11   Force curves of steel disks with uncoated and DLC-coated surfaces before and after steel/steel and DLC/DLC contacts tests.

1.5   CONCLUSION

In conventional applications, cyano-based ILs have generally demonstrated poor tribological performance. However, the number of reports detailing improved performance of these ILs is increasing. In this study, cyano-based ILs demonstrated ultralow friction on ta-C/ta-C contacts coated rubbing pairs. For successful application of cyano-based ILs as lubricants, consideration of both the IL structure and the chemistry of the sliding surface is crucial. Overall, it can be concluded that cyano-based ILs represent a novel class of environmentally friendly lubricants with great potential.

REFERENCES

1. T. F. Stocker, D. Qin, G.-K. Plattner, M. M. B. Tignor, S. K. Allen, J. Boschung, A. Nauels, Y. Xiam V. Bex, and P. M. Midgley, *Climate Change 2013 The Physical Science Basis*, Cambridge University Press, Cambridge, UK, (2016).
2. M. D. Bermudez, A. E. Jiménez, J. Sanes, and F. J. Carrion, Ionic liquids as advanced lubricant fluids, *Molecules*, 14, 2888–2908 (2009).
3. F. Zhou, Y. Liang, and W. Liu, Ionic liquid lubricants: Designed chemistry for engineering applications, *Chem. Soc. Rev.*, 38, 2590–2599 (2009).

4. I. Minami, Ionic liquids in tribology, *Molecules*, 14, 2286–2305 (2009).
5. M. Palacio and B. Bhushan, A review of ionic liquids for green molecular lubrication in nanotechnology, *Tribol. Lett.*, 40, 247–268 (2010).
6. E. Schlücker and P. Waserscheid, Ionic liquids in mechanical engineering, *Chemie Ingeniuer Technik*, 83, 1476–1484 (2011).
7. C. F. Ye, W. M. Liu, Y. X. Chen, and L. G. Yu, Room-temperature ionic liquids: A novel versatile lubricant, *Chem. Commun.*, 21, 2244–2245 (2001).
8. Z. Mu, F. Zhou, S. Zhang, Y. Liang, and W. Liu, Effect of the functional groups in ionic liquid molecules on the friction and wear behavior of aluminum alloy in lubricated aluminum-on-steel contact, *Tribol. Int.*, 38, 725–731 (2005).
9. A. E. Jiménez, M. D. Bermúdez, P. Iglesias, F. J. Carrión, and G. Martínez-Nicolás, 1-n-alkyl-3-methylimidazolium ionic liquids as neat lubricants and lubricant additives in steel-aluminium contacts, *Wear*, 260, 766–782 (2006).
10. J. Qu, P.J. Blau, S. Dai, H. Luo, and H.M. Meyer III, Ionic liquids as novel lubricants and additives for diesel engine applications, *Tribol. Lett.*, 35, 181–189 (2009).
11. J. S. Wikes, A short history of ionic liquids—From molten salts to neoteric solvents, *Green Chem.*, 4, 73–80 (2002).
12. H. Kondo, J. Seto, K. Ozawa, and S. Haga, Novel lubricants for magnetic thin film media, *J. Magn. Soc. Jpn.*, 13, 89, 213–218 (1989).
13. H. Kondo, A. Seki, H. Watanabe, and J. Seto, Frictional properties of novel lubricants for magnetic thin film media, *IEEE Trans. Magn.*, 26, 2691–2693 (1990).
14. H. Kondo, A. Seki, and A. Kita, Comparison of an amide and amine salt as friction modifiers for a magnetic thin film medium, *Tribol. Trans.*, 37, 99–104 (1994).
15. W. Liu, C. Ye, Q. Gong, H. Wang, and P. Wang, Tribological performance of room-temperature ionic liquids as lubricant, *Tribol. Lett.*, 13, 81–85 (2002).
16. A. Suzuki, Y. Shinka, and M. Masuko, Tribological characteristics of imidazolium-based room temperature ionic liquids under high vacuum, *Tribol. Lett.*, 27, 307–313 (2007).
17. Y. Kondo, S. Yagi, T. Koyama, R. Tsuboi, and S. Sasaki, Lubricity and corrosiveness of ionic liquids for steel-on-steel sliding contacts, *J. Eng. Tribol.*, 226, 991–1006 (2011).
18. C. A. Wamser, Hydrolysis of fluoboric acid in aqueous solution, *J. Am. Chem. Soc.*, 70, 1209–1215 (1948).
19. R. P. Swatloski, J. D. Holbrey, and R. D. Rogers, Ionic liquids are not always green: Hydrolysis of 1-butyl-3-methylimidazolium hexafluorophosphates, *Green Chem.*, 5, 361–363 (2003).
20. J. Arias-Pardilla, T. Espinosa, and M. D. Bermudez, Ionic liquids in surface protection, *Electrochemistry in Ionic Liquids*, Springer, Berlin, Germany, Vol. 2, pp. 533–561 (2015).
21. V. Totolin, I. Minami, C. Gabler, J. Brenner, and N. Dörr, Lubrication mechanism of phosphonium phosphate ionic liquid additive in alkylborane-imidazole complexes, *Tribol. Lett.*, 53, 421–432 (2014).
22. A. E. Somers, S. M. Biddulph, P. C. Howlett, J. Sun, D. R. MacFarlane, and M. Forsyth, A comparison of phosphorus and fluorine containing IL lubricants for steel on aluminium, *Phys. Chem. Chem. Phys.*, 14, 8224–8231 (2012).
23. M. Mahrova, F. Pagano, V. Pejakovic, A. Valea, M. Kalin, A. Igartua, and E. Tojo, Pyridinium based dicationic ionic liquids as base lubricants or lubricant additives, *Tribol. Int.*, 82, 245–254 (2015).
24. V. Pejaković, M. Kronberger, and M. Kalin, Influence of temperature of tribological behaviour of ionic liquids as lubricants and lubricant additives, *Lubrication Sci.*, 26, 107–115 (2014).

25. H. Okubo, S. Watanabe, C. Tadokoro, and S. Sasaki, Effects of concentration of zinc dialkyldithiophosphate on the tribological properties of tetrahedral amorphous carbon films in presence of organic friction modifiers, *Tribol. Int.*, 94, 446–457 (2016).
26. H. Okubo, S. Watanabe, C. Tadokoro, and S. Sasaki, Effects of structure of zinc dialkyldithiophosphates on tribological properties of tetrahedral amorphous carbon film under boundary lubrication, *Tribol. Int.*, 98, 26–40 (2016).
27. H. Okubo, C. Tadokoro, and S. Sasaki, Tribological properties of a tetrahedral amorphous carbon (ta-C) film under boundary lubrication in the presence of organic friction modifiers and zinc dialkyldithiophosphate (ZDDP), *Wear*, 332–333, 1293–1302 (2015).
28. Y. Kondo, T. Koyama, R. Tsuboi, M. Nakano, K. Miyake, and S. Sasaki, Tribological performance of halogen-free ionic liquids as lubricants of hard coatings and ceramics, *Tribol. Lett.*, 51, 243–249 (2013).
29. H. Okubo, S. Kawada, S. Watanabe, and S. Sasaki, Tribological performance of halogen-free ionic liquids in steel-steel and DLC-DLC contacts, *Tribol. Trans.*, 61, 1–9 (2017).

2 Reduction and Control of Friction in Hydrogels

Tetsuo Yamaguchi

CONTENTS

2.1 INTRODUCTION

Hydrogels are soft elastic materials composed of sparsely crosslinked polymers and a large amount of water [1], as schematically shown in Figure 2.1. Corresponding to such a specific structure of hydrogels, several striking features, such as flexibility, optical transparency, water retention (swelling) ability, stimuli (pH, temperature, electric field) response, and so on, are provided. Owing to the similarities in structures and properties to living tissues, hydrogels are expected to be biocompatible materials or replacements for damaged parts of the human body, such as artificial articular skins, corneas, and cartilages [2].

In this mini-review, we focus on the frictional properties of hydrogels [3–6]. As mentioned above, hydrogels are elastic solids containing an abundant amount of water. Such wet structures contribute to suppressing direct contact between solid components (polymer network and counterpart material/substrate) and to reducing friction. Furthermore, if there are "third" materials (e.g., surfactants) in the system and if they are adsorbed onto the frictional interface, they also contribute to reducing friction.

In Section 2.2, we will show a few examples of low-friction hydrogels and discuss the basic properties and the mechanisms behind them. In Section 2.3, we will introduce the friction of hydrogels in the presence of surfactants. In Section 2.4, we will briefly explain how one can control the friction of hydrogels by applying electric fields.

2.2 LOW FRICTION OF HYDROGELS

Friction is the phenomenon under relative motion between two elastic bodies in contact, and the frictional force originates from interactions between them [7,8]. Since hydrogels are elastic materials containing a large amount of water inside the polymer

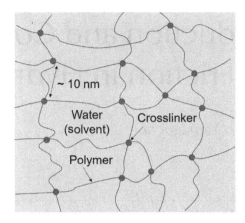

FIGURE 2.1   Schematic of microstructure of hydrogels.

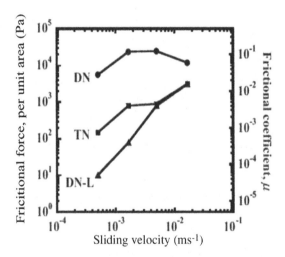

FIGURE 2.2   Friction-sliding velocity curves for three different types of hydrogels: double-network gel (DN), triple-network gel (TN), and double-network gel with linear polymers (DN-L). (From Kaneko, D. et al., *Adv. Mater.*, 17, 535–538, 2005.)

networks, the direct interaction between polymer and counterpart substrate is suppressed and the resulting friction becomes extremely low in some conditions [9,10]. As an example, friction coefficients for double-network (DN) hydrogels and their families sliding against a glass substrate are shown in Figure 2.2 [10].

A double-network hydrogel consists of two interpenetrated crosslinked polymer networks: negatively charged poly(2-acrylamido-2methylpropane sulfonic acid) (PAMPS) and neutral poly(acrylamide) (PAAm) networks (see Figure 2.3). While

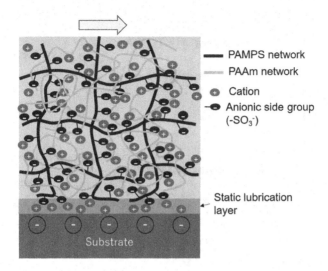

FIGURE 2.3 Microstructure of a double-network (DN) hydrogel. Electric charges in the gel cause osmotic repulsion against the negatively charged glass substrate, resulting in forming a static lubrication layer and leading to extremely low friction.

a DN hydrogel exhibits high mechanical strength (~10 MPa) [11], the friction coefficient is not so small ($\mu \sim 10^{-1}$) because the second network of PAAm has stickiness against a glass substrate (counterpart). However, addition of linear PAMPS chains into the DN hydrogel leads to enhancing osmotic repulsion (not direct electrostatic repulsion between negative charges but effective repulsion via dilution of counterions with water at the interface) and results in extremely low friction as small as ~10^{-4} to 10^{-5} [10]. The introduction of repulsive interaction between polymer and counterpart substrate is one of the key techniques to obtain low friction in a water environment.

Then we discuss a second example exhibiting low friction in hydrogels. Figure 2.4 shows friction coefficients for three different types of poly(vinyl alcohol) (PVA) hydrogels (electrically neutral hydrogels), prepared in three different manners: freeze-thawing, cast-drying, and hybrid methods (details are described in ref. [12]). Among these, the hybrid PVA hydrogel exhibits the lowest friction ($\mu \sim 0.01$) over the whole sliding distance. The mechanism responsible for such low friction has been discussed in the biomechanics community and is called biphasic lubrication [13–19]: once the normal load is applied onto the gel, the stress is supported not only by elastic rebound stress of the polymer network but also by hydrostatic pressure generated due to the fluid component, as schematically shown in Figure 2.5. This hydrostatic pressure of the fluid diminishes the elastic contact stress and effectively reduces friction between the gel and the counterpart.

The biphasic lubrication is a transient process: the hydrostatic pressure causes internal fluid flow sideways and vanishes when the flow stops. However, since the

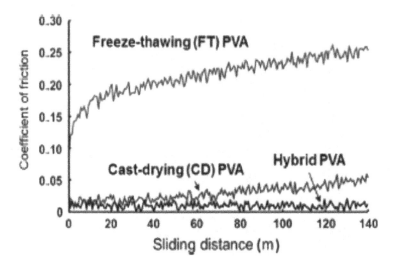

FIGURE 2.4 Friction coefficients of three different types of poly(vinyl alcohol) (PVA) hydrogels against a glass substrate. (From Murakami, T. et al., *Tribol. Int.*, 89, 19–26, 2015.)

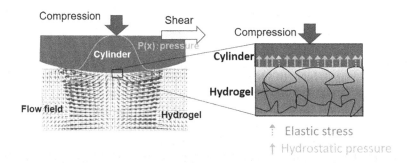

FIGURE 2.5 Schematic of biphasic lubrication mechanism. A part of the load is supported by hydrostatic pressure of the fluid, contributing to reduction in elastic contact stress and resulting in low friction.

pore size of the gel network is in general sufficiently small (diameter ~ nm), it takes a long time for water to flow from the high-pressure region to the low-pressure region inside the network. Furthermore, even if the load-supporting mechanism by the fluid stops, water can be quickly supplied from the gel surface once the counterpart is detached. This recovery mechanism works well, for example, during walking or running (repeat loading and unloading) in knee joints.

It is important to note that the above two mechanisms (the former/latter originate from osmotic/hydrostatic effects, respectively) are completely independent and complementary. It is possible to design low-friction hydrogels by combining these two mechanisms.

2.3 FRICTION IN THE PRESENCE OF SURFACTANTS

Surfactants are known to play an essential role in the friction of hydrogels by adsorption onto the hydrogel surfaces. For example, the effects of proteins in artificial articular cartilages have been discussed by Yarimitsu et al. [20], reporting that adsorption of albumin and γ-globulin forms boundary films and reduces friction and wear of PVA hydrogels. Kamada et al. [21] studied friction between poly(2-acrylamido-2-methylpropanesulfonic acid sodium salt) (PNaAMPS) hydrogel and hydrophilic/hydrophobic substrates in the presence of different types of surfactants. As is clearly seen in Figure 1 of Kamada et al. [21], friction is reduced with an increasing amount of surfactants at smaller sliding velocities. This indicates that the gel friction is dominated by the boundary lubrication at smaller sliding velocities and by fluid lubrication at larger velocities.

2.4 CONTROL OF FRICTION

Finally, we discuss the control of friction in hydrogels. We conducted a series of friction control experiments using hydrogels: (i) friction experiments between a polyelectrolyte hydrogel and an electrode [22], (ii) friction experiments between a neutral hydrogel and an electrode with ionic surfactants [23], and (iii) preparation of neutral hydrogel with addition of a small amount of polyelectrolyte monomer on an electrode [24]. Here we discuss only the second example, friction experiments between a neutral hydrogel and an electrode with ionic surfactants. Figure 2.6a shows our setup: we adhered a cylindrical poly(acrylamide) (PAAm) hydrogel onto the top plate and rotated at a constant speed under a given vertical displacement, and then measured the frictional torque against an electrode in surfactant solution. Figure 2.6b is a typical example of the voltage response. As the positive voltage is applied onto the bottom electrode, the frictional torque decreased and was recovered to the original value when it was switched off. The frictional behavior is explained by the dynamic adsorption of SDS surfactant, as schematically described in Figure 2.6c. Thus, one can actively and easily control friction by adding a small amount of an ionic surfactant and by applying electric voltage.

2.5 SUMMARY

In this mini-review, we have discussed the friction of hydrogels by focusing on reduction and control of friction. In terms of low friction, not only the electric charge–induced osmotic repulsion but also the biphasic lubrication play essential roles. As for the control of friction, the applied voltage and electric charges in the constituent molecules drive "active" surface modification and reduction in friction. Other than the mechanisms presented here, the effect of confinement might be one of the friction reduction mechanisms [25]. We believe that these mechanisms contribute to basic understanding of the frictional behavior of hydrogels as well as development of new stimuli-responsive materials or switching devices.

(a)   (b)

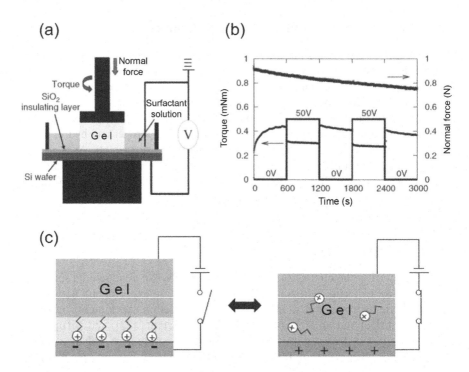

FIGURE 2.6 (a) Schematic of experimental setup, (b) frictional torque (left axis) and normal force (right axis) responses with 10 mmol/L SDS solution under applied voltage (0V/50V), and (c) schematic for the torque response. (From Takata, M. et al., *J. Phys. Soc. Jpn.*, 79, 063602, 2010.)

ACKNOWLEDGMENTS

The author acknowledges M. Doi, M. Takata, R. Suzuki, T. Murakami, Y. Sawae, N. Sakai, K. Nakashima, and S. Yarimitsu for useful discussions and helpful comments. This work was supported by JSPS Kakenhi JP16H06478, "Science of Slow Earthquakes: Physical Modeling."

REFERENCES

1. S. J. Buwalda, K. W. M. Boere, P. J. Dijkstra, J. Feijen, T. Vermonden and W. E. Hennink, "Hydrogels in a historical perspective: From simple networks to smart materials," *J. Control. Release* **190**, 254–273 (2014).
2. E. Caló and V. V. Khutoryanskiy, "Biomedical applications of hydrogels: A review of patents and commercial products," *European Polymer J.* **65**, 252–267 (2015).
3. J. P. Gong, "Friction and lubrication of hydrogels: Its richness and complexity," *Soft Matter* **2**, 544–552 (2006).
4. T. Baumberger, C. Caroli and O. Ronsin, "Self-healing slip pulses along a gel/glass interface," *Phys. Rev. Lett.* **88**, 75509 (2002).
5. T. Baumberger, C. Caroli and O. Ronsin, "Self-healing slip pulses and the friction of gelatin gels," *Eur. Phys. J. E* **11**, 85–93 (2003).

6. T. Nitta, H. Kato, H. Haga, K. Nemoto and K. Kawabata, "Static friction of agar gels: Formation of contact junctions at frictional interface," *J. Phys. Soc. Jpn.* **74**, 2875–2879 (2005).

7. E. Rabinowicz, *Friction and Wear of Materials*, 2nd ed., John Wiley & Sons, New York (2013).

8. B. N. J. Persson, *Sliding Friction: Physical Principles and Applications*, 2nd ed., Springer, Berlin, Germany (1998).

9. J. P. Gong, T. Kurokawa, T. Narita, G. Kagata, Y. Osada, G. Nishimura and M. Kinjo, "Synthesis of hydrogels with extremely low surface," *J. Am. Chem. Soc.* **123**, 5582–5583 (2001).

10. D. Kaneko, T. Tada, T. Kurokawa, J. P. Gong and Y. Osada, "Mechanically strong hydrogels with ultra-low frictional coefficients," *Adv. Mater.* **17**, 535–538 (2005).

11. J. P. Gong, Y. Katsuyama, T. Kurokawa and Y. Osada, "Double-network hydrogels with extremely high mechanical strength," *Adv. Mater.* **15**, 1155–1158 (2003).

12. T. Murakami, N. Sakai, T. Yamaguchi, S. Yarimitsu, K. Nakashima, Y. Sawae and A. Suzuki, "Evaluation of a superior lubrication mechanism with biphasic hydrogels for artificial cartilage," *Tribol. Int.* **89**, 19–26 (2015).

13. V. C. Mow, S. C. Kuei, W. M. Lai and C. G. Armstrong, "Biphasic creep and stress relaxation of articular cartilage in compression: Theory and experiments," *J. Biomech. Eng.* **102**, 73–84 (1980).

14. G. A. Ateshian, "The role of interstitial fluid pressurization in articular cartilage lubrication," *J. Biomech.* **42**, 1163–1176 (2009).

15. N. Sakai, Y. Hagihara, T. Furusawa, N. Hosoda, Y. Sawae and T. Murakami, "Analysis of biphasic lubrication of articular cartilage loaded by cylindrical indenter," *Tribol. Int.* **46**, 225–236 (2012).

16. T. Murakami, S. Yarimitsu, K. Nakashima, T. Yamaguchi, Y. Sawae, N. Sakai and A. Suzuki, "Superior lubricity in articular cartilage and artificial hydrogel cartilage," *Proc. IMechE Part J: J. Eng. Tribol.* **228**, 1099–1111 (2014).

17. T. Murakami, S. Yarimitsu, K. Nakashima, N. Sakai, T. Yamaguchi, Y. Sawae and A. Suzuki, "Biphasic and boundary lubrication mechanisms in artificial hydrogel cartilage: A review," *Proc. IMechE Part H: J. Eng. Med.* **229**, 864–878 (2015).

18. T. Murakami, S. Yarimitsu, N. Sakai, K. Nakashima, T. Yamaguchi, Y. Sawae and A. Suzuki, "Superior lubrication mechanism in poly (vinyl alcohol) hybrid gel as artificial cartilage," *Proc. IMechE Part J: J. Eng. Tribol.* **231**, 1160–1170 (2017).

19. T. Yamaguchi, R. Sato and Y. Sawae, "Propagation of fatigue cracks in friction of brittle hydrogels," *Gels* **4**, 53 (2018).

20. S. Yarimitsu, K. Nakashima, Y. Sawae and T. Murakami, "Study on the mechanisms of wear reduction of artificial cartilage through in situ observation on forming protein boundary film," *Tribol. Online* **2**, 114–119 (2007).

21. K. Kamada, H. Furukawa, T. Kurokawa, T. Tada, T. Tominaga, Y. Nakano and J. P. Gong, "Surfactant-induced friction reduction for hydrogels in the boundary lubrication regime," *J. Phys. Condens. Matter* **23**, 284107 (2011).

22. M. Takata, T. Yamaguchi, J. P. Gong and M. Doi, "Electric field effect on the sliding friction of a charged gel," *J. Phys. Soc. Jpn.* **78**, 084602 (2009).

23. M. Takata, T. Yamaguchi and M. Doi, "Friction control of a gel by electric field in ionic surfactant solution," *J. Phys. Soc. Jpn.* **79**, 063602 (2010).

24. R. Suzuki, T. Yamaguchi and M. Doi, "Frictional property of hydrogels prepared under electric fields," *J. Phys. Soc. Jpn.* **82**, 124803 (2013).

25. D. Kaneko, M. Oshikawa, T. Yamaguchi, J. P. Gong and M. Doi, "Friction coefficient between rubber and solid substrate -Effect of rubber thickness-," *J. Phys. Soc. Jpn.* **76**, 043601 (2007).

3 Biosurfactants
Properties and Applications

Dina A. Ismail, Mona A. Youssif,
and Nabel A. Negm

CONTENTS

3.1 INTRODUCTION

Surfactants are amphiphilic molecules consisting of hydrophilic and lipophilic domains. The former can be nonionic, cationic, anionic, or amphoteric. The lipophilic domain usually consists of hydrocarbon species [1,2]. Common nonionic surfactants include ethoxylates, ethylene and propylene oxide copolymers, and sorbitan esters. Monovalent carboxylic acid salts, ester sulphonates, or sulphates are anionics, whereas quaternary ammonium salts are typical cationic surfactants. Since both the hydrophilic and lipophilic groups reside within the same molecule, surfactants tend to partition preferentially at the interfaces between fluid phases with different polarities such as at oil and water or air/water interfaces. Unique properties of surfactant molecules are the reduction of interfacial energy (interfacial tension) and surface tension through the formation of an ordered molecular film at interface. These properties are essential requirements for industrial applications such as emulsification, foaming, detergency, wetting, and phase dispersion or solubilization.

Many biological molecules are amphiphilic and partition preferentially at interfaces. Microbial compounds that exhibit particularly high surface activity and emulsifying activity are classified as biosurfactants. Biosurfactants, also known as microbial surfactants, are an emerging and diverse class of biomolecules that are commercially applied in different industries including the cosmetics industry. Biosurfactants are surface-active compounds produced by microorganisms such as yeast, fungi, and bacteria.

Most biosurfactants are either anionic or nonionic; and a few are cationics with amine groups. The lipophilic part is mainly long-chain hydrocarbon, which may be attached to fatty acids, hydroxyl fatty acids, or α-alkyl-β-hydroxy fatty acids. The hydrophilic part can be carbohydrates, amino acids, cyclic peptides, phosphates, carboxylic acids, or alcohols [3].

Biosurfactants are mainly grouped in terms of their chemical composition and microbial (biological) origin. The main groups include glycolipids, lipopeptides, lipoproteins, fatty acids, phospholipids, and neutral lipids [4–7]. Biosurfactants can be also classified based on their molecular weights (MW) into low-MW surfactants (called biosurfactants) and high-MW surfactants (called bioemulsans) [8,9]. Biosurfactants such as glycolipids, lipopeptides, phospholipids, and proteins have lower surface and interfacial tension. On the other hand, bioemulsans such as polymers of polysaccharides, lipoproteins, and small molecules surfactants are effective emulsion stabilizing agents, that is, they stabilize oil-in-water emulsions [8–10]. Biosurfactants are synthesized by different microorganisms in the presence of organic carbon sources. Their biological synthesis is mainly influenced by the composition of the nutrients, medium and culture conditions [4,11]. Nutritional carbon sources for the microorganisms include hydrocarbons, carbohydrates, vegetable oils and oil wastes, olive oil mill effluent, lactic whey and distiller wastes, starchy substrates, and industrial and/or municipal wastewater, under aerobic conditions [4,7,12].

The most important characteristics of biosurfactants are their excellent environmental properties such biodegradability and lower toxicity compared to synthetic surfactants. They are also effective at extreme temperature, pH, and salinity, and are ecologically safe [4,12]. These properties have allowed their application for safe remediation of inorganic toxins such as heavy metals [12,13] and organic toxins such as hydrocarbons [14]. Also, their ability to reduce the interfacial tension at oil/water interface and their fast biodegradation allows them to break down oil/water emulsions in petroleum processing [15], as well as their applicability in enhanced oil recovery [16,17].

Moreover, biosurfactants are used in the food processing industry, in health care, and in cosmetics industries [4]. These properties of biosurfactants have spurred a large number of investigations, which have resulted in the identification of new microorganisms that produce natural surfactants, determination of their structure, and discovery of new sources of carbon to enhance the production of biosurfactants [18].

3.2   BIOSURFACTANT-PRODUCING MICROORGANISMS

In the microbial production of biosurfactants, microorganisms use a set of carbon sources and different energy sources for growth. The variation of carbon sources and their combination with insoluble substances facilitates intracellular diffusion of the different components and consequently, production of different biological substances [19–21]. Microorganisms (yeasts, bacteria, and some filamentous fungi) are capable of producing biosurfactants with different molecular structures and surface activities. Table 3.1 shows the different types of microorganisms and the correspondingly produced biosurfactants.

TABLE 3.1
Microorganisms and the Main Classes of Biosurfactants They Produce

Microorganisms	Biosurfactant Class	References
Acinetobacter calcoaceticus	Glycolipids	[43–47]
Alcanivorax borkumensis		
Arthrobacter paraffineus		
Arthrobacter sp.		
Candida antartica		
Candida apicola		
Candida batistae		
Candida bogoriensis		
Candida bombicola		
Candida ishiwadae		
Candida lipolytica		
Lactobacillus fermentum		
Nocardia sp.		
Pseudomonas aeruginosa		
Pseudomonas sp.		
Rhodococcus erythropolis		
Rhodotorula glutinus		
Rhodotorula graminus		
Serratia marcescens		
Tsukamurella sp.		
Ustilago maydis		
Acinetobacter calcoaceticus	Polymeric Surfactants	[61,62]
Acinetobacter calcoaceticus		
Acinetobacter calcoaceticus		
Acinetobacter calcoaceticus		
Bacillus stearothermophilus		
Candida lipolytica		
Candida utilis		
Halomonas eurihalina		
Mycobacterium thermoautotrophium		
Sphingomonas paucimobilis		
Acinetobacter sp.	Lipopeptides	[54,55]
Bacillus licheniformis		
Bacillus pumilus		
Bacillus subtilis		
Candida lipolytica		
Gluconobacter cerinus		
Pseudomonas fluorescens		
Serratia marcescens		
Streptomyces sioyaensis		
Thiobacillus thiooxidans		

(Continued)

TABLE 3.1 (*Continued*)

Microorganisms and the Main Classes of Biosurfactants They Produce

Microorganisms	Biosurfactant Class	References
Arthrobacter paraffineus	Fatty Acids	[56]
Capnocytophaga sp.		
Corynebacterium insidibasseosum		
Corynebacterium lepus		
Nocardia erythropolis		
Penicillium spiculisporum		
Talaramyces trachyspermus		
Acinetobacter calcoaceticus	Small Molecule Surfactant	[65]
Cyanobacteria		
Pseudomonas marginalis		
Acinetobacter sp.	Phospholipids	[57]
Aspergillus		
Corynebacterium lepus		

3.2.1 BACTERIAL BIOSURFACTANTS

Microorganisms make use of a wide range of organic compounds as a source of carbon and energy for their growth. If the carbon source is insoluble in water, like a hydrocarbon; the microorganisms make it possible to diffuse into the cell by producing a variety of substances, the biosurfactants. Some bacterial strains and yeasts excrete ionic surfactants which emulsify the hydrocarbon substances present in the growth medium. A few examples of this group of biosurfactants are rhamnolipids that are produced by different *Pseudomonas* spp. [22] or sophorolipids that are produced by several *Torulopsis* spp. [23].

Some microorganisms are able to alter the structure of their cell wall by producing nonionic or lipopolysaccharide surfactants in their cell wall. Some examples of such organisms are *Rhodococcus erythropolis* and various *Mycobacterium* spp., *Arthrobacter* spp., which produce the nonionic trehalose dimycolate (Figure 3.1) [24–25]. Lipopolysaccharides, such as emulsans, are produced by *Acinetobacter* spp. [25] and lipoproteins that are produced by *Bacillus subtilis* [26].

3.2.2 FUNGAL BIOSURFACTANTS

Although the field of biosurfactants production by bacterial species is well explored, relatively fewer fungi are known to produce biosurfactants. Among the fungi, *Candida bombicola* [27], *Candida lipolytica* [28], *Candida ishiwadae* [29], *Candida batistae* [30], *Aspergillus ustus* [31], and *Trichosporon ashii* [32] have been explored. Many of these fungal strains are known to produce biosurfactants from low-cost raw materials. The major type of biosurfactant produced by these strains is sophorolipids (glycolipids). *Candida lipolytica* produces cell wall–bound lipopolysaccharides when it is growing on n-alkane hydrocarbons [33].

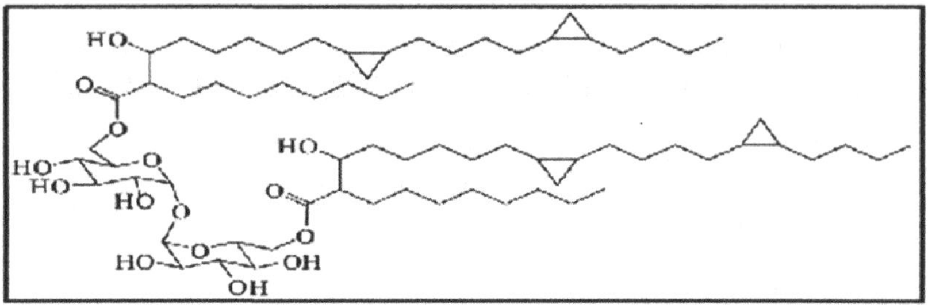

FIGURE 3.1 Chemical structure of trehalose dimycolate. (From Cooper, D.G. et al., *Appl. Environ. Microbiol.*, 42, 408–412, 1981; Casas, J.A. et al., *Enzyme Microb. Technol.*, 21, 221–229, 1997; Sarubbo, L.A. et al., *Curr. Microbiol.*, 54, 68–73, 2007.)

3.2.3 YEAST BIOSURFACTANTS

Biosurfactants can be produced by yeasts, either extracellularly or attached to parts of the cell, predominantly during their growth on water-immiscible substrates. However, some yeasts may produce biosurfactants in the presence of different types of substrates, such as carbohydrates. The use of different carbon sources changes the structure of the biosurfactant produced and, consequently, its properties. These changes may be welcomed when some properties are required for particular applications [34]. There are a number of studies in biosurfactant production involving the optimization of their physicochemical properties [35,36]. The composition and characteristics of biosurfactants are also reported to be influenced by the nature of the nitrogen source as well as the presence of iron, magnesium, manganese, phosphorus, and sulfur.

3.3 FACTORS AFFECTING BIOSURFACTANT PRODUCTION

The composition and emulsifying activity of a biosurfactant depends not only on the producer bacterial strain but also on the culture conditions, such as nature of nutrients including carbon and nitrogen source, C/N ratio, chemical and physical conditions such as temperature, aeration, divalent cations, and pH. These factors influence the amount of biosurfactants produced and the types of polymers produced [37].

3.3.1 CARBON SOURCES

The quality and quantity of biosurfactant production are influenced by the nature of the carbon substrate [38]. Diesel, crude oil, glucose, sucrose, and glycerol have been reported to be good sources of carbon substrates for biosurfactant production [39].

3.3.2 NITROGEN SOURCES

Nitrogen is important for biosurfactant production medium because it is essential for microbial growth as protein and enzyme synthesis depends on it. Different nitrogen compounds have been used for the production of biosurfactants such as urea peptone, yeast extract, ammonium sulphate, ammonium nitrate, sodium nitrate, meat extract, and malt extracts. Though yeast extract is the most used nitrogen source for biosurfactant production, its concentration in the medium depends on the organism and culture medium. Ammonium salts and urea are the preferred nitrogen sources for biosurfactant production with *Arthrobacter paraffineus*, whereas nitrate supports the maximum surfactant production in *P. aeruginosa* [40].

3.3.3 ENVIRONMENTAL FACTORS

Environmental factors are extremely important to the yield and characteristics of the biosurfactants produced by the different microorganisms. To obtain large quantities of biosurfactants, it is necessary to optimize the biosynthesis conditions such as temperature, pH, aeration, or agitation speed of the cultivation medium. Most biosurfactant productions are performed in the temperature range of 25°C–30°C [39]. The effect of pH on biosurfactants production was studied by Zinjarde and Pant [41], who reported that the optimum biosurfactants production was achieved at pH of 8, which is the natural pH of sea water.

3.3.4 AERATION AND AGITATION

Aeration and agitation are important factors that influence the production of biosurfactants since both facilitate oxygen transfer from the outer atmosphere to the cultivation medium. They may also be linked to the physiological function of microbial emulsifier. It has been suggested that the production of bioemulsifiers can enhance the solubilization of insoluble substrates and consequently facilitate nutrients transport to microorganisms. Adamczak and Bednarski [40] observed that the highest surfactant production amount (45.5 g/L) was obtained at an air flow rate of 1 vessel volume per minute and when the dissolved oxygen concentration maintained at 50% of saturation.

3.3.5 SALT CONCENTRATION

Salt concentration of a particular medium has a significant effect on biosurfactant production as the cellular activities of microorganisms are affected by the medium. Nevertheless, contrary observations were made for some biosurfactant products which were not affected by concentration of up to 10%, although slight reductions in the critical micelle concentration (cmc) were detected [39].

3.4 CHEMICAL CLASSIFICATIONS OF BIOSURFACTANTS

Many microorganisms have the ability to produce molecules with surface activity. Two main types of surface-active compounds are produced by microorganisms: biosurfactants and bioemulsifiers. Biosurfactants significantly reduce the air-water surface tension while bioemulsifiers do not reduce the surface tension as much but stabilize oil-in-water emulsions [42]. Several biosurfactants produced by microorganisms are discussed below.

3.4.1 GLYCOLIPIDS

Generally, most known biosurfactants in nature are glycolipid. Their structure includes carbohydrates combined with long-chain aliphatic acids or hydroxyl fatty acids, which are further linked with either ester or an ether group. Among the glycolipids, the best known are the rhamnolipids, trehaloselipids, and sophorolipids (Figure 3.2).

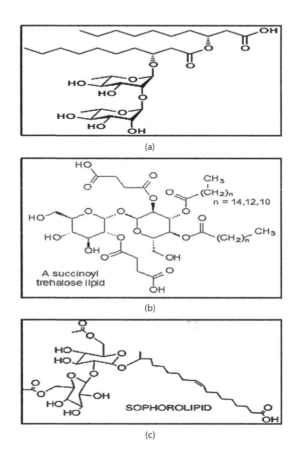

FIGURE 3.2 Some structures of biosurfactants: (a) rhamnolipid, (b) trehaloselipid, and (c) sophorolipid. (From Karanth, N.G.K. et al., *Current Sci.*, 77, 116–126, 1999; Jarvis, F.G. and Johnson, M.J., *J. Am. Chem. Soc.*, 71, 4124–4126, 1949; Lang, S. and Wagner, F., Structure and properties of biosurfactants, in *Biosurfactants and Biotechnology*, N. Kosaric, W.L. Cairns and N.C.C. Gray (Eds.), Marcel Dekker, New York, pp. 21–47, 1987; Li, Z. et al., *Appl. Environ. Microbiol.*, 48, 610–617, 1984; Ristau, E. and Wagner, F., *Biotechnol. Lett.*, 5, 95–100, 1983; Göbbert, U. et al., *Biotechnol. Lett.*, 6, 225–230, 1984; Inoue, S. and Itoh, S., *Biotechnol. Lett.*, 4, 3–8, 1982; Cooper, D. G. and Paddock, D. A., *Appl. Environ. Microbiol.*, 46, 1426–1429, 1983; Tulloch, P. et al., *J. Chem. Soc. Chem. Commun.*, 12, 584–586, 1967; Hommel, R.K. et al., *Appl. Microbiol. Biotechnol.*, 26, 199–205, 1987.)

3.4.2 Rhamnolipids

A rhamnolipids (Figure 3.2a) consists of rhamnose molecules that are linked to one or two molecules of β-hydroxyl decanoic acid. One hydroxyl group of the hydroxyl decanoic acid is involved in glycosidic linkage with the reducing end of the rhamnose disaccharide; the OH group of the second hydroxyl decanoic is involved in ester linkage [43]. Production of rhamnose-containing glycolipids was first observed during cultivation of *Pseudomonas aeruginosa* [44].

3.4.3 Trehaloselipids

Trehaloselipid (Figure 3.2b) is a nonreducing disaccharide in which the two glucose units are linked in an α,α-1,1-glycosidic linkage. It is the basic component of the cell wall glycolipids in *Mycobacteria* and *Corynebacteria*. Several structures of trehaloselipids were reported [45,46]. Trehaloselipids were subsequently isolated from *Rhodococcus erythropolis* [47].

3.4.4 Sophorolipids

Many species of yeast, such as *Torulopsis bombicola* [48,49], *T. petrophilum* [50] and *T. apicola* [51], produce mostly sophorolipids (Figure 3.2c). Sophorolipids are dimeric carbohydrates sophorose that are linked to a long-chain hydroxy fatty acid. At least six to nine different structures of lipophilic sophorosides are found in a mixture [52]. Although sophorolipids can lower surface and interfacial tension, they are not effective as emulsifying agents [53].

3.4.5 Lipopeptides and Lipoprotein

Lipopeptides and lipoprotein biosurfactants in general consist of a large number of cyclic lipopetides linked to a fatty acid. Examples include decapeptide antibiotics (gramicidins) and lipopeptide antibiotics (polymyxins), which possess remarkable surface-active properties (Figure 3.3). Lipid-carrying particles are collectively known as lipoproteins (Figure 3.3b). Lipoproteins are classified according to their densities. The names of these particles, from least to most dense, are very-low-density lipoproteins (VLDL), intermediate-density lipoproteins (IDL), low-density lipoproteins (LDL), and high-density lipoproteins (HDL). Several bacteria are known to produce these antibiotic-like molecules. The cyclic lipopeptide surfactant, produced by *Bacillus subtilus*, is one of the most surface-active biosurfactants [54]. Cyclic lipopeptide surfactant (Figure 3.4) lowers the surface tension of water to 27.9 mN/m at a concentration of 0.005% by weight [55]. Furthermore, it possesses antibacterial, antiviral, antifungal, antimycoplasma, and hemolytic activities. This lipopeptide consists of a hexapeptide lactonised linked to hydroxy fatty acid.

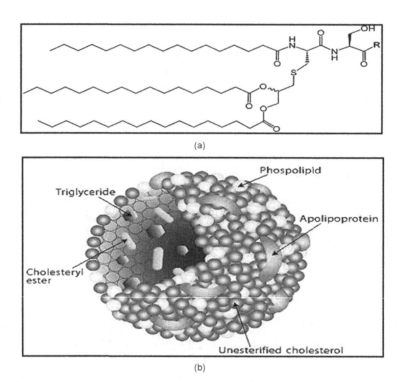

(a)

(b)

FIGURE 3.3 Chemical structures of (a) lipopeptides, and (b) lipoprotein; biosurfactants with antibiotic properties. (From Ron, E.Z. and Rosenberg, E., *Environ. Microbiol.*, 3, 229–236, 2001.)

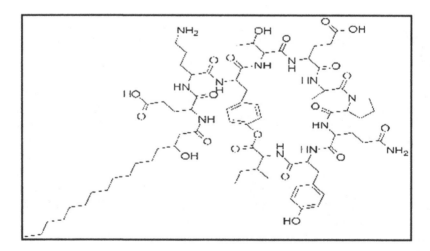

FIGURE 3.4 Structure of cyclic lipopeptide biosurfactant with antibacterial and antiviral properties. (From Arima, K. et al., *Biochem. Biophys. Res. Commun.*, 31, 488–494, 1968.)

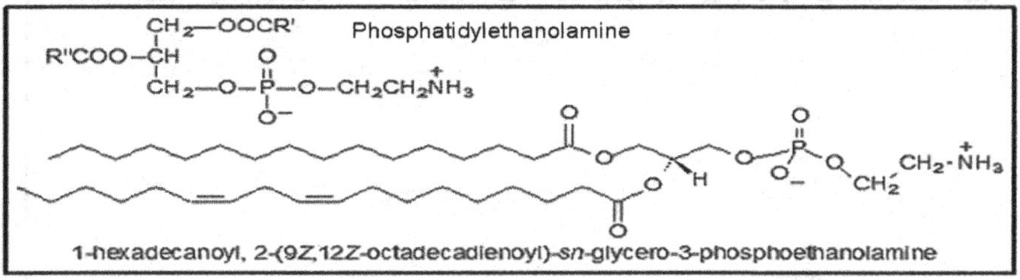

FIGURE 3.5 Structure of phosphatidyl ethanolamine biosurfactant. (From Niels, W. et al., *Biochim. Biophys. Acta*, 1831, 652–62, 2012.)

3.4.6 FATTY ACIDS, PHOSPHOLIPIDS, AND NEUTRAL LIPIDS

Different types of fatty acids and phospholipid surfactants are produced by several bacteria and yeast during growth on n-alkanes [56–58]. When *R. erythropolis* grows on n-alkanes, it produces a substance known as phosphatidyl ethanolamine (Figure 3.5). Phosphatidylethanolamine is a class of phospholipids found in biological membranes [59], and is synthesized by the addition of cytidine diphosphate-ethanolamine to diglycerides, releasing cytidine monophosphate. Phosphatidylethanolamine decreases the interfacial tension between water and hexadecane to less than 1 mN/m and has a cmc of 30 mg/liter [60].

3.4.7 POLYMERIC BIOSURFACTANTS

The best-studied polymeric biosurfactants are those from several well-known components such as emulsan, liposan, mannoprotein (Figure 3.6), and other polysaccharide-protein complexes. Emulsan, synthesized by *Acinetobacter calcoaceticus*, consists of a heteropolysaccharide backbone covalently linked to fatty acids [61]. Another example is liposan, an extracellular water-soluble emulsifier synthesized by the yeast *Yarrowia lipolytica* [62], which is composed of 83% carbohydrate and 17% protein.

3.4.8 PARTICULATE BIOSURFACTANTS

These particulate biosurfactants are also high-MW emulsifiers or biosurfactants. There are two types of biosurfactants in this group as shown in Figure 3.7. Vesicles or whole microbial cell acts as biosurfactants. Extracellular membrane vesicles partition hydrocarbons in an aqueous phase to form a stable microemulsion, which plays an important role in alkane uptake by microbial cells [63,64]. Vesicles of *Acinetobacter* sp. strain HO1-N with a diameter of 20–50 nm and a density of 1.158 g/cm^3 are composed of protein, phospholipid, and lipopolysaccharide [65].

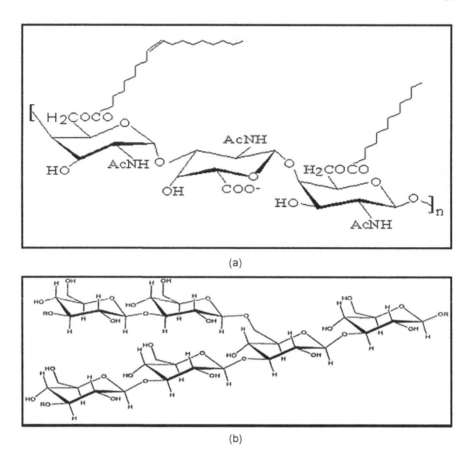

(a)

(b)

FIGURE 3.6 Structure of polymeric surfactants: (a) emulsan and (b) mannoprotein. (From Rosenberg, E. et al., *Appl. Environ. Microbiol.*, 54, 323–326, 1988.)

FIGURE 3.7 Types of particulate biosurfactant.

3.5 EXTRACTION OF BIOSURFACTANTS

In many biotechnological processes, the downstream processing operations consume up to 60% of the total production cost [7]. Due to the lower purity requirements and economic considerations, most biosurfactant preparations would have to involve whole-cell or spent culture broths. In addition, biosurfactant activity may be affected by impurities present in these preparations. Biosurfactant extraction depends mainly on its ionic charge, water solubility, and location (intracellular, extracellular, or cell bound) with respect to the cell surface.

Most widely used techniques involve extractions with solvents such as: chloroform-methanol, dichloromethane-methanol, butanol, ethyl acetate, pentane, hexane, acetic acid, ether, and so on. Several well-known examples of biosurfactants recovered by solvent extraction are

1. trehaloselipids of *Mycobacterium* spp, and *Arthrobacter paraffineus*,
2. trehalose corynomycolates and tetraesters of *Rhodococcus erythropolis*,
3. mono-, di-, and pentasaccharide lipids of *A. paraffineus* and *Nocardia corynebacterioids*,
4. cellobiolipids of *Ustilago* spp.,
5. sophorolipids of several yeast species,
6. liposan from *C. lipolytica*, and
7. rhamnolipids of *Pseudomonas* spp. [7].

Glycolipids produced by *T. bombicola*, *T. petrophilum*, and *T. apicola* are extracted with chilled ethyl acetate after adsorption on charcoal. Biosurfactant from *P. aeruginosa* has also been recovered in a similar way, except that extraction was carried out in acetone. Both the glycolipid produced by *Ustilago zeae* and the mannosyl erythritol lipid produced by *Candida* species are sedimented as heavy oils upon centrifugation and then extracted with either ethanol or methanol [7].

Glycolipids from *P. aeruginosa* and *U. zeae* have also been recovered by acid precipitation at low temperature. Other glycolipids from a mixed microbial population and rhamnolipids from both *P. aeruginosa* and *C. lipolytica* have been recovered by acidification followed by extraction in chloroform-methanol solvent. A cell-bound bioemulsifier [66] has been extracted from *Scerevisiae* at 121°C in a buffer containing potassium metabisulfite, followed by precipitation in an ethanol-acetic acid mixture.

Ammonium sulfate precipitation has been effectively used for isolating emulsan and biodispersan from *Acinetobacter calcoaceticus*, as well as a bioemulsifier from an unidentified gram-negative bacterium. Surfactin and surfactin-like biosurfactants produced by *B. subtilis* and *B. licheniformis*, respectively, have been recovered by acid precipitation, whereas other biosurfactants from *Pseudomonas* spp., *Endomycopsis lipolytica* and *C. tropicalis*, and *Debaryomyces polymorphus* were best recovered by acetone precipitation [7].

3.6 BIOLOGICAL ACTION OF BIOSURFACTANTS IN MICROORGANISMS

Biosurfactants are produced by a variety of microorganisms; they are secreted either extracellularly or are attached to parts of cells predominantly during growth on water immiscible substrates [7]. The main biological role of biosurfactants is to permit microorganisms to grow on water-immiscible substrates by reducing the surface tension at the phase boundary, thus making the substrate more readily available for uptake and metabolism, though the molecular mechanisms related to the uptake of their substrates are still neither clear nor fully understood [7]. Ron and Rosenberg [54] suggested that, in limited area for microorganisms growth, biomass increases arithmetically rather than exponentially, and the evidence that emulsification is a natural process brought about by extracellular agents is indirect and there are certain conceptual difficulties in understanding how emulsification can provide an (evolutionary) advantage for the microorganism producing the emulsifiers. Another biological role of biosurfactants is their antimicrobial activity toward various microorganisms. In addition, biosurfactants have been shown to be involved in cell adhesion, which imparts high stability under hostile environmental conditions, and in cell desorption when organisms need to find new habitats for survival [7].

3.7 BIOSURFACTANTS AND SYNTHETIC SURFACTANTS

In this chapter, synthetic surfactant refers to surfactants not produced by biological organisms such as bacteria, yeast, and fungi. Comparing biosurfactants and synthetic surfactants brings a distinctive challenge of homogenizing two different kinds of materials that share an analogous function but differ considerably relative to their origins. These divergent origins have a great influence on various properties such as toxicity, environmental stability, degradation pattern in soil, and production cost to the environment.

Microbial biosurfactants are produced by various microorganisms, such as bacteria, yeast, and fungi, and have surface-active properties. Biosurfactant-mediated interactions are facilitated by the amphiphilic behavior of the surfactants from the hydrophilic and lipophilic moieties. The wide range of solubility, surface tension, and interfacial tension (IFT) reduction, along with detergency, wetting, and foaming efficiency, make biosurfactants more or less appropriate for sustainable application [67].

Similar properties are obtained from synthetic surfactants derived from petrochemical and other sources [67]. Synthetic surfactants have been broadly developed for various industrial applications, including for many cleaning products because of their detergent action and application in surface cleaning formulations. The present scenario for industrial formulation and sustainability has encouraged active attention to biosurfactant research as a potential alternate. Also, microbial biosurfactants are being commercially derived from agro-industrial feedstock. In their degradation, biosurfactants have considerably less negative impact on the environment, tolerate high concentrations of salt, and have very high stability at higher temperatures compared to synthetic surfactants. This suggests that biosurfactants are more effective surfactants for numerous industrial applications [67,68].

3.8 NATURE AND PROPERTIES OF BIOSURFACTANTS

Surface-active characteristics of biosurfactants are usually close to those of synthetic surfactants. However, biosurfactants also have the following properties: low toxicity, biodegradability by environmental microorganisms, effectiveness in a wide range of acidity and salinity, and ability to be biosynthesized using relatively cheap or low polluting substrates.

3.8.1 SURFACE AND INTERFACIAL TENSIONS OF BIOSURFACTANT SOLUTIONS

The forces between the liquid molecules are responsible for the phenomenon known as surface tension. The efficacy of a microbial surfactant is estimated by its ability to reduce the surface tension of the growth medium. An effective microbial surfactant can decrease the surface tension of water from 72 to 35 mN/m, so the ability of a biosurfactant to reduce surface tension by 38 mN/m means it should be regarded as a potentially effective biosurfactant [69].

3.8.2 CRITICAL MICELLE CONCENTRATION

Figure 3.8 shows an important property of biosurfactant molecules used in industrial and environmental applications. Basically surfactant molecules have two different moieties with distinct properties. The moiety that is sensitive to polar solutions such as

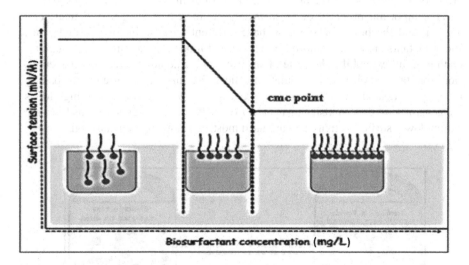

FIGURE 3.8 Simplified surface tension—concentration profile of a biosurfactant in an aqueous medium.

water is the hydrophilic component, whereas the moiety that has attraction for nonpolar solutions such as hydrocarbons is the lipophilic component. The amphiphilic molecules display diverse behavior in water. Such biological mixtures are known as "surface-active agents" or biosurfactants [67]. Efficient biosurfactants have low cmc values and require lower concentrations to decrease the surface tension than synthetic surface-active agents [70]. At low concentrations, biosurfactant molecules are adsorbed at the solution interface and lower the surface tension of the aqueous medium. As the surface gets crowded and saturated by the adsorbed biosurfactant molecules, the remaining biosurfactant molecules spontaneously aggregate in the solution to form micelles.

The biosurfactant mixtures form micelles because of the lipophilic interaction between the lipophilic chains and the van der Waals bonding between the hydrophilic groups. The micelles are arranged to capture the lipophilic molecules in their emulsification with water. The cmc values of biosurfactants are typically 10- to 40-fold lower than synthetic surfactants. Biosurfactants with their lower cmc can achieve high emulsification efficiency for varied industrial and environmental processes [71].

3.8.3 EMULSIFICATION BY BIOSURFACTANTS

Emulsions are intrinsically immiscible two-phase systems (Figure 3.9). Based on the surface tension concept, emulsification takes place by reducing the IFT, and the biosurfactants act as bioemulsifiers. Bioemulsifiers are biological mixtures derived from the microorganisms [72]. Since 1982, more than 250 patents have been issued to industrial formulations with biosurfactants predominantly in food formulations, cosmetic preparations, and pharmaceuticals processing. Effective emulsification property is vital for biosurfactants for use in various industrial formulations [67].

The important property of biosurfactants is to reduce surface and interfacial tension, and the biosurfactants also have different purpose in food formulations. Biosurfactants are used to improve the texture of food preparations such as agglomeration of fat, extend the shelf life of sugar-based formulations, and stabilize emulsion from fat- and oil-based formulations. Biosurfactants are also used in the baking industry for delaying the fermentation process [73]. Biosurfactants with high MWs are generally used as emulsifying agents. Low-MW biosurfactants, such as Liposan, exhibit lower surface activity and are used mostly to emulsify vegetable oils.

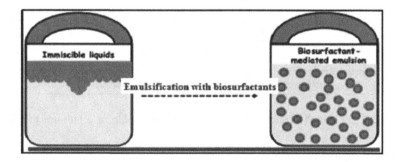

FIGURE 3.9 Simplified representation of emulsification by biosurfactant.

There are some studies about the application of rhamnolipids to improve the properties of various creams and frozen preparations. Very few efforts have been reported about investigating emulsions prepared from biosurfactants for food processing [74].

3.8.4 STABILITY OF BIOSURFACTANTS

The applications of biosurfactants are largely dependent on their performance and stability at different environmental conditions, such as temperature and pH [74]. In the application of biosurfactants in emulsifying edible oils in food formulations, they should be stable in various temperature ranges. Therefore, biosurfactants are used as an active ingredient in preparations such as oral suspensions [75]. The stability of microbial surfactants in various pH and temperature ranges is beneficial for industrial and environmental applications.

3.9 APPLICATION OF BIOSURFACTANTS IN THE PETROLEUM INDUSTRY

In the petroleum industry, biosurfactants have been applied effectively for the exploration of heavy oil, offering advantages over their synthetic counterparts throughout the entire petroleum processing chain (extraction, transportation, and storage). Biosurfactants are used in microbial-enhanced oil recovery, cleaning contaminated vessels and facilitating the transportation of heavy crude oil via pipeline [76]. Table 3.2 represents a list of biosurfactant applications in the oil industry [76].

TABLE 3.2
Common Applications of Biosurfactants in the Petroleum Industry

Petroleum Production	Applications
Extraction	Reservoir wettability modification
	Oil viscosity reduction
	Drilling mud
	Paraffin/asphalt deposition control
	Enhanced oil displacement
	Oil viscosity reduction
Transportation	Oil viscosity reduction
	Oil emulsion stabilization
	Paraffin/asphalt deposition
Oil tank/container cleaning	Oil viscosity reduction
	Oily sludge emulsification
	Hydrocarbon dispersion
Oil waste treatment	Solubilization and mobilization

Source: Luna, J.M. et al. *Chem. Eng. Trans.*, 27, 67–72, 2012.

3.9.1 Extraction of Crude Oil from Reservoirs

Several enhanced oil recovery processes are currently employed worldwide using thermal, chemical, and physical treatments [77]. However, these processes are very expensive as well as environmentally harmful. Thus, the search for cost-effective, eco-friendly alternatives to chemical- and thermal-enhanced oil recovery is necessary. A number of biotechnology-based processes have been proposed to increase oil production [78]. Biosurfactants can be applied in this field to enhance the mobilization of hydrocarbons, thereby enhancing the recovery of crude oil from reservoirs in a process called microbial-enhanced oil recovery (MEOR) [79].

MEOR uses strategies such as the injection and subsequent in situ spreading of microorganisms that produce biosurfactants into the reservoir, and the injection of nutrients into the reservoir to stimulate the growth of microorganisms that produce biosurfactants, or further production of biosurfactants in reactors and subsequent injection into the reservoir [80]. These processes enhance oil recovery from a nearly depleted reservoir, thereby extending the life of the reservoir. MEOR is less expensive compared to chemically enhanced oil recovery, as microorganisms produce efficient products from low-cost raw materials substrates [81].

3.9.2 Transport of Crude Oil by Pipelines

Crude oil often needs to be transported over long distances from extraction fields to refineries. The transportation of heavy and extra-heavy crude oil requires operations that limit its economic viability. These problems are low fluidity due to the high viscosity and the asphaltene content in heavy crude oil, which leads to the deposition of asphaltenes and/or paraffins as well as a drop in pressure that causes plugging problems in the pipeline [79]. Asphaltenes precipitate in the pipelines, and they combine with ferric ions in the acidic conditions to form a solid sediment called asphaltene mud. Asphaltene mud deposits in petroleum pipelines and decreases the flow of crude oil. Organic solvents such as toluene and xylene are used to dissolve this type of mud, which increases the production cost and generates highly toxic wastes [75].

A promising technology involving the production of a stable oil-in-water emulsion that facilitates oil mobility has been recently developed. Biosurfactant-based emulsifiers are particularly suitable for this application. Bioemulsifiers are effective in stabilizing oil-in-water emulsions. Due to the high functionality of the biosurfactants (bioemulsifiers), these molecules tightly bind to oil droplets and form an effective barrier that prevents drop coalescence or coagulation [79]. Emulsan is the most powerful bioemulsifier and has potential applications in the petroleum industry, including the formation of heavy oil-water emulsions for viscosity reduction during pipeline transport [75,79].

3.9.3 Storage Tank Cleaning

Crude oil is stored in storage tanks, and the maintenance of these tanks requires periodic cleaning. However, waste and heavy oil fractions that sediment at the bottom and on the walls of these storage tanks are highly viscous and become solid deposits that cannot be removed with conventional pumping. The removal of these materials requires washing with solvents and manual cleaning, which is a hazardous, time-consuming, labor intensive, and expensive procedure, as cleaning operations may include hot water spraying, solvent liquefaction, and subsequent land waste disposal [79,82].

The use of biosurfactants provides an alternative cleaning procedure to decrease the viscosity of sludge and oil deposits through the formation of oil-in-water emulsions, consequently facilitating the pumping of wastes. Furthermore, this allows for the recovery of crude oil when the emulsion is broken [82]. Dispersion activity was greater than that achieved with a synthetic surfactant. The biosurfactant remained stable for at least three weeks. The effect of the biosurfactant matrix obtained from *P. aeruginosa* [83] was evaluated for the cleaning of oil-contaminated storage tanks. The oil-in-water emulsions recovered from the storage tanks were mostly stable for 50 minutes and then the emulsions are broken into two phases, water and oil. This indicates that the biosurfactant in the sterilized supernatant of *P. aeruginosa* can be used directly for cleaning oil storage tanks and other vessels used for the transportation and storage of crude oil. The potential of biosurfactant isolated from *Pseudomonas cepacia* has also been tested [84] and found to provide a cleaning rate of 80%, which suggests the applicability of this biosurfactant in the cleaning of oil storage tanks.

3.9.4 REMOVAL OF OIL AND PETROLEUM CONTAMINATION

Recent research findings confirmed the effects of biosurfactants on hydrocarbon biodegradation by increasing the microbial accessibility to insoluble substrates and thus enhancing their biodegradation [85,86]. Various experiments showed the effects of biosurfactants on hydrocarbons in enhancing their water solubility and increasing the displacement of oily substances from soil particles. Thus, biosurfactants increase the apparent solubility of these organic compounds at concentrations above their cmc, which enhances their availability for microbial uptake [87]. For these reasons, inclusion of biosurfactants in a bioremediation treatment of a hydrocarbon-polluted environment is promising in facilitating their production by microorganisms [88].

Many of the biosurfactants known today have been investigated to examine their potential technical applications. Most of these applications involve their efficiency in bioremediation, dispersion of oil spills, and enhanced oil recovery. *Alcanivorax* and *Cycloclasticus* bacterial genera are highly specialized hydrocarbon degraders in marine environments. *Alcanivorax borkumensis* utilizes aliphatic hydrocarbons as its main carbon source for growth and produces an anionic glycolipid biosurfactant and thus the potential of *Alcanivorax* strains during bioremediation of hydrocarbon pollution in marine habitats has been studied [89].

Several species of *P. aeruginosa* and *B. subtilis* produce rhamnolipid, a commonly isolated glycolipid biosurfactant and surfactant—a lipoprotein type biosurfactant—and these two biosurfactants increase the solubility [90] and the bioavailability of petrochemical mixtures and also stimulate original microorganisms for enhanced biodegradation of diesel-contaminated soil. *Gordonia* bacteria grow on aliphatic hydrocarbons as the sole carbon source and produce Bioemulsan which effectively degrades crude oil, polycyclic aromatic hydrocarbons (PAHs) and other recalcitrant branched hydrocarbons from contaminated soils. The rate of biodegradation is mainly dependent on the physicochemical properties of the biosurfactants [91].

3.10 BIOREMEDIATION OF TOXIC POLLUTANTS

Bioremediation involves the acceleration of natural biodegradative processes in contaminated environments by enhancing the availability of the materials (e.g., nutrients and oxygen), conditions (e.g., pH and moisture content) and prevailing microorganisms. The addition of emulsifiers is advantageous when bacterial growth is slow (e.g., at cold temperatures or in the presence of high concentrations of pollutants) or when the pollutants consist of compounds that are difficult to degrade such as PAHs.

Bioemulsifiers can be applied as an additive to stimulate the bioremediation process, and with advanced genetic technologies, it is expected that the increase in bioemulsifier concentration during bioremediation would be achieved by the addition of bacteria that produces bioemulsifiers. Recently, this approach has been used successfully in the cleaning of oil pipes. Cultures of *A. radioresistens* [92] which produce bioemulsifier but are unable to use hydrocarbons as a carbon source were added to a mixture of oil-degrading bacteria to enhance oil bioremediation.

Persistent organic pollutants found in oil containing wastewater and sediments, such as PAHs, are lipophilic in nature and thus water solubility of PAHs normally decreases, with increasing number of rings in molecular structure. This property induces a low bioavailability of these organic compounds which is a crucial factor in the biodegradation of PAHs. The water solubility of some PAHs can be improved by several folds by addition of biosurfactants owing to their amphipathic structure [93]. In addition, most hydrocarbons exist in strongly adsorbed form when they are introduced into soils. Thus, their removal efficiency can be limited at low concentrations. However, addition of biosurfactants to the system enhances the bioavailability of low solubility and highly sorptive compounds [94].

3.11 USE OF BIOSURFACTANTS AS FLOATING AGENTS

Flotation is the only processing method in which the three phases of air, liquid, and solid are involved simultaneously [95]. In this regard, flotation reagents include collectors, surface modifiers, frothers, and pH-regulators. These allow for controlling physical and chemical conditions of solid, air, and liquid phases, respectively [95]. Frothers are heteropolar surface-active compounds containing a polar group and a hydrocarbon radical, capable of adsorbing at the air-water interface [95]. By the application of a two-stage separation process, termed sorptive flotation (the first stage consists of the metal ions sorption onto an appropriate sorbent material, followed by successive flotation as the second stage), the biologically produced surface-active agents Surfactin-105 and Lichenysin-A (Figure 3.10) were applied as flotation collectors for the separation of metal-loaded sorbents [96].

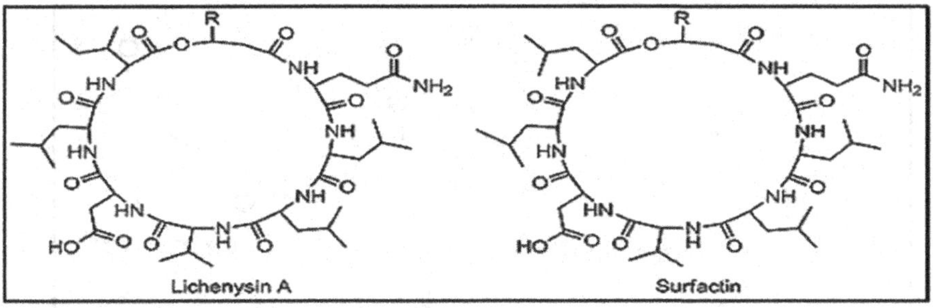

FIGURE 3.10 Chemical structures of Lichenysin-A and Surfactin-105 biosurfactants used as flotation agents for metal removal. (From Zouboulis, A.I. et al., *Mineral Eng.*, 16, 1231–1236, 2003.)

3.12 BIOSURFACTANT AS CORROSION INHIBITORS

A corrosion inhibitor is a chemical substance which, when added in small concentrations to corrosive solutions, minimizes or prevents corrosion of metals [97]. Corrosion inhibitors are used to protect metals, including temporary protection during storage or transport as well as localized protection, required, for example, to prevent corrosion that may result from accumulation of small amounts of corrosive medium [97]. Previous studies have shown that the adsorption of surface-active compounds on stainless steel surfaces plays a major role in decreasing or preventing of corrosion. Corrosion inhibitors are additives used to protect steel pipelines in the petroleum industry as well as in the food and pharmaceutical industries [98,99]. Recently, biosurfactants were investigated as corrosion inhibitors [99]. Stainless steel corrosion was delayed in the presence of a lipopeptide biosurfactant produced by a *P. fluorescens* strain [100]. A surfactant-like compound produced by *Bacillus* sp. was shown to be active against sulfate-reducing bacteria, which is the major bacterial group responsible for biogenic souring and biocorrosion in petroleum reservoirs. This enabled its use for sulfate-reducing bacteria control in the petroleum industry as a promising alternative to synthetic biocides [101].

3.13 APPLICATION OF BIOSURFACTANTS IN AGRICULTURE

The solubility of biohazardous chemical compounds such as PAH by surfactants as mobilizing agents increases the apparent solubility of lipophilic organic contaminants (LOCs) in the aqueous medium. Also, surfactants help microbes to adsorb on soil particles, thus decreasing the diffusion path between the site of adsorption and site of bio-uptake by the microorganisms [102]. In agriculture, biosurfactants are used for hydrophilization of heavy soils to obtain good wettability and to achieve even distribution of fertilizer in the soil. Biosurfactants also prevent the caking of certain fertilizers during storage and promote spreading and penetration of the toxicants in pesticides [102].

The rhamnolipid biosurfactant, mostly produced by the genus *Pseudomonas*, is known to possess potent antimicrobial activity. Furthermore, no adverse effects on

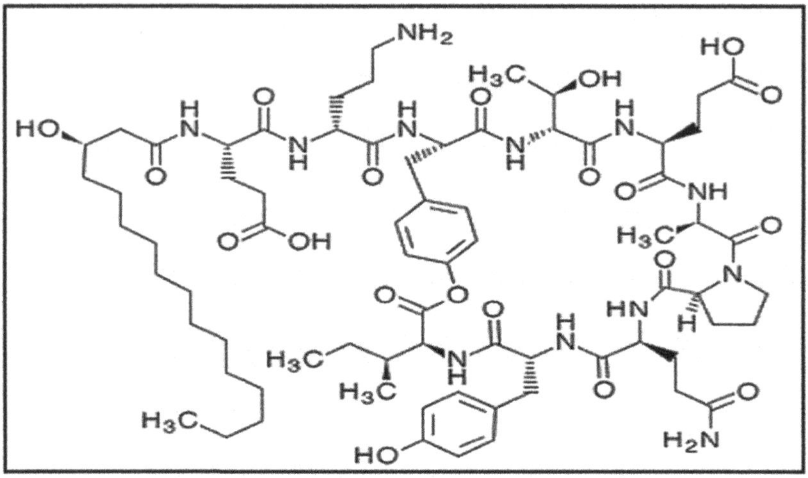

FIGURE 3.11 Chemical structure of Fengycin, biosurfactant with antifungal property for biocontrol of plant disease. (From Kachholz, T. and Schlingmann, M., Possible food and agricultural application of microbial surfactants: An assessment, in *Biosurfactant and Biotechnology*, N. Kosaric, W.L. Cairns and N.C.C. Grey (Eds.), Marcel Dekker, New York, pp. 183–208, 1987.)

humans or the environment are anticipated from aggregate exposure to rhamnolipid biosurfactant. Fengycin (Figure 3.11) was reported to possess antifungal activity and therefore may be employed in the biocontrol of plant diseases [103].

3.14 BIOSURFACTANTS AS BIOPESTICIDES

Conventional arthropod control strategy involves applications of broad-spectrum chemicals and pesticides, which often can produce undesirable effects. Further, emergence of pesticide-resistant insect population as well as rising prices of new chemical pesticides have stimulated the search for new eco-friendly vector control tools. Lipopeptide biosurfactants produced by several bacteria exhibit insecticidal activity against the fruit fly *Drosophila melanogaster* and hence are promising for use as biopesticides [104].

3.15 APPLICATIONS OF BIOSURFACTANTS IN COMMERCIAL LAUNDRY DETERGENTS

Almost all surfactants used in modern-day commercial laundry detergents are synthetic and can exert toxicity to freshwater living organisms. Growing public awareness about the environmental hazards and risks associated with synthetic surfactants has stimulated the search for natural, eco-friendly substitutes for synthetic surfactants in laundry detergents. Biosurfactants such as cyclic lipopeptide (CLP) are stable over a wide pH range (7–12) as well as temperature (40°C–80°C) range, with no considerable change or loss of their surface-active property. They also show good emulsion formation capability with vegetable oils and excellent compatibility and stability with commercial laundry detergents, thus favoring their inclusion in laundry detergent formulations [105].

3.16 APPLICATION OF BIOSURFACTANTS IN MEDICINE

Mukherjee et al. [106] elucidated a wide range of applications of biosurfactants in medicine. These include antimicrobial activity, anticancer activity, anti-adhesion agents, immunological adjuvants, antiviral activity, gene delivery, and many more.

3.16.1 ANTIMICROBIAL ACTIVITY

The diverse structures of biosurfactants confer them with the ability to display versatile performance [107]. Several biosurfactants have strong antibacterial [108], antifungal, and antiviral capabilities. These surfactants play the role of anti-adhesion agents to pathogens making them useful for treating many diseases as well as their use as therapeutic and probiotic agents. A good example is the biosurfactant produced by marine *B. circulans* that displays potent antimicrobial activity against gram-positive and gram-negative pathogens and semi-pathogenic microbial strains including multidrug-resistant (MDR) strains [109].

3.16.2 ANTI-ADHESION AGENTS

Biosurfactants have been found to inhibit the adhesion of pathogenic organisms to solid surfaces or to infection sites. Rodrigues et al. [110] established that precoating vinyl urethral catheters by running the surfactin solution through them before inoculation with media reduced the amount of biofilm formed by *Salmonella typhimurium, Salmonella enterica, E. coli* and *Proteus mirabilis.* Krishnaswamy et al. [109] reported that pretreatment of silicone rubber with *S. thermophilus* biosurfactant reduced the adhesion of *C. albicans* by 85%. Also, the biosurfactants from *L. fermentum* and *L. acidophilus* adsorbed on glass reduced the number of adhering pathogenic cells of *Enterococcus faecalis* by 77%.

3.16.3 IMMUNOLOGICAL ADJUVANTS

Bacterial lipopeptides create potent nontoxic, nonpyrogenic immunological adjuvants when mixed with conventional antigens. An improvement in the humoral human response was demonstrated with low molecular mass antigens Iturin AL (Figure 3.12) and herbicolin A (Figure 3.13) [108].

3.16.4 ANTIVIRAL ACTIVITY

Antibiotic effects and inhibition of growth of human immunodeficiency virus (HIV) in leucocytes by biosurfactants have been cited in the literature [109–111]. Furthermore, Krishnaswamy et al. [109] reported that due to the increased incidence of HIV in

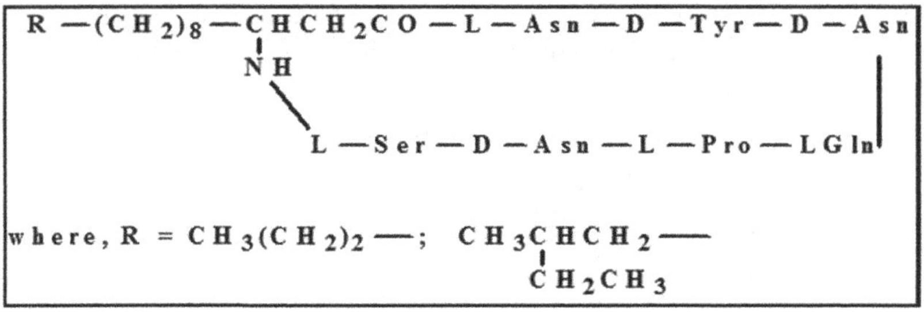

R —(CH₂)₈—CHCH₂CO —L—Asn —D —Tyr—D —Asn

$$R-(CH_2)_8-\underset{\underset{}{\overset{\displaystyle |}{NH}}}{CH}CH_2CO-L-Asn-D-Tyr-D-Asn$$

$$L-Ser-D-Asn-L-Pro-LGln$$

$$where, R = CH_3(CH_2)_2-; \quad CH_3\underset{\underset{CH_2CH_3}{\overset{\displaystyle |}{}}}{CH}CH_2-$$

FIGURE 3.12 Chemical structure of the biosurfactant Iturin AL, which has been found to display an immune-enhancing capability. (From Gharaei-Fathabad, E., *Am. J. Drug Discov. Devel.*, 1, 58–69, 2011.)

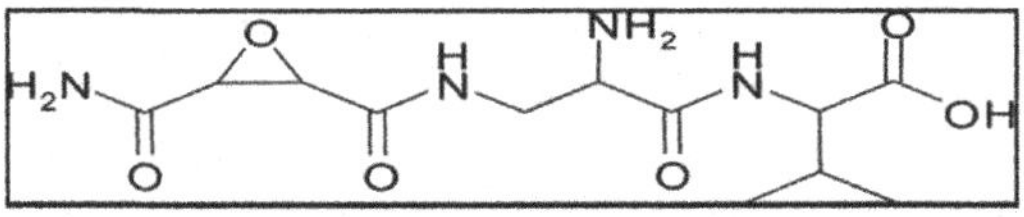

FIGURE 3.13 Chemical structure of the biosurfactant herbicolin A, reported to possess an immune-enhancing capability. (From Gharaei-Fathabad, E., *Am. J. Drug Discov. Devel.*, 1, 58–69, 2011.)

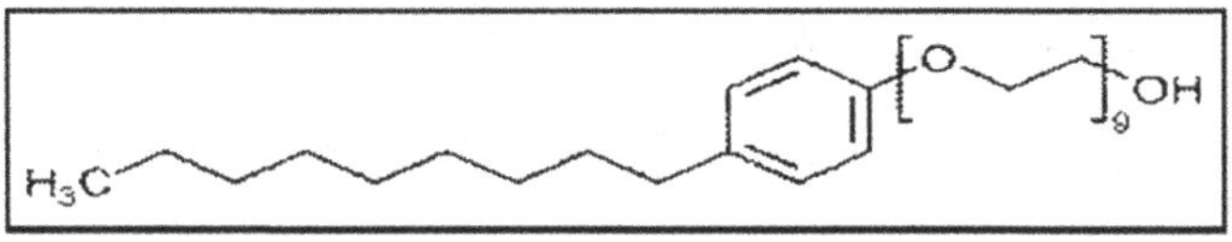

FIGURE 3.14 Chemical structure of the biosurfactant nonoxynol–9 surfactant, reported to have a potent virucidal property. (From Zhao, Z. et al., *Bioresour. Technol.*, 101, 292–297, 2010.)

women, there is the need for a female-controlled, efficacious, and safe topical vaginal microbiocide. Sophorolipid surfactant from *C. bombicola*, and its structural analogues such as the sophorolipid diacetate ethyl ester, has been reported [112] to be the most potent spermicidal and virucidal agent. It was also reported [113] that this substance has a virucidal capability similar to nonoxynol–9 (Figure 3.14) against human semen.

3.16.5 GENE DELIVERY

Gharaei-Fathabad [108] stated that the establishment of an efficient and safe method for introducing exogenous nucleotides into mammalian cells is critical for basic sciences and clinical applications.

3.16.6 OTHER ADVANTAGES AND APPLICATIONS OF BIOSURFACTANTS IN MEDICINE

The use of surfactants as agents for stimulating stem fibroblast metabolism and immunomodulatory action has been reported. However, it has also been reported that the deficiency of a pulmonary surfactant, a phospholipid protein complex, is responsible for the failure of respiration in premature infants. However, isolation of its genes from its protein molecules and cloning it in bacteria has made possible its medical application through production by fermentation [109].

3.17   FOOD INDUSTRY

Biosurfactants have various functions in the food industry, apart from their role as agents that decrease surface and interfacial tensions and facilitate the formation and stabilization of emulsions. Example of applications include control of the aggregation of fat globules, stabilization of aerated systems, improvement of texture and shelf life of products containing starch, modification of rheological properties of wheat dough, and improvement of constancy and texture of fat-based products [114]. In bakery and ice cream formulations, biosurfactants act by controlling the consistency, slowing staling and solubilizing the flavoring oils and helping during cooking of fats and oil. Improvement in the stability of dough and the texture and conservation of bakery products is obtained by the addition of rhamnolipid biosurfactants [115].

3.18   THE ROLE OF BIOSURFACTANTS IN THE COSMETIC INDUSTRY

Biosurfactants are used in the same way as synthetic surfactants in the cosmetic industry. This includes detergency, emulsification, de-emulsification, wetting, foaming, dispersion, solubilization of lipophilic substances, or modification of surfaces [116–118].

The emulsifier function is probably the most important property of biosurfactants in the formulation of cosmetics because emulsions have considerable advantages over other types of preparations. Emulsions are easy to apply, relatively inexpensive (due to their high water content), and simultaneously allow for the use of lipo-soluble and hydro-soluble substances. It is important to note that low-MW biosurfactants reduce the surface tension at the air/water interface as well as the interfacial tension at oil/water (O/W) interface, while high-MW biosurfactants are more effective in stabilizing O/W emulsions [117].

Several authors [119,120] have discussed that most challenging problems related to the use of synthetic surfactants in cosmetics formulations is their potential to cause skin irritation and allergic reactions. Synthetic surfactants can interact with proteins, remove lipids from the epidermal surface by disorganizing the intercellular structure of such lipids, and also affect living cells in the skin. These harmful effects can potentially be avoided or reduced by using biosurfactants, for example, such as those produced from lactic acid bacteria [121–123]. Some of these biosurfactants can also be found in agro-industrial streams, such as corn steep liquor, due to the spontaneous fermentation that occurs during the refining process of corn [124–126].

Biosurfactants are usually more biodegradable and eco-friendly than synthetic surfactants. However, they have to follow the same regulations and restrictions as their synthetic counterparts in order to be incorporated into cosmetic formulations.

3.19   TRIBOLOGICAL APPLICATION OF BIOSURFACTANTS

Biosurfactants can be considered as valuable candidates for tribological applications. Biosurfactants can be used in several systems including metalworking fluids during metal processing [127]. Biosurfactants are used as anticorrosives, biocides, and emulsifiers [128] during formulation of the metalworking fluids. Furthermore, heavy metals pollution in metalworking fluids is a serious environmental issue and various technologies have been discovered for their cleanup from environment. The use of biosurfactants for this purpose was found to be an eco-friendly approach and also an alternative to conventional complex remediation systems.

3.20   SUMMARY AND CONCLUSIONS

This chapter discusses the potential roles and applications of biosurfactants, mainly focusing on bacteria-, fungi-, and yeast-based biosurfactants and their potential application areas, such as food and food-processing industries and environmental biotechnology. The optimization of the biosurfactants production process is the key factor to improving yield and reducing costs. Estimates show that the utilization of renewable sources as substrates may reduce up to 30% of the production costs. Factors that influence biosurfactant production, such as nitrogen source, carbon source, pH, temperature, agitation, and aeration, are also relevant for optimization of production and thus on the production cost.

REFERENCES

1. X. Mao, R. Jiang, W. Xiao and J. Yu, "Use of surfactants for the remediation of contaminated soils: A review," *J. Hazard. Mater.*, 285, 419–435 (2015).
2. R. C. F. S. Silva, D. G. Almeida, J. M. Luna, R. D. Rufino, V. A. Santos and L. A. Sarubbo, "Applications of biosurfactants in the petroleum industry and the remediation of oil spills," *Int. J. Mol. Sci.*, 15, 12523–12542 (2014).
3. C. N. Mulligan, R. N. Yong and B. F. Gibbs, "Surfactant- enhanced remediation of contaminated soil: A review," *Eng. Geol.*, 60, 371–380 (2001).
4. J. D. Desai and I. M. Banat, "Microbial production of surfactants and their commercial potential," *Microbiol. Mol. Biol. Rev.*, 61, 47–64 (1997).
5. N. Kosaric, "Biosurfactants and their application for soil bioremediation," *Food Technol. Biotechnol.*, 39, 295–304 (2001).
6. K. S. M. Rahman and E. Gakpe, "Production, characterization and applications of biosurfactants," *Biotechnology*, 7, 360–370 (2008).
7. K. K. Gautam and V. K. Tyagi, "Microbial surfactants: A review," *J. Oleo Sci.*, 55, 155–166 (2006).
8. T. R. Neu, "Significance of bacterial surface-active compounds in interaction of bacteria with interfaces," *Microbiol. Rev.*, 60, 151–166 (1996).
9. E. Rosenberg and E. Z. Ron, "High- and low-molecular- mass microbial surfactants," *Appl. Microbiol. Biotechnol.*, 52, 154–162 (1999).
10. A. Perfumo, T. J. P. Smyth, R. Marchant and I. M. Banat, "Production and roles of biosurfactants and bioemulsifiers in accessing lipophilic substrates," In: *Handbook of Hydrocarbon and Lipid Microbiology*, K. N. Timmis (Eds.), pp. 1501–1512, Springer, Berlin, Germany (2010).

11. A. Franzetti, P. Caredda, C. Ruggeri, P. La Colla, E. Tamburini, M. Papacchini and G. Bestetti, "Potential applications of surface active compounds by *Gordonia sp. strain BS29* in soil remediation technologies," *Chemosphere*, 75, 801–807 (2009).

12. N. Kosaric, "Biosurfactants in industry," *Pure Appl. Chem.*, 64, 1731–1737 (1992).

13. A. I. Zouboulis, K. A. Matis, N. K. Lazaridis and P. N. Golyshin, "The use of biosurfactant in flotation: Application for the removal of metal ions," *Miner. Eng.*, 16, 1231–1236 (2003).

14. A. Franzetti, P. Di Gennaro, G. Bestetti, M. Lasagni, D. Pitea and E. Collina, "Selection of surfactants for enhancing diesel hydrocarbons-contaminated media bioremediation," *J. Hazard. Mater.*, 152, 1309–1316 (2008).

15. C. N. Mulligan, "Environmental applications for biosurfactants," *Environ. Pollut.*, 133, 183–198 (2005).

16. G. A. Plaza, K. Lukasik, J. Wypych, G. Nalecz-Jawecki, C. Berry and R. L. Brigmon, "Biodegradation of crude oil and distillation products by biosurfactant-producing bacteria," *Pol. J. Environ. Stud.*, 17, 87–94 (2008).

17. H. R. Abdolhamid, M. A. R. Sadiq Al-Baghdadi and A. K. El Hinshiri, "Evaluation of bio-surfactants enhancement on bioremediation process efficiency for crude oil at oilfield: Strategic study," *Ovidus Univ. Ann. Chem.*, 20, 25–30 (2009).

18. A. M. Shete, G. Wadhawa, I. M. Banat and B. A. Chopade, "Mapping of patents on bioemulsifiers and biosurfactant: A review," *J. Sci. Ind. Res.*, 65, 91–115 (2006).

19. M. Deleu and M. Paquot, "From renewable vegetables resources to microorganisms: New trends in surfactants," *C R Chimie*, 7, 641–646 (2004).

20. R. Marchant, S. Funston, C. Uzoigwe, P. K. S. M. Rahman and I. M. Banat, "Production of biosurfactants from nonpathogenic bacteria, In: *Biosurfactants*," R. Sen (Eds.), pp. 73–81, Taylor & Francis Group, New York (2014).

21. J. Chakraborty and S. Das, "Biosurfactant-based bioremediation of toxic metals," In: *Microbial Biodegradation and Bioremediation*, S. Das, Eds., pp. 167–201, Elsevier, Rourkela Odisha, India (2014).

22. M. M. Burger, L. Glaser and R. M. Burton, "The enzymatic synthesis of a rhamnose-containing glycolipid by extracts of *Pseudomonas aeruginosa*," *J. Biol. Chem.*, 238, 2595–2602 (1963).

23. A. J. Cutler and R. J. Light, "Regulation of hydroxydocosanoic acid sophoroside production in *Candida bogoriensis* by the levels of glucose and yeast extract in the growth medium," *J. Biol. Chem.*, 254, 1944–1950 (1979).

24. E. Ristau and F. Wanger, "Formation of novel anionic trehalose tetraesters from *Rhodococcus erythropolis* under growth limiting conditions," *Biotechnol. Lett.* 5, 95–100 (1983).

25. A. Kretschmer, H. Bock and F. Wagner, "Chemical and physical characterization of interfacial-active lipids from *Rhodococcus erythropolis* grown on n-alkanes," *Appl. Environ. Microbiol.*, 44, 864–870 (1982).

26. D. G. Cooper, C.R. Macdonald, S. J. B. Duff and N. Kosaric, "Enhanced production of surfactin from *Bacillus subtilis* by continuous product removal and metal cation additions," *Appl. Environ. Microbiol.*, 42, 408–412 (1981).

27. J. A. Casas, S. G. de Lara and F. Garcia-Ochoa, "Optimization of a synthetic medium for *Candida bombicola* growth using factorial design of experiments," *Enzyme Microb. Technol.*, 21, 221–229 (1997).

28. L. A. Sarubbo, C. B. B. Farias and G. M. Campos-Takaki, "Co-Utilization of canola oil and glucose on the production of a surfactant by *Candida lipolytica*," *Curr. Microbiol.*, 54, 68–73 (2007).

29. B. Thanomsub, T. Watcharachaipong, K. Chotelersak, P. Arunrattiyakorn, T. Nitoda and H. Kanzaki, "Monoacylglycerols: Glycolipid biosurfactants produced by a thermotolerant yeast, *Candida ishiwadae*," *J. Appl. Microbiol.*, 96, 588–592 (2004).

30. M. Konishi, T. Fukuoka, T. Morita, T. Imura and D. Kitamoto, "Production of new types of sophorolipids by *Candida batistae*," *J. Oleo Sci.*, 57, 359–369 (2008).

31. C. S. Alejandro, H. S. Humberto and J. F. Maria, "Production of glycolipids with antimicrobial activity by *Ustilago maydis* FBD12 in submerged culture," *Afr. J. Microbiol. Res.*, 5, 2512–2523 (2011).

32. P. Chandran and N. Das, "Biosurfactant production and diesel oil degradation by yeast species *Trichosporon asahii* isolated from petroleum hydrocarbon contaminated soil," *Int. J. Eng. Sci. Technol.*, 2, 6942–6953 (2010).

33. R. D. Rufino, L. A. Sarubbo and G. M. Campos-Takaki, "Enhancement of stability of biosurfactant produced by *Candida lipolytica* using industrial residue as substrate," *World J. Microbiol. Biotechnol.*, 23, 729–734 (2007).

34. D. G. Cooper, "Biosurfactants," *Microbiol. Sci.*, 3, 145–149 (1986).

35. P. F. F. Amaral, M. Lehocky and J. M. da Silva, "Production and characterization of a bioemulsifier from *Yarrowia lipolytica*," *Proc. Biochem.*, 41, 1894–1898 (2006).

36. L. A. Sarubbo, J. M. Luna and G. M. Campos-Takaki, "Production and stability studies of the bioemulsifiers obtained from a new strain of *Candida glabrata*," *Eletron. J. Biotechnol.*, 9, 400–406 (2006).

37. A. Salihu, I. Abdulkadir and M. N. Almustapha, "An investigation for potential development of biosurfactants," *Microbiol. Mol. Biol. Rev.*, 3, 111–117 (2009).

38. P. K. S. M. Rahman and E. Gakpe, "Production, characterization and application of biosurfactants—Review," *Biotechnology*, 7, 360–370 (2008).

39. J. D. Desai and I. M. Banat, "Microbial production of surfactants and their commercial potential," *Microbiol. Mol. Biol. Rev.*, 61, 47–64 (1997).

40. M. Adamczak and W. Bednarski, "Influence of medium composition and aeration on the synthesis of biosurfactants produced by *Candida antartica*," *Biotechnol. Lett.*, 22, 313–316 (2000).

41. S. S. Zinjarde and A. Pant, "Emulsifier from a tropical marine yeast *Yarrowia lipolytica*," *J. Basic Microbiol.*, 42, 67–73 (2002).

42. R. D. Rufino, L. A. Sarubbo and G. M. Campos-Takaki, "Enhancement of stability of biosurfactant produced by *Candida lipolytica* using industrial residue as substrate," *World J. Microbiol. Biotechnol.*, 23, 729–734 (2007).

43. N. G. K. Karanth, P. G. Deo and N. K. Veenanadig, "Microbial production of biosurfactants and their importance," *Current Sci.*, 77, 116–126 (1999).

44. F. G. Jarvis and M. J. Johnson, "A glycolipid produced by *Pseudomonas aeruginosa*," *J. Am. Chem. Soc.*, 71, 4124–4126 (1949).

45. S. Lang and F. Wagner, "Structure and properties of biosurfactants," In: *Biosurfactants and Biotechnology*, N. Kosaric, W. L. Cairns and N. C. C. Gray (Eds.), pp. 21–47, Marcel Dekker, New York (1987).

46. Z. Li, S. Lang, F. Wagner, L. Witte and V. Wray, "Formation and identification of interfacial-active glycolipids from resting microbial cells," *Appl. Environ. Microbiol.*, 48, 610–617 (1984).

47. E. Ristau and F. Wagner, "Formation of novel anionic trehalose tetraesters from *Rhodococcus erythropolis* under growth-limiting conditions," *Biotechnol. Lett.*, 5, 95–100 (1983).

48. U. Göbbert, S. Lang and F. Wagner, "Sophoroselipids formation by resting cells of *Torulopsis bombicola*," *Biotechnol. Lett.*, 6, 225–230 (1984).

49. S. Inoue and S. Itoh, "Sorphorolipids from *Torulopsis bombicola* as microbial surfactants in alkane fermentation," *Biotechnol. Lett.*, 4, 3–8 (1982).

50. D. G. Cooper and D. A. Paddock, "*Torulopsis petrophilum* and surface activity," *Appl. Environ. Microbiol.*, 46, 1426–1429 (1983).

51. P. Tulloch, A. Hill and J. F. T. Spencer, "A new type of macrocyclic lactone from *Torulopsis apicola*," *J. Chem. Soc. Chem. Commun.*, 12, 584–586 (1967).

52. R. K. Hommel, O. Stuwer, W. Stuber, D. Haferburg and H. P. Kleber, "Production of water-soluble surface-active exolipids by *Torulopsis apicola*," *Appl. Microbiol. Biotechnol.*, 26, 199–205 (1987).

53. D. G. Cooper and D. A. Paddock, "Production of a biosurfactant from *Torulopsis bombicola*," *Appl. Environ. Microbiol.*, 47, 173–176 (1984).

54. E. Z. Ron and E. Rosenberg, "Natural roles of biosurfactants," *Environ. Microbiol.*, 3, 229–236 (2001).

55. K. Arima, A. Kakinuma and G. Tamura, "Surfactin, a crystalline peptide surfactant produced by *Bacillus subtilis*: Isolation, characterization and its inhibition of fibrin clot formation," *Biochem. Biophys. Res. Commun.*, 31, 488–494 (1968).

56. C. Asselineau and J. Asselineau, "Trehalose containing glycolipids," *Prog. Chem. Fats Lipids*, 16, 59–99 (1978).

57. M. C. Cirigliano and G. M. Carman, "Purification and characterization of liposan, a bioemulsifier from *Candida lipolytica*," *Appl. Environ. Microbiol.*, 50, 846–850 (1985).

58. D. G. Cooper, J. E. Zajic and D. F. Gerson, "Production of surface-active lipids by *Corynebacterium lepus*," *Appl. Environ. Microbiol.*, 37, 4–10 (1979).

59. N. Wellner, T. Diep, C. Janfelt and H. S. Hansen, "N-acylation of phosphatidyl ethanolamine and its biological functions in mammals," *Biochim. Biophys. Acta*, 1831, 652–662 (2012).

60. Z. Y. Li, S. Lang, F. Wagner, L. Witte and V. Wray, "Formation and identification of interfacial-active glycolipids from resting microbial cells," *Appl. Environ. Microbiol.*, 48, 610–617 (1984).

61. E. Rosenberg, C. Rubinovitz, R. Legmann and E. Z. Ron, "Purification and chemical-properties of *Acinetobacter calcoaceticus* A2 biodispersan," *Appl. Environ. Microbiol.*, 54, 323–326 (1988).

62. M. C. Cirigliano and G. M. Carman, "Isolation of a bioemulsifier from *Candida lipolytica*," *Appl. Environ. Microbiol.*, 48, 747–750 (1984).

63. S. A. Monteiro, G. L. Sassaki, L. M. de Souza, J. A. Meira, J. M. de Araújo, D. A. Mitchell, L. P. Ramos and N. Krieger, "Molecular and structural characterization of the biosurfactant produced by *Pseudomonas aeruginosa* DAUPE 614," *Chem. Phys. Lipids*, 147, 1–13 (2007).

64. S. Mukherjee, P. Das and R. Sen, "Towards commercial production of microbial surfactants," *Trends Biotechnol.*, 24, 509–515 (2006).

65. O. Kappeli and W. R. Finnerty, "Partition of alkane by an extracellular vesicle derived from hexadecane-grown Acinetobacter," *J. Bacteriol.*, 140, 707–712 (1979).

66. D. R. Cameron, D. G. Cooper and R. J. Neufield, "The Mannoprotein of *Saccharomyces Cerevisiae* is an effective bioemulsifier," *Appl. Environ. Microbiol.*, 54, 1420–1425 (1988).

67. B. S. Saharan, R. K. Sahu and D. Sharma, "A review on biosurfactants: Fermentation, current developments and perspectives," *Genetic Eng. Biotechnol. J.*, 1, 1–14 (2011).

68. I. M. Banat, S. K. Satpute, S. S. Cameotra, R. Patil and N. V. Nyayanit, "Cost effective technologies and renewable substrates for biosurfactants' production," *Front. Microbiol.*, 5, 697–702 (2014).

69. E. J. Silva, N. M. P. R. Silva, R. D. Rufino, J. M. Luna, R. O. Silva and L. A. Sarubbo, "Characterization of a biosurfactant produced by *Pseudomonas cepacia* in the presence of industrial wastes and its application in the biodegradation of hydrophobic compounds in soil," *Colloids Surf. B*, 117, 36–41 (2014).

70. S. George and K. Jayachandran, "Analysis of rhamnolipid biosurfactants produced through submerged fermentation using orange fruit peelings as sole carbon source," *Appl. Biochem. Biotechnol.*, 158, 694–705 (2009).

71. K. Muthusamy, S. Gopalakrishnan, T. K. Ravi and P. Sivachidambaram, "Biosurfactants: Properties, commercial production and application," *Curr. Sci.*, 94, 736–748 (2008).

72. T. R. Neu and K. Poralla, "Emulsifying agents from bacteria isolated during screening for cells with hydrophobic surfaces," *Appl. Microbiol. Biotechnol.*, 32, 521–525 (1990).

73. M. Nitschke and S. G. V. Costa, "Biosurfactants in food industry," *Trends Food Sci. Technol.*, 18, 252–259 (2007).

74. D. Sharma, B. S. Saharan, N. Chauhan, S. Procha and S. Lal, "Isolation and functional characterization of novel biosurfactant produced by *Enterococcus faecium*," *Springer Plus*, 4, 1–14 (2015).

75. D. Sharma, B. S. Saharan, N. Chauhan, A. Bansal and S. Procha, "Production and structural characterization of *Lactobacillus helveticus* derived biosurfactant," *Sci. World J.*, 2014, Article ID 493548 (2014).

76. J. M. Luna, R. D. Rufino, G. M. Campos-Takakia and L. A. Sarubbo, "Properties of the biosurfactant produced by *Candida sphaerica* cultivated in low-cost substrates," *Chem. Eng. Trans.*, 27, 67–72 (2012).

77. H. Al-Sulaimani, S. Joshi, Y. Al-Wahaibi, S. N. Al-Bahry, A. Elshafie and A. Al-Bemani, "Microbial biotechnology for enhancing oil recovery: Current developments and future prospects," *Biotechnol. Bioinf. Bioeng. J.*, 1, 147–158 (2011).

78. S. Sun, Z. Zhang, Y. Luo, W. Zhong, M. Xiao, W. Yi, L. Yub and P. Fu, "Exopolysaccharide production by a genetically engineered *Enterobacter cloacae* strain for microbial enhanced oil recovery," *Bioresour. Technol.*, 102, 6153–6158 (2011).

79. A. Perfumo, I. Rancich and I. M. Banat, "Possibilities and challenges for biosurfactants use in petroleum industry," *Adv. Exp. Med. Biol.*, 672, 135–145 (2010).

80. S. N. Al-Bahry, Y. M. Al-Wahaibi, A. E. Elshafie, A. S. Al-Bemani, S. J. Joshi, H. S. Al-akhmari and H. S. Al-Sulaimani, "Biosurfactant production by *Bacillus subtilis* B20 using date molasses and its possible application in enhanced oil recovery," *Int. Biodeterior. Biodegrad.*, 81, 141–146 (2013).

81. P. Sarafzadeh, A. Niazi, V. Oboodi, M. Ravanbakhsh, A. Z. Hezave, S. Shahab Ayatollahi and S. Raeissi, "Investigating the efficiency of MEOR processes using *Enterobacter cloacae* and *Bacillus stearothermophilus* (biosurfactant-producing strains) in carbonated reservoirs," *J. Pet. Sci. Eng.*, 113, 46–53 (2014).

82. T. Matsui, T. Namihira, T. Mitsuta and H. Saeki, "Removal of oil tank bottom sludge by novel biosurfactant," *J. Jpn. Pet. Inst.*, 55, 138–141 (2012).

83. A. Diab and S. G. El Din, "Application of the biosurfactants produced by *Bacillus* spp. and *Pseudomonas aeruginosa* isolated from the rhizosphere soil of an Egyptian salt marsh plant for the cleaning of oil-contaminated vessels and enhancing the biodegradation of oily sludge," *Afr. J. Environ. Sci. Technol.*, 7, 671–679 (2013).

84. N. M. P. Rocha e Silva, R. D. Rufino, J. M. Luna, V. A. Santos and L. A. Sarubbo, "Screening of *Pseudomonas* species for biosurfactant production using low-cost substrates," *Biocatal. Agric. Biotechnol.*, 3, 132–139 (2013).

85. Y. Zhang and R. M. Miller, "Enhanced octadecane dispersion and biodegradation by a *Pseudomonas rhamnolipid* surfactant (biosurfactant)," *Appl. Environ. Microbiol.*, 58, 3276–3282 (1992).

86. W. P. Hunt, K. G. Robinson and M. M. Ghosh, "The role of biosurfactants in biotic degradation of hydrophobic organic compounds," In: *Hydrocarbon Bioremediation*, R. E. Hinchee, R. E. Hoeppel and R. N. Miller (Eds.), pp. 318–322, CRC Press, Boca Raton, FL, (1994).

87. M. W. Chang, T. P. Holoman and H. Yi, "Molecular characterization of surfactant-driven microbial community changes in anaerobic phenanthrene-degrading cultures under methanogenic conditions," *Biotechnol. Lett.*, 30, 1595–1601 (2008).

88. C. Calvo, M. Manzanera, G. A. Silva-Castro, I. Uad and J. Gonzalez-Lopez, "Application of bioemulsifiers in soil oil bioremediation processes: Future prospects," *Sci. Total Environ.*, 407, 3634–3640 (2009).

89. N. L. Olivera, M. L. Nievas, M. Lozada, G. del Prado, H. M. Dionisi and F. Sineriz, "Isolation and characterization of biosurfactant-producing *Alcanivorax* strains: Hydrocarbon accession strategies and alkane hydroxylase gene analysis," *Res. Microbiol.*, 160, 19–26 (2009).

90. L. M. Whang, P. W. Liu, C. C. Ma and S. S. Cheng, "Application of biosurfactants, rhamnolipid and surfactin, for enhanced biodegradation of diesel-contaminated water and soil," *J. Hazard. Mater.*, 151, 155–163 (2008).

91. A. Franzetti, G. Bestetti, P. Caredda, P. La Colla and E. Tamburini, "Surface-active compounds and their role in the access to hydrocarbons in *Gordonia* strains," *Microbiol. Ecol.*, 63, 238–248 (2008).

92. S. Navon-Venezia, Z. Zosim, A. Gottlieb, R. Legmann, S. Carmeli, E. Z. Ron and E. R. Alasan, "A new bioemulsifier from *Acinetobacter radioresistens*," *Appl. Environ. Microbiol.*, 61, 3240–3244 (1995).

93. H. Yin, J. Qiang, Y. Jia, J. Ye, H. Peng, H. M. Qin, N. Zhang and B. Y. He, "Characteristics of biosurfactant produced by *Pseudomonas aeruginosa* S6 isolated from oil-containing wastewater," *Proc. Biochem.*, 44, 302–308 (2009).

94. K. H. Shin, K. W. Kim and E. A. Seagren, "Combined effects of pH and biosurfactant addition on solubilization and biodegradation of phenanthrene," *Appl. Microbiol. Biotechnol.*, 65, 336–343 (2004).

95. H. Khoshdast and A. Sam, "Flotation frothers: Review of their classifications, properties and preparation," *Open Miner. Proc. J.*, 4, 25–44 (2011).

96. A. I. Zouboulis, K. A. Matis, N. K. Lazaridis and P. N. Golyshin, "The use of biosurfactants in flotation: Application for the removal of metal ions," *Miner. Eng.*, 16, 1231–1236 (2003).

97. S. P. Canmet, Materials Technology Laboratory Ottawa. "Microbial Degradation of Materials: General Processes," In: *Uhlig's Corrosion Handbook*, 2nd edn., R. Winston Revie (Ed.), Ottawa, ON (2000).

98. M. A. Hegazy, M. Abdallah and H. Ahmed, "Novel cationic Gemini surfactants as corrosion inhibitors for carbon steel pipelines," *Corros. Sci.*, 52, 2897–2904 (2010).

99. R. Zuo, "Biofilms: Strategies for metal corrosion inhibition employing microorganisms," *Appl. Microbiol. Biotechnol.*, 76, 1245–1253 (2007).

100. C. Dagbert, T. Meylheuc and M. N. Bellon-Fontaine, "Pit formation on stainless steel surfaces pre-treated with biosurfactants produced by *Pseudomonas fluorescens*," *Electrochim. Acta*, 54, 35–40 (2008).

101. E. Korenblum, L. V. de Araujo, C. R. Guimaraes, L. M. de Souza, G. Sassari, F. Abreu, M. Nitschke et al., "Purification and characterization of a surfactin-like molecule produced by *Bacillus sp.* H_2O-1 and its antagonistic effect against sulfate reducing bacteria," *BMC Microbiol.*, 12, 252–261 (2012).

102. R. S. Makkar and K. J. Rockne, "Comparison of synthetic surfactants and biosurfactants in enhancing biodegradation of polycyclic aromatic hydrocarbon," *Environ. Toxicol. Chem.*, 22, 2280–2292 (2003).

103. T. Kachholz and M. Schlingmann, "Possible food and agricultural application of microbial surfactants: An assessment," In: *Biosurfactant and Biotechnology*, N. Kosaric, W. L. Cairns and N. C. C. Grey (Eds.), pp. 183–208, Marcel Dekker, New York (1987).

104. C. N. Mulligan, "Recent advances in the environmental applications of biosurfactants," *Curr. Opin. Colloid Interface Sci.*, 14, 372–378 (2009).

105. K. Das and A. K. Mukherjee, "Crude petroleum-oil biodegradation efficiency of *Bacillus subtilis* and *Pseudomonas aeruginosa* strains isolated from petroleum oil contaminated soil from North-East India," *Bioresour. Technol.*, 98, 1339–1345 (2007).

106. S. Mukherjee, P. Das and R. Sen, "Towards commercial production of microbial surfactants," *Trends Biotechnol.*, 24, 509–515 (2006).

107. Z. Zhao, Q. Wang, K. Wang, K. Brain, C. Liu and Y. Gu, "Study of the antifungal activity of *Bacillus vallismortis* ZZ185 *in vitro* and identification of its antifungal components," *Bioresour. Technol.*, 101, 292–297 (2010).

108. E. Gharaei-Fathabad, "Biosurfactants in pharmaceutical industry: A mini review," *Am. J. Drug Discov. Devel.*, 1, 58–69 (2011).

109. M. Krishnaswamy, G. Subbuchettiar, T. K. Ravi and S. Panchaksharam, "Biosurfactants properties, commercial production and application," *Current Sci.*, 94, 736–747 (2008).

110. L. Rodrigues, I. M. Banat, J. Teixeira and R. Oliveira, "Biosurfactant: Potential applications in medicine," *J. Antimicrob. Chemother.*, 57, 609–618 (2006).

111. J. D. Desai and I. M. Banat, "Microbial production of surfactants and their commercial potential," *Microbiol. Mol. Biol. Rev.*, 61, 47–64 (1997).

112. R. M. Mann and J. R. Bidwell, "The acute toxicity of agricultural surfactants to the tadpoles of four Australian and two exotic frogs," *Environ. Pollut.*, 114, 195–205 (2001).

113. A. A. Al-Jabri, M. D. Wigg, E. Elias, R. Lambkin, C. O. Mills and J. S. Oxford, *In-vitro* anti-HIV-1 virucidal activity of tyrosine-conjugated tri- and dihydroxy bile salt derivatives," *J. Antimicrob. Chemother.*, 45, 617–621 (2000).

114. I. M. Banat, R. S. Makkar and S. S. Cameotra, "Potential commercial applications of microbial surfactants, *Appl. Microbiol. Biotechnol.*, 53, 495–508 (2000).

115. I. P. H. Van Haesendonck and E. C. A. Vanzeveren, *Rhamnolipids in Bakery Products.* W. O. 2004/040984, International Application Patent (PCT), Washington, DC (2004).

116. R. Marchant and I. M. Banat, "Biosurfactants: A sustainable replacement for chemical surfactants?," *Biotechnol. Lett.*, 34, 1597–1605 (2012).

117. I. M. Banat, A. Franzetti, I. Gandolfi, G. Bestetti, M. G. Martinotti, L. Fracchia, T. J. Smyth and R. Marchant, "Microbial biosurfactants production, applications and future potential," *Appl. Microbiol. Biotechnol.*, 87, 427–444 (2010).

118. K. Holmberg, "Natural surfactants," *Curr. Opin. Colloid Interface Sci.*, 6, 148–159 (2001).

119. T. Bujak, T. Wasilewski and Z. Nizioł-Łukaszewska, "Role of macromolecules in the safety of use of body wash cosmetics," *Colloids Surf. B*, 135, 497–503 (2015).

120. G. Lu and D. J. Moore, "Study of surfactant-skin interactions by skin impedance measurements," *Int. J. Cosmetic Sci.*, 34, 74–80 (2012).

121. C. Duarte, E. J. Gudina, F. L. Cristovao and L. R. Rodrigues, "Effects of biosurfactants on the viability and proliferation of human breast cancer cells," *AMB Express*, 4, 40–45 (2014).

122. E. J. Gudiña, V. Rangarajan, R. Sen and L. R. Rodrigues, "Potential therapeutic applications of biosurfactants," *Trends Pharmacol. Sci.*, 34, 667–675 (2013).

123. L. Rodrigues, I. M. Banat, J. Teixeira and R. Oliveira, "Biosurfactants: Potential applications in medicine," *J. Antimicrob. Chemother.*, 57, 609–618 (2006).

124. X. Vecino, L. Barbosa-Pereira, R. Devesa-Rey, J. M. Cruz and A. B. Moldes, "Study of the surfactant properties of aqueous stream from the corn milling industry," *J. Agric. Food Chem.*, 62, 5451–5457 (2014).

125. X. Vecino, R. Devesa-Rey, J. M. Cruz and A. B. Moldes, "Procedimiento de separacion de los surfactantes presentes de licores de maiz yusos," Patent ES-2435324 B2 (2014).

126. X. Vecino, L. Barbosa-Pereira, R. Devesa-Rey, J. M. Cruz and A. B. Moldes, "Optimization of liquid–liquid extraction of biosurfactants from corn steep liquor," *Bioproc. Biosyst. Eng.* doi:10.1007/s00449-015-1404-9 (2015).

127. M. Fakruddin, "Biosurfactant: Production and application," *J. Pet. Environ. Biotechnol.*, 3, 124–135 (2012).

128. L. V. Araujo, D. M. G. Freire and M. Nitschke, "Biosurfactants: Properties, anticorrosion, antibiofilm, antimicrobial," *Quim. Nova*, 36, 848–858 (2013).

4 Microbial Lipids for Potential Tribological Applications—An Overview*

Daniel K.Y. Solaiman, Richard D. Ashby, and Girma Biresaw

CONTENTS

4.1 INTRODUCTION

Historically, vegetable oils and animal fats (VOAFs) were used as lubricating agents primarily to reduce friction and wear in various operations. It was even proposed that Egyptians in ancient times used olive oil and tallow as lubricants to maneuver the massive boulders for constructing the pyramids and to grease the axles of their chariots [1]. With the dawn of the petroleum age, mineral oils began to replace VOAFs as the preferred lubricants in industrial operations and applications. Mineral oils possess physical and chemical properties that are better suited for industrial lubrication purposes than the VOAFs. These desired properties include the ability to withstand high temperature (e.g., high thermal stability and resistance to oxidation), and to remain fluid in cold environments (i.e., low cloud point and low freezing point) [2]. Over the years, mineral oil–based lubricants have grown into a business with an estimated global market of ~15 million tons valued at ~US $70 billion. Aside from the base oils, various chemical compounds have also been developed and blended as additives in the formulation of industrial lubricants to impart or improve their desired properties. With the attention

---

* Mention of trade names or commercial products in this article is solely for the purpose of providing specific information and does not imply recommendation or endorsement by the U.S. Department of Agriculture. USDA is an equal opportunity provider and employer.

of the industry now turning to renewable and sustainable natural resources, biobased feedstocks have become the focus of research to produce biobased lubricant base oils and additives. VOAFs as a group are classified as renewable and sustainable feedstocks. Over the years, we have seen and continue to see many exciting advancements and new inventions of biobased lubricants and additives from VOAFs [3–8]. Another equally valuable source to harvest renewable and sustainable feedstocks for the manufacturing of biobased lubricants and additives is the microbial world. Microorganisms produce a myriad of natural products, some of which are well-suited for development into bio-based lubricants and additives. Notable among these are the hydroxy fatty acids and the triacylglycerol (TAG) compounds that are biosynthesized by bacteria, yeast, and algae. Unlike crops and livestock that provide the VOAFs, the fermentation-based biorefineries producing the microbial bioproducts are less susceptible to unexpected climate conditions (i.e., drought, flood, storm, etc.). In this respect, microbial bioproducts from the biorefineries could serve to buffer the impact of sudden adverse climates on the supply chain of biobased lubricants and additives. Furthermore, valuable rangeland and arable land need not be preempted by these biorefineries. In certain cases, such as with some algal oils, the theoretical product yields of lipid products per acre of land could substantially exceed those of oilseed crops, reaching as high as 30,000 L (equivalent to ~200 barrels) of oil per hectare of land [9]. The perceived effects on food supplies are therefore minimal. In this mini-review, a quick survey of the biosynthesis and the potential uses in tribology of microbial fatty acids and triacylglycerols are presented.

4.2   ALGAL LIPIDS AND HYDROXY FATTY ACIDS

Algae are a diverse group of single-cell or multicellular aquatic organisms that share an ability to perform photosynthesis to assimilate CO_2. They range from the bacteria-like cyanobacteria to the meter-long sea kelps [10]. In this article, we focus only on the microalgal group of microorganisms that have the metabolic capability to produce TAGs and/or hydroxy fatty acids useful in tribological applications. Microalgae are a group of very diverse aquatic organisms covering tens of thousands of species. A heightened interest in these organisms occurred during the oil crisis of the 1970s, which prompted the US Department of Energy to initiate the landmark Aquatic Species Program (ASP) research undertaken to evaluate microalgae as a potential source of renewable oils for energy uses [11]. At the height of the program, a collection of some 3,000 species of microalgae was assembled and characterized with particular regard to their growth properties and lipid content. Of these thousands of species, only 25 strains were reported to be viable in the collection [12]. The majority of these 25 species belong to the classes of green algae (i.e., Chlorophyceae and Prasinophyceae) and diatom (i.e., Bacillariophyceae), and they were screened for their high potential in producing TAGs in commercially viable yields. It would not be an understatement to say that the ASP was the standard-bearer for subsequent research on algal oil over the next two decades. In Table 4.1, we have listed several representative strains of microalgae that belong to different taxonomic classes studied in the ASP project. During the years since the conclusion of ASP, researchers worldwide continued to base the selection of strains for their studies on the microalgae first surveyed in that project. A survey of recent research literature showed that the lipid contents (% cell dry weight [CDW] or % biomass) of these

TABLE 4.1
National Renewable Energy Laboratory Aquatic Species Program (NREL ASP) Existing Species (as of 2009)

Species	Strain	SERI Strain Designations	Class	Representative Values of Lipid Contents (% CDW) and the Major Fatty Acids
Ankistrodesmus falcatus 91–1	Pyramid Lake 91–1	ANKIS1	Chlorophyceae	59.6%; C16:0, C18:1, C18:2, C18:3 [14]
Chaetoceros muelleri	NM-6	CHAET6	Bacillariophyceae	43.4%; C16:1, C16:0,
Chaetoceros muelleri subsalsum	—	CHAET59	Bacillariophyceae	C14:0 [16]
Chlorella sp.	S01	CHLOR1	Chlorophyceae	15.9%; C18:1, C18:2, C16:0 [13]
Chlorella ellipsoidea	BL-6	CHLOR2	Chlorophyceae	32%; C18:2, C16:0, C18:3 [17]
Cyclotella cryptica	DI-35	CYCLO1	Bacillariophyceae	42.1%; C14:0, C16:1, C16:0 [18,19]
Ellipsoidon sp.	70–01	ELLIP1	Eustigmatophyceae	n.a.
Franceia sp.	LCC-1, ASU 0146	FRANC1	Chlorophyceae	n.a.
Monoraphidium sp.	Mom's Ranch	MONOR1	Chlorophyceae	43.5%; C16:0,
Monoraphidium sp.	—	MONOR2	Chlorophyceae	C18:1 [20]
Nannochloris sp.	HB44	NANNO2	Chlorophyceae	16.2%; C18:1, C16:0, C18:2 [13]
Nannochloropsis salina	GSBSTICHO	NANNP1	Eustigmatophyceae	53.0%; C16:0, C18:1 [21]
Nannochloropsis sp.	Nanno-Q	NANNP2	Eustigmatophyceae	19.4%; C16:0, C16:1, C14:0, C18:3 [22]
Navicula saprophila	—	NAVIC2	Bacillariophyceae	42.5%; C16:0, C16:1,
Navicula saprophila	BB 260	NAVIC7	Bacillariophyceae	C18:0 [18,23]
Navicula saprophila	—	NAVIC24	Bacillariophyceae	
Nitzschia pusilla monoensis	Mono Lake	NITZS1	Bacillariophyceae	47.2%; C16:0, C14:0, C12:0 [18,24]
Nitzschia dissipata	SB-307	NITZS13	Bacillariophyceae	
Nitzschia communis	S-16	NITZS14	Bacillariophyceae	
Oocystis pusilla	Walker Lake	OOCYS1	Chlorophyceae	18.1%; C16:0, C18:1, C18:3 [25]
Phaeodactylum tricornutum	TFX-1	PHAEO1	Bacillariophyceae	32.2%; C16:1, C16:0 [26]
Phaeodactylum tricornutum	BB	PHAEO2	Bacillariophyceae	
Pleurochrysis dentata	—	PLEUR1	Prymnesiophyceae	25.9%; [27]
Tetraselmis suecica	—	TETRA1	Prasinophyceae	29%; C16:0, C18:1 [28]
Tetraselmis sp.	HB47	TETRA4	Prasinophyceae	

organisms ranged widely from 15.9% in a *Chlorella* sp. [13] to 59.6% in *Ankistrodesmus falcutus* [14]. The diatoms (i.e., Bacillariophyceae), however, uniformly exhibited relatively high lipid contents in the 40%–55% range (Table 4.1). A quick literature survey of the fatty acid profiles of the species in Table 4.1 also revealed that the prominent fatty acid components of the microalgal lipids were commonly palmitic acid (C16:0), oleic acid (C18:1), and myristic acid (C14:0), though other unsaturated and saturated fatty acids in the C12–C22 range also showed up as minor components. However, it must be further noted that the microalgae belong to the class of Bacillariophyceae that almost uniformly contain shorter-chain (C12–C16) and saturated fatty acids as the predominant components of their lipids. Beer et al. [15] surveyed microalgae to specifically identify strains that could biosynthesize lipids having a high C18:1 fatty acid content. Table 4.2 lists representative strains that were found to produce lipids in concentrations up to 73% of the CDW with C18:1 compositions as high as 61% (i.e., *C. protothecoides*). It is noteworthy that none of these species belong to the diatoms (i.e., Bacillariophyceae). It could be concluded that depending on the application, the microalgal strain selected for lipid production must contain the appropriate fatty acid profiles [15].

In the laboratory setting, numerous microalgal culture growth parameters have been extensively examined for their influence on lipid yields and fatty acid compositions. There is no single set of parameters, however, that is universally applicable to increase the lipid yield and specific fatty acid composition in the diverse strains of microalgae. For the diatoms, for example, limitation of the availability of a silicone supplement plays an important role in influencing cell growth rates, lipid contents, and fatty acid compositions [29–31], but it has no apparent effect on the culturing of the microalgae in the green algae class. Conversely, while wavelength and flux of light irradiation can dramatically alter cell growth characteristic and lipid production in the green

TABLE 4.2
Total Lipid and Oleic Acid Contents of Selected Microalgae

Species	Lipid Content (% CDW)[a]	C18:1 (% w/w Lipid Composition)[b]	Class
Ankistrodesmus braunii	73	54	Chlorophyceae
Chlorella protothecoides	58	61	Trebouxiophyta
Neochloris oleoabundans	54	36	Chlorophyceae
Pleurochrysis carterae	50	<10	Prymnesiophyceae
Nannochloropsis spp.	41	<12	Eustigmatophyceae
Tetraselmis suecica	25	16	Chlorodendrophyceae
Dunaliella tertiolecta	23	<10	Chlorophyceae

Source: Beer, T. et al., Biodiesel from Algae, in: Regional Forum on Bioenergy Sector Development: Challenges, Opportunities, and Way Forward, Report from the United Nations Economic and Social Commission for Asia and the Pacific-Asian and Pacific Centre for Agricultural Engineering and Machinery, (ESCAP-APCAEM), pp 4–12, 2008.

Note: Microalgae with a high potential for biodiesel production.

[a] The upper limit of the lipid content.

[b] The upper limit of %-lipid composition unless otherwise indicated.

algae [21,22,32], these factors often do not affect the cell growth and lipid production of the diatoms in a significant manner. The controlling of the culturing parameters is even more complicated at the pilot- and production-scale bioprocesses. In general, there are two types of bioreactors to carry out large- or production-scale microalgal bioprocesses, that is, the open bioreactors and the closed fermentor systems. Each type of bioreactor has its own advantages and disadvantages. Granata [33] recently surveyed research results published over the last 60 years on the culturing and lipid production by microalgae in different bioreactor configurations. The survey showed that for the open bioreactor system, the most popular configurations are the pond and the raceway flume systems. A larger number of fermentor configurations are possible with the closed bioreactor system; these include the tube (vertical, horizontal, and helical), the flat plate, and other miscellaneous types of reactor vessels. Granata [33] classified all of these open and closed bioreactor systems into four groups based on the lipid contents of the culture and the productivity rates. Based on techno-economic considerations, it was concluded that intermediate-volume bioreactors with high surface-to-volume ratios were desirable for algal lipid production. For example, a closed bioreactor system consisting of 33 units of 200-L fermentors with an algal production rate of 4 g L^{-1} Day^{-1} and lipid contents of at least 50% of the CDW would be adequate for economic production of algal oils. Hess et al. [32] cited the commercial offering of algal oil-based surfactants, namely, Dehyton® AO 45 (BASF, Florham Park, NJ, USA), as a successful example of economical microalgal lipid production in a closed bioreactor system. In terms of the present discussion, it is thus economically feasible to obtain microalgal lipids suitable for tribological applications by adopting the existing bioreactor technologies for biofuels and biochemical production.

4.3   BACTERIAL LIPIDS AND HYDROXY FATTY ACIDS

Unlike the algal lipids, research activities on the production of industrial lipids using the prokaryotic bacteria only intensified starting in the mid-1990s. Many bacteria were previously studied for their production of a specific class of lipidic biodegradable polymers called poly(hydroxyalkanoates) (PHAs) [34,35]. The occurrence of intracellular non-PHA lipids was first reported in Mycobacterium [36], and then in Streptomyces [37,38], and Acinetobacter [39,40]. Alvarez and colleagues [41,42] subsequently reported in detail the high-level accumulation of non-PHA lipid in the inclusion bodies of gram-positive *Rhocococcus opacus* and *Rhodococcus ruber.* They reported that *R. opacus* strain PD630 produced lipids at a high level of 78%–87% of CDW when grown on gluconate or olive oil as the carbon source [41]. Furthermore, they determined that 98% (w/w) of the lipid produced by *R. opacus* were neutral lipids having palmitic acid (C16:0), oleic acid (C18:1), and margaric acid (C17:0) as the predominant fatty acid components [41]. This high lipid content rivals the lipid content of many oleaginous microalgae, leading to a heightened interest in the research community to identify and study additional oleaginous bacteria. Table 4.3 lists a few representative bacteria that have been shown to produce lipids. A few noteworthy observations could be deduced from the tabulation. The majority of these microorganisms belonged to the gram-positive group of bacteria that possess a thick polysaccharide cell wall. Since the lipid inclusion bodies in which the lipids are stored are intracellular, this could pose potential challenges and add to the costs of breaking apart or lysing the bacterial cells in order to harvest the lipid. A second observation was that the fatty acid composition of the lipids in these bacteria consisted of C16- to C18-chain length species commonly found in VOAFs that were investigated for applications in lubricant formulations [43–46]. An important observation was that the high lipid contents occurred in the *Rhodococcus* and the "blue-green algae" (i.e., Cyanobacterium and Synechococcus) species, which helped to focus further research and development efforts on potential industrial products from these bacteria. In fact, Table 4.3 shows that about one-third of the listed bacteria belong to the *Rhodococcus* species. Simple substrates such as hexadecane, gluconate, and dodecane could be used by the *Rhodococcus* bacteria to synthesize lipids reaching contents of 43%–84% of the CDW (Table 4.3). Low-cost waste streams such as Kraft hardwood pulp, biomass

TABLE 4.3
Representative Bacteria and Their Lipid Contents

Bacterium	Carbon Sources	Lipid Content (% CDW) and Major Fatty Acids	References
Gram-positive			
Bacillus subtilis	nutrient broth (nitrogen-deficient)	33.4%; C18:1, C16:0, C18:0	[51]
Dietzia maris	hexadecane	19.2%; C16:0, C16:1	[52]
Gordonia amarae	gluconate	6.1%; C18:1, C16:0	[52]
	hexadecane	5.1%; C16:0, C16:1	
Gordonia sp. DG	dodecane	69%; C18:3, 18:0, C18:1	[53]
Mycobacterium avium	palmitic acid	5%; C16:0, C16:1, C18:1	[36]
Nocardia asteroides	gluconate	12.2%; C16:0, C18:1	[52]
Nocardia corallina	valerate	23.9%; C18:1, C17:1, C17:0, C16:0	[54]
Nocardia globerula	gluconate	18.6%; C16:0, C18:1, C18:0	[52]
Nocardia restricta	gluconate	19.3%; C16:0, C18:1, C18:0	[52]
Rhodococcus erythropolis	hexadecane	43.4%; C16:0, C16:1	[52]
Rhodococcus fascians	hexadecane	12.9%; C16:0, C16:1, C18:1	[52]
Rhodococcus opacus	gluconate	76%; C16:0, C18:1	[50]
Rhodococcus opacus	dodecane	84%; 18:0, C18:1	[53]
Rhodococcus opacus Xsp8[a]	KHP[b]	45.8%; C16:0, C18:1, C17:1	[55]
Rhodococcus opacus DSM 43205	BGW[b]	62.8%; C16:0, C18:0, C20:0	[56]
Rhodococcus opacus PD630	dairy wastewater	51%; C18:0, C16:0	[57]
Rhodococcus ruber	hexadecane	26.0%; C16:0, C14:0, C16:1	[58]
Rhodococcus sp. strain 20	hexadecane	8.1%; C16:0, C16:1	[52]
	gluconate	7.6%; C16:0, C18:1, C18:0	
Streptomyces coelicolor	glucose	3.8%[c]; C17:0, C18:0	[37]
Streptomyces lividans	glucose	4.6%[c]; C17:0, C16:0, C15:0	
Streptomyces albus	glucose	2.2%[c]; C18:0	
Streptomyces griseus	glucose	2.8%[c]; C18:0	

(Continued)

TABLE 4.3 (Continued)
Representative Bacteria and Their Lipid Contents

Bacterium	Carbon Sources	Lipid Content (% CDW) and Major Fatty Acids	References
Gram-negative			
Alcanivorax borkumensis	pyruvate	23.2%; C16:0, C18:1, C16:1	[59]
Acinetobacter sp. strain 211	olive oil	25.0%; C18:1, C16:1; C16:0	[54]
Pseudomonas aeruginosa strain 44T1	olive oil	38.0%; C18:1, C16:0	[60]
Rhodobacter sphaeroides	lactate	35.0%; C18:1, C16:0, C18:0	[61]
Rhodopseudomonas palustris	malic acid	39.0%; (n.a.)	[62]
"Blue-green algae"			
Cyanobacterium aponinum	BG-11 medium (CO_2)	45.0%; C18:0, C16:0	[63]
Synechococcus sp. HSO1	BG-11 medium + ostrich oil (CO_2)	32.0%; C16:1, C18:2, C17:0	[64]

Source: Qadeer, S. et al., *J. Cleaner Prod.*, 168, 917–928, 2017; Alvarez, H. and Steinbüchel, A., *Appl. Microbiol. Biotechnol.*, 60, 367–376, 2002.

Note: n.a.—not available.

[a] Genetically or metabolically engineered strain.

[b] KHP: Kraft hardwood pulp (unbleached); BGW: biomass gasification wastewater.

[c] Values were calculated based on data in [37].

gasification wastewater, and dairy wastewater were equally suitable to achieve high lipid yields in the *Rhodococcus* species (Table 4.3). The commercial feasibility of using a *Rhodococcus* species, namely, *R. opacus* strain PD630, to produce industrial lipid was tested at a 500-L pilot plant scale [47]. Cells with 38.4% CDW of lipids in the form of TAGs were obtained at a high cell density of 18.4 g CDW/L, effectively yielding TAG at a volumetric productivity of ~3.5 kg (~4 L at an average density of 0.9 kg/L) in the process. Metabolic and/or genetic engineering could be used to greatly increase the lipid productivity of the producing bacteria [48]. A few commercial ventures, notably the former LS9 Inc. (San Carlos, CA), to produce bacterial oils using metabolically engineered strains were founded in the United States in the mid-2000s. Unfortunately, the volatile petroleum geopolitical and economic conditions were not favorable to commercial success of these companies targeting the huge-volume transportation fuel market, leading to the opportunity to direct the marketing focus to higher-value intermediate-volume industrial products including the biolubricants.

4.4   YEAST GLYCOLIPIDS AND HYDROXY FATTY ACIDS

Biobased-hydroxy fatty acids such as ricinoleic acid and lesquerolic acid that are derived from plants make ideal substrates for biolubricant synthesis and/or additives [65,66]. However, these hydroxy fatty acids suffer from the lack of a cropping system in the United States to grow the seeds, as well as from the presence of potentially toxic byproducts within the seed oil [67]. Aside from TAGs, some yeasts (and certain *Pseudomonas* bacteria) produce glycolipids such as sophorolipids (SLs) and mannosylerythritol lipids (MELs) often in high product yields. For example, *Starmerella bombicola* (formerly *Candida bombicola*) synthesizes extracellular SLs in volumetric yields reportedly as high as 400 g/L under appropriate fermentation conditions [68]. The potential tribological applications of microbial glycolipids (MGLs) and derivatives have been recently reviewed [69]. In this section we will highlight the tribological applications of estolides that could be synthesized using SLs as starting materials. SL molecules are composed of a sophorose sugar as well as a hydroxy fatty acid tail. Typically, the hydroxy group on the fatty acid tail of SLs that are synthesized by *S. bombicola* is located at the ω-1 position, which is relatively unique among natural fatty acids. Since SLs have been extensively studied and production protocols optimized owing to their application potential as biosurfactants, there are advantages in utilizing the hydroxy fatty acids from SLs in biolubrication applications. These advantages include the relative ease of producing the ω-1 hydroxy fatty acids and the ease of converting these molecules into other derivatives suitable for lubrication applications [70]. Estolides (intermolecular ester derivatives of fatty acids that are generally composed of at least two fatty acid molecules) are typically considered as effective substitutes/additives in biobased functional fluids [71]. Zerkowski et al. [72] described an exquisite reaction strategy to synthesize structured estolides using as starting materials the 17-hydroxy stearic acid and 17-hydroxy oleic acid that were first obtained through hydrolysis of SLs from *S. bombicola* fermentation. The reaction strategy entailed the adoption of ester-forming reactions such as carbodiimide coupling and a modified Yamaguchi symmetrical anhydride method, which together allowed for the control of the length and the sequence of the resultant structured estolides. Ashby et al. [73] further demonstrated the synthesis of biobased estolides from the hydroxy fatty acids derived from SLs. In that study [73], SLs were produced via fermentation using either oleic acid or linoleic acid as the hydrophobic carbon source. These fatty acids were chosen in order to produce molecules containing higher concentrations of unsaturated ω-1 hydroxy fatty acids [74]. The fatty acids were then liberated from the sophorose sugar through acid-catalyzed alcoholysis and used to synthesize both unsaturated and epoxy estolides; these compounds were shown to be effective plasticizers for biopolymer films.

4.5 PROSPECTIVE OF MICROBIAL LIPIDS IN TRIBOLOGICAL APPLICATIONS

The overview of the production of microbial lipids presented in this mini-review showed that many microalgae and bacterial species could biosynthesize glycolipids with fatty acid profiles that are suitable for biolubricant and other tribological applications. The key techno-economic hurdle remains the production of glycolipids at competitive cost and in yields high enough to compete with current petroleum-based products. Metabolic/genetic engineering approaches could help tackle this hurdle and also lead to the production of fatty acid profiles better suited for tribological applications. The competition from the petro-based mineral oils remains an uncertain and tough challenge to the VOAFs as well as to the microbial-based biolubricants.

Unlike vegetable oils, which contain fatty acids (or lipids) and glycerol in their structures, MGLs comprise various combinations of fatty acids and sugars in their structures (Figure 4.1). While there is a great deal of interest in the potential lubricant application of the lipid component, the effect of the unique structure of MGLs on lubrication has yet to be explored in a meaningful way. Recently, based on published properties, a list of potential application areas for MGLs in lubricant formulations was outlined [69]. According to this outline, glycolipid application areas can be categorized into two classes: (a) applications where, based on previously known properties, MGLs, without further chemical or enzymatic modifications, can be used as lubricant additives, such as biocides, emulsifiers, and friction modifiers; and (b) applications where chemically and/or enzymatically modified MGLs and/or corresponding lipids and sugars can be developed as biobased base oils or any number of biobased lubricant additives.

A recent literature search showed a total of only three publications in which the glycolipids (i.e., SL and rhamnolipids [RLs]) were used in both types of applications discussed above. A Chinese patent invention disclosure by Ren (2015) [75] discusses

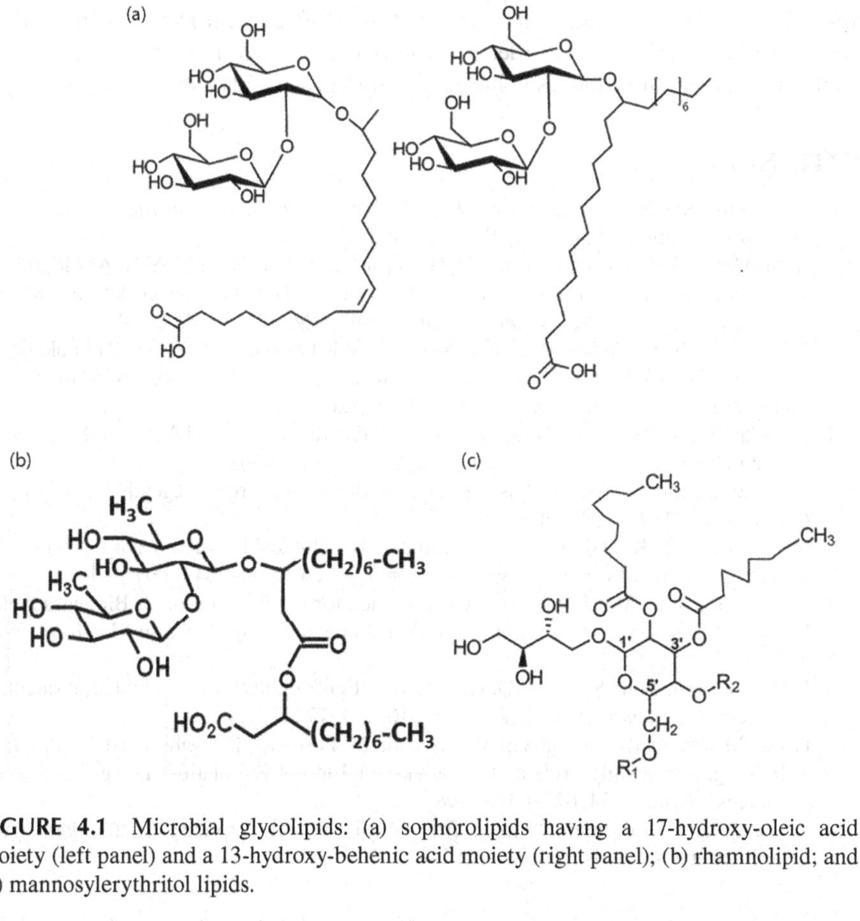

FIGURE 4.1 Microbial glycolipids: (a) sophorolipids having a 17-hydroxy-oleic acid moiety (left panel) and a 13-hydroxy-behenic acid moiety (right panel); (b) rhamnolipid; and (c) mannosylerythritol lipids.

an electric hammer cylinder lubricating grease formulation based on jojoba oil, with a soap/surfactant thickener mixture that includes 3–5 parts by weight of RL. It is claimed that the grease displayed a number of superior properties including excellent antiwear, load-carrying, stability, as well as resistance to wear, corrosion, and adhesion. Another Chinese patent invention disclosure by Chen (2015) [76] discusses a high-performance water-based metal-cutting fluid, with a surfactant/emulsifier package that includes 1%–10% (w/w) of SL. It is claimed that this high-performance fluid displayed excellent cleaning, antiseptic, cooling, lubricity, hard-water resistance, and storage stability. Sturms et al. [77] investigated the friction modifier additive properties of SL, two ω-1 hydroxy fatty acid methyl esters (FAMEs) synthesized from the SL, olive oil, and olive oil FAMEs. Blends with 1% and 5% (w/w) of additives in an API Group III base oil (NEXBASE 3043), were tested on a reciprocating ball-on-flat tribometer at room temperature, using steel/steel and SiC/Al ball flat friction surfaces. For the test conditions (load, stroke length, speed, contact pressure, number of test cycles) applied in the study, the coefficient of friction of the blends (1%–5%), relative to the pure base oil, was slightly lower for the SiC-Al test, but showed no change for steel/steel test. The wear depth tests showed lower wear values for all the 5% blends relative to the pure base oil, with the wear depth increasing in the order: olive oil ~ SL FAMEs < olive oil FAMEs < SL. A great deal more investigations will need to be carried out in order to build our knowledge base and successfully develop cost-competitive biobased lubricants from glycolipids and their derivatives.

REFERENCES

1. F. D. Gunstone, *The Chemistry of Oils and Fats: Sources, Composition, Properties, and Uses.* Blackwell Publishing, Boca Raton, FL (2004).
2. H. M. Mobarak, E. Niza Mohamad, H. H. Masjuki, M. A. Kalam, K. A. H. Al Mahmud, M. Habibullah and A. M. Ashraful, "The Prospects of Biolubricants as Alternatives in Automotive Applications," *Renew. Sustain. Energy Rev.*, 33, 34–43 (2014).
3. S. M. Alves, B. S. Barros, M. F. Trajano, K. S. B. Ribeiro and E. Moura, "Tribological Behavior of Vegetable Oil-Based Lubricants with Nanoparticles of Oxides in Boundary Lubrication Conditions," *Tribol. Int.*, 65, 28–36 (2013).
4. S. Asadauskas, J. M. Perez and J. L. Duda, "Oxidative Stability and Antiwear Properties of High Oleic Vegetable Oils," *Lubr. Eng.*, 52, 877–82 (1996).
5. S. Z. Erhan and S. Asadauskas, "Lubricant Basestocks from Vegetable Oils," *Ind. Crops Prod.*, 11, 277–282 (2000).
6. N. H. Jayadas, P. K. Nair and G. Ajithkumar, "Tribological Evaluations of Coconut Oil as an Environment-Friendly Lubricant," *Tribol. Int.*, 40, 350–354 (2007).
7. P. S. Lathi and B. Mattiasson, "Green Approach for the Preparation of Biodegradable Lubricant Basestock from Epoxidized Vegetable Oil," *Appl. Catal. Environ.*, 69, 207–212 (2007).
8. P. Nagendramma and S. Kaul, "Development of Ecofriendly/Biodegradable Lubricants: An Overview," *Renew. Sust. Energy Rev.*, 16, 764–774 (2012).
9. Q. Hu, M. Sommerfeld, E. Jarvis, M. Ghirardi, M. Posewitz, M. Seibert and A. Darzins, "Microalgal Triacylglycerols as Feedstocks for Biofuel Production: Perspectives and Advances," *Plant J.*, 54, 621–639 (2008).
10. A. Dean and J. Pittman, "Lipids from Algae: Novel Applications and Powerful Quantification Methods," *Inform*, 26, 14–17 (2015).

11. J. Sheehan, T. Dunahay, J. Benemann and P. Roessler, "A Look Back at the U.S. Department of Energy's Aquatic Species Program: Biodiesel from Algae," NREL/ TP-580-24190 (1998).

12. E. P. Knoshaug, R. Sestric, E. Jarvis, Y.-C. Chou, P. T. Pienkos and A. Darzins, "Current Status of the Department of Energy's Aquatic Species Program Lipid-Focused Algae Collection," *Abstract of the 31st Symposium on Biotechnology for Fuels and Chemicals.* NREL/PO-510-45788 (2009).

13. V. Andruleviciute, V. Makareviciene, V. Skorupskaite and M. Gumbyte, "Biomass and Oil Content of *Chlorella* sp., *Haematococcus* sp., *Nannochloris* sp. and *Scenedesmus* sp. under Mixotrophic Growth Conditions in the Presence of Technical Glycerol," *J. Appl. Phycol.*, 26, 83–90 (2014).

14. P. Singh, A. Guldhe, S. Kumari, I. Rawat and F. Bux, "Investigation of Combined Effect of Nitrogen, Phosphorus and Iron on Lipid Productivity of Microalgae *Ankistrodesmus Falcatus* KJ671624 using Response Surface Methodology," *Biochem. Eng. J.*, 94, 22–29 (2015).

15. T. Beer, D. Batten, J. Volkman, G. Dunstan and S Blackburn, "Biodiesel from Algae," in: Regional Forum on Bioenergy Sector Development: Challenges, Opportunities, and Way Forward, Report from the United Nations Economic and Social Commission for Asia and the Pacific-Asian and Pacific Centre for Agricultural Engineering and Machinery (ESCAP-APCAEM), pp 4–12 (2008).

16. X.-W. Wang, J.-R. Liang, C.-S. Luo, C.-P. Chen and Y.-H. Gao, "Biomass, Total Lipid Production, and Fatty Acid Composition of the Marine Diatom *Chaetoceros Muelleri* in Response to Different CO_2 Levels," *Bioresource Technol.*, 161, 124–130 (2014).

17. R. A. I. Abou-Shanab, J.-H. Hwang, Y. Cho, B. Min and B.-H. Jeon, "Characterization of Microalgal Species Isolated from Fresh Water Bodies as a Potential Source for Biodiesel Production," *Appl. Energy*, 88, 3300–3306 (2011).

18. M. G. Tadros and J. R. Johansen, "Physiological Characterization of Six Lipid-Producing Diatoms from the Southeastern United States," *J. Phycol.*, 24, 445–452 (1988).

19. S. Sriharan, D. Bagga and M. Nawaz, "The Effects of Nutrients and Temperature on Biomass, Growth, Lipid Production, and Fatty Acid Composition of *Cyclotella cryptica* Reimann, Lewin, and Guillard," *Appl. Biochem. Biotechnol.*, 28, 317–326 (1991).

20. Q. He, H. Yang, L. Wu and C. Hu, "Effect of Light Intensity on Physiological Changes, Carbon Allocation and Neutral Lipid Accumulation in Oleaginous Microalgae," *Bioresource Technol.*, 191, 219–228 (2015).

21. C. H. Ra, P. Sirisuk, J.-H. Jung, G.-T. Jeong and S.-K. Kim, "Effects of Light-Emitting Diode (LED) with a Mixture of Wavelengths on the Growth and Lipid Content of Microalgae," *Bioprocess Biosyst. Eng.*, 41, 457–465 (2018).

22. P. Das, W. Lei, S. S. Aziz and J. P. Obbard, "Enhanced Algae Growth in Both Phototrophic and Mixotrophic Culture under Blue Light," *Bioresource Technol.*, 102, 3883–3887 (2011).

23. X.-L. Li, T. K. Marella, L. Tao, R. Li, A. Tiwari and G. Li, "Optimization of Growth Conditions and Fatty Acid Analysis for Three Freshwater Diatom Isolates," *Phycol. Res.*, 65, 177–187 (2017).

24. M. M. Joseph, K. R. Renjith, G. John, S. M. Nair and N. Chandramohanakumar, "Biodiesel Prospective of Five Diatom Strains using Growth Parameters and Fatty Acid Profiles," *Biofuels*, 8, 81–89 (2017).

25. S. Vidyashankar, K. S. VenuGopal, G. V. Swarnalatha, M. D. Kavitha, V. S. Chauhan, R. Ravi, A. K. Bansal, R. Singh, A. Pande, G. A. Ravishankar and R. Sarada, "Characterization of Fatty Acids and Hydrocarbons of *Chlorophycean* Microalgae Towards Their Use as Biofuel Source," *Biomass Bioenergy*, 77, 75–91 (2015).

26. M. Yang, Y. Fan, P.-C. Wu, Y.-D. Chu, P.-L. Shen, S. Xue and Z.-Y. Chi, "An Extended Approach to Quantify Triacylglycerol in Microalgae by Characteristic Fatty Acids," *Front. Plant Sci.*, 8, Article no. 1949 (2017).

27. N. R. Moheimani, A. Isdepsky, J. Lisec, E. Raes and M. A. Borowitzka, "*Coccolithophorid* Algae Culture in Closed Photobioreactors," *Biotechnol. Bioeng.*, 108, 2078–2087 (2011).

28. P. Bondioli, L. Della Bella, G. Rivolta, G. Chini Zittelli, N. Bassi, L. Rodolfi, D. Casini, M. Prussi, D. Chiaramonti and M. R. Tredici, "Oil Production by The Marine Microalgae *Nannochloropsis* sp. F&M-M24 and *Tetraselmis suecica* F&M-M33," *Bioresource Technol.*, 114, 567–572 (2012).

29. X. Xiang, A. Ozkan, O. Chiriboga, N. Chotyakul and C. Kelly, "Techno-Economic Analysis of Glucosamine and Lipid Production from Marine Diatom *Cyclotella* sp.," *Bioresource Technol.*, 244, 1480–1488 (2017).

30. J. Feng, J. Cheng, R. Cheng, C. Zhang, J. Zhou and K. Cen, "Screening the Diatom *Nitzschia* sp. Re-Mutated by ^{137}Cs-γ Irradiation and Optimizing Growth Conditions to Increase Lipid Productivity," *J. Appl. Phycol.*, 27, 661–672 (2015).

31. Y. Maeda, D. Nojima, T. Yoshino and T. Tanaka, 'Structure and Properties of Oil Bodies in Diatoms," *Philos. Trans. Royal Soc. B*, 372, Article no. 20160408 (2017).

32. S. K. Hess, B. Lepetit, P. G. Kroth and S. Mecking, "Production of Chemicals from Microalgae Lipids—Status and Perspectives," *Eur. J. Lipid Sci. Technol.*, 120, Article no. 1700152 (2018).

33. T. Granata, "Dependency of Microalgal Production on Biomass and the Relationship to Yield and Bioreactor Scale-up for Biofuels: A Statistical Analysis of 60+ Years of Algal Bioreactor Data," *Bioenergy Res.*, 10, 267–287 (2017).

34. A. Steinbüchel and S. Hein, "Biochemical and Molecular Basis of Microbial Synthesis of Polyhydroxyalkanoates in Microorganisms," *Adv. Biochem. Eng. Biotechnol.*, 71, 81–123 (2001).

35. G.-Q. Chen and X.-R. Jiang, "Engineering Microorganisms for Improving Polyhydroxyalkanoate Biosynthesis," *Curr. Opin. Biotechnol.*, 53, 20–25 (2018).

36. L. Barksdale and K. S. Kim, "Mycobacterium," *Bacteriol. Rev.*, 41, 217–372 (1977).

37. E. R. Olukoshi and N. M. Packter, "Importance of Stored Triacylglycerols in *Streptomyces*: Possible Carbon Source for Antibiotics," *Microbiology*, 140, 931–943 (1994).

38. N. M. Packter and E. R. Olukoshi, "Ultrastructural Studies of Neutral Lipid Localisation in *Streptomyces*," *Arch. Microbiol.*, 164, 420–427 (1995).

39. R. A. Makula, P. J. Lockwood and W. R. Finnerty, "Comparative Analysis of the Lipids of *Acinetobacter* Species Grown on Hexadecane," *J. Bacteriol.*, 121, 250–258 (1975).

40. C. C. L. Scott and W. R. Finnerty, "Characterization of Intracytoplasmic Hydrocarbon Inclusions from the Hydrocarbon Oxidizing *Acinetobacter* Species HO1 N," *J. Bacteriol.*, 127, 481–489 (1975).

41. H. M. Alvarez, F. Mayer, D. Fabritius and A. Steinbüchel, "Formation of Intracytoplasmic Lipid Inclusions by *Rhodococcus opacus* Strain PD630," *Arch. Microbiol.*, 165, 377–386 (1996).

42. R. Kalscheuer, M. Wältermann, H. Alvarez and A. Steinbüchel, "Preparative Isolation of Lipid Inclusions from *Rhodococcus opacus* and *Rhodococcus ruber* and Identification of Granule-Associated Proteins," *Arch. Microbiol.*, 177, 20–28 (2002).

43. A. E., Atabani, A. S. Silitonga, H. C. Ong, T. M. I. Mahlia, H. H. Masjuki and H. Fayaz, "Non-Edible Vegetable Oils: A Critical Evaluation of Oil Extraction, Fatty Acid Compositions, Biodiesel Production, Characteristics, Engine Performance and Emissions Production," *Renew. Sust. Energ Rev.*, 18, 211–245 (2013).

44. L. A. Quinchia, M. A., Delgado, T. Reddyhoff, C. Gallegos and H. A. Spikes, "Tribological Studies of Potential Oil-Based Lubricants Containing Environmentally Friendly Viscosity Modifiers," *Tribol. Int.* 69, 110–117 (2014).

45. A. Adhvaryu, S. Z. Erhan and J. M. Perez, "Tribological Studies of Thermally and Chemically Modified Vegetable Oils for Use as Environmentally Friendly Lubricants," *Wear*, 257, 359–367 (2004).

46. G. Biresaw, "Elastohydrodynamic Properties of Seed Oils," *J. Am. Oil Chem. Soc.*, 83, 559–566 (2006).

47. I. Voss and A. Steinbüchel, "High Cell Density Cultivation of *Rhodococcus opacus* for Lipid Production at a Pilot-Plant Scale," *Appl. Microbiol. Biotechnol.*, 55, 547–555 (2001).

48. M.-H. Liang and J.-G. Jiang, "Advancing Oleaginous Microorganisms to Produce Lipid via Metabolic Engineering Technology," *Prog. Lipid Res.*, 52, 395–408 (2013).

49. S. Qadeer, A. Khalid, S. Mahmood, M. Anjum and Z. Ahmad, "Utilizing Oleaginous Bacteria and Fungi for Cleaner Energy Production," *J. Clean. Prod.*, 168, 917–928 (2017).

50. H. Alvarez and A. Steinbüchel, "Triacylglycerols in Prokaryotic Microorganisms," *Appl. Microbiol. Biotechnol.*, 60, 367–376 (2002).

51. S. Patnayak and A. Sree, "Screening of Bacterial Associates of Marine Sponges for Single Cell Oil and PUFA," *Lett. Appl. Microbiol.*, 40, 358–363 (2005).

52. H. M. Alvare, "Relationship Between β-Oxidation Pathway and the Hydrocarbon-Degrading Profile in *Actinomycetes* Bacteria," *Int. Biodeterior. Biodegradation*, 52, 35–42 (2003).

53. M. K. Gouda, S. H. Omar and L. M. Aouad, "Single Cell Oil Production by *Gordonia* sp. DG using Agro-Industrial Wastes," *World J. Microbiol. Biotechnol.*, 24, 1703–1711 (2008).

54. H. M. Alvarez, O. H. Pucci and A. Steinbüchel, "Lipid Storage Compounds in Marine Bacteria," *Appl. Microbiol. Biotechnol.*, 47, 132–139 (1997).

55. K. Kurosawa, S. J. Wewetzer and A. J. Sinskey, "Engineering Xylose Metabolism in Triacylglycerol-Producing *Rhodococcus Opacus* for Lignocellulosic Fuel Production," *Biotechnol. Biofuels*, 6, Article no. 134 (2013).

56. L. Goswami, M. M. T. Namboodiri, R. V. Kumar, K. Pakshirajan and G. Pugazhenthi, "Biodiesel Production Potential of Oleaginous *Rhodococcus Opacus* Grown on Biomass Gasification Wastewater," *Renew. Energ.*, 105, 400–406 (2017).

57. S. Kumar, N. Gupta and K. Pakshirajan, "Simultaneous Lipid Production and Dairy Wastewater Treatment using *Rhodococcus opacus* in a Batch Bioreactor for Potential Biodiesel Application," *J. Environ. Chem. Eng.*, 3, 1630–1636 (2015).

58. H. M. Alvarez, R. Kalscheuer and A. Steinbuchel, "Accumulation of Storage Lipids in Species of *Rhodococcus* and *Nocardia* and Effect of Inhibitors and Polyethylene Glycol," *Fett/Lipid*, 99, 239–246 (1997).

59. R. Kalscheuer, T. Stöveken, U. Malkus, R. Reichelt, P. N. Golyshin, J. S. Sabirova, M. Ferrer, K. N. Timmis and A. Steinbüchel, "Analysis of Storage Lipid Accumulation in *Alcanivorax borkumensis*: Evidence for Alternative Triacylglycerol Biosynthesis Routes in Bacteria," *J. Bacteriol.*, 189, 918–928 (2007).

60. C. de Andrés, M. J. Espuny, M. Robert, M. E. Mercadé, A. Manresa and J. Guinea, "Cellular Lipid Accumulation by *Pseudomonas aeruginosa* 44T1," *Appl. Microbiol. Biotechnol.*, 35, 813–816 (1991).

61. D.-H. Kim, J.-H. Lee, Y. Hwang, S. Kang and M.-S. Kim, "Continuous Cultivation of Photosynthetic Bacteria for Fatty Acids Production," *Bioresource Technol.*, 148, 277–282 (2013).

62. P. Carlozzi, A. Buccioni, S. Minieri, B. Pushparaj, R. Piccardi, A. Ena and C. Pintucci, "Production of Bio-Fuels (Hydrogen and Lipids) Through a Photofermentation Process," *Bioresource Technol.*, 101, 3115–3120 (2010).

63. S. E. Karatay and G. Dönmez, "Microbial Oil Production from Thermophile Cyanobacteria for Biodiesel Production," *Appl. Energy*, 88, 3632–3635 (2011).

64. S. Modiri, H. Sharafi, L. Alidoust, H. Hajfarajollah, O. Haghighi, A. Azarivand, Z. Zamanzadeh, H. S. Zahiri, H. Vali and K. A. Noghabi, "Lipid Production and Mixotrophic Growth Features of Cyanobacterial Strains Isolated from Various Aquatic Sites," *Microbiology*, 161, 662–673 (2015).

65. S. C. Cermak, K. B. Brandon and T. A. Isbell, "Synthesis and Physical Properties of Estolides from Lesquerella and Castor Fatty Acid Esters," *Ind. Crops Prod.*, 23, 54–64 (2006).

66. T. A. Isbell, B. A. Lowery, S. S. DeKeyser, M. L. Winchell and S. C. Cermak, "Physical Properties of Triglyceride Estolides from Lesquerella and Castor Oils," *Ind. Crops Prod.*, 23, 256–263 (2006).

67. D. M. Schieltz, L. G. McWilliams, Z. Kuklenyik, S. M. Prezioso, A. J. Carter, Y. M. Williamson, S. C. McGrath, S. A. Morse and J. R. Barr, "Quantification of Ricin, RCA and Comparison of Enzymatic Activity in 18 *Ricinus Communis* Cultivars by Isotope Dilution Mass Spectrometry," *Toxicon*, 95, 72–83 (2015).

68. H.-J. Daniel, M. Reuss and C. Syldatk, "Production of Sophorolipids in High Concentration from Deproteinized Whey and Rapeseed Oil in a Two Stage Fed Batch Process using *Candida bombicola* ATCC 22214 and *Cryptococcus curvatus* ATCC 20509," *Biotechnol. Lett.*, 20, 1153–1156 (1998).

69. D. K. Y. Solaiman, R. D. Ashby and G. Biresaw, "Biosynthesis and Derivatization of Microbial Glycolipids and Their Potential Application in Tribology," in: *Surfactants in Tribology*, Volume 5, G. Biresaw and K. L. Mittal (Eds.), pp. 263–288, Taylor & Francis Group, Boca Raton, FL. (2017).

70. A. Nunez, R. Ashby, T. A. Foglia and D. K. Y. Solaiman, "Analysis and Characterization of Sophorolipids by Liquid Chromatography with Atmospheric Pressure Chemical Ionization," *Chromatographia*, 53, 673–677 (2001).

71. S. Cermak and T. Isbell, "Estolides—The Next Biobased Functional Fluid," *Inform*, 15, 515–517 (2004).

72. J. A. Zerkowski, A. Nuñez and D. K. Y. Solaiman, "Structured Estolides: Control of Length and Sequence," *J. Am. Oil Chem. Soc.*, 85, 277–284 (2008).

73. R. D. Ashby, D. K. Y. Solaiman, C.-K. Liu, G. Strahan and N. Latona, "Sophorolipid-Derived Unsaturated and Epoxy Fatty Acid Estolides as Plasticizers for Poly(3-Hydroxybutyrate)," *J. Am. Oil Chem. Soc.*, 93, 347–358 (2016).

74. R. D. Ashby, D. K. Y. Solaiman and T. A. Foglia, "Property Control of Sophorolipids: Influence of Fatty Acid Substrate and Blending," *Biotechnol. Lett.*, 30, 1093–1100 (2008).

75. X. Ren, "Electric Hammer Gas Cylinder Lubricating Grease and Preparation Method Thereof," Chinese Patent Invention Disclosure, CN 104449987 A 20150325 (2015).

76. X. Chen, "High-Performance Metal Cutting Fluid," Chinese Patent Invention Disclosure CN 104403772 A 20150311 (2015).

77. R. Sturms, D. White, K. L. Vickerman, T. Hattery, S. Sundararajan, B. J. Nikolau and S. Garg, "Lubricant Properties of ω-1 Hydroxy Branched Fatty Acid-Containing Natural and Synthetic Lipids," *Tribol. Lett.*, 65, Article no. 99 (2017).

Section II

Adhesion, Wetting, and Their Relevance to Tribology

5 Moving Contact Line of Droplets on Structured Surfaces

Some Problems Relevant to Tribology

Ya-Pu Zhao and Zhanlong Wang

CONTENTS

5.1 INTRODUCTION

In recent years, droplet wetting and spreading have received significant interest from the scientific community due to their broad applications in, for example, self-assembly [1,2], droplet movement controlling [3,4], water directional transport [5,6], water harvesting [7,8], self-clean surfaces [9,10], printing, and anti-icing [11,12]. For these droplets, the roughness of a surface can significantly alter the behavior of fluids moving over that surface [13], for example, the leaves of plants where micrometer-scale bumps on the leaves lead to super-hydrophobic behavior [14], desert beetles who use hydrophilic patches on their back to collect dew [8], and butterfly wings that are patterned anisotropically to promote directional runoff [15]. On some special surfaces like nepenthes, droplets are easily moved and gnats are prone to slip off. These specially structured surfaces provide effective methods to control the droplet movement and help to fabricate ultrasmooth surfaces [6]. Based on these phenomena, surfaces with regular arrays of chemical patches [16–18] and posts [19–22] are fabricated to provide special functions. Previous works have shown the possibilities of manufacturing multi-scale surface patterns [23–25].

On patterned surfaces, the motion of drops is generally complex due to the contact line pinning and contact angle hysteresis. Surfaces with pillars often generate larger contact angles for partially wetting liquids and faster velocities for totally wetting liquids. On anisotropically patterned surfaces, the droplet is elongated in the direction parallel to texture, and it has an elliptic drop shape and different spreading behaviors [26,27]. On these surfaces, drop volume generally affects the drop anisotropy and contact angle hysteresis to some extent [21]. On chemically [16,28,29] nanostructured surfaces, there is an anisotropy in the average value of the contact angle and contact angle hysteresis. Elongated drop shapes were also obtained by Chen et al. [30] and Chung et al. [31] for hydrophobic and hydrophilic grooved surfaces, respectively.

Drops condensing on grooved surfaces also show similar elongated drop shapes during growth when the surface is hydrophilic [32]. At the nanometer scale, the anisotropy still persists [33]. The anisotropic behavior depends on the size of surface structure, and has a nonlinear relation with the surface morphology. For the dynamics of drops sliding on chemically striped and hydrophobic grooved surfaces, the sliding angles are considerably larger for drops moving perpendicular to the stripes [17,22]. In nature, the use of these specific wettability patterned surfaces helps many animals and plants in arid environments to collect humidity more efficiently to obtain water. In this review, we summarize the latest developments in wetting and electrowetting (EW) on structured surfaces, including pillar-arrayed and corrugated surfaces. The apparent contact angle, spreading scaling law, and EW state transformation are explained in detail.

5.2 THEORETICAL BACKGROUND

Wetting is a natural phenomenon, depicting the contact situation between a liquid and a solid. Droplet wetting situation can be characterized by contact angle. When the contact angle is 0°, the liquid wets the solid surface completely. When the

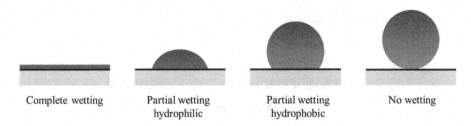

FIGURE 5.1 The cases of droplets wetting a solid surface. Droplets wet the solid surface completely, partially (hydrophilic), partially (hydrophobic), and not at all in the figures from left to right.

contact angle is larger than 0°, less than 180°, the liquid partially wets the solid surface. When the contact angle equals 180°, the liquid does not wet the solid surface. For the cases of partial wetting, the solid surfaces are hydrophilic when the contact angle is less than 90°, and hydrophobic when the contact angle is larger than 90° (Figure 5.1).

For a droplet deposited on a solid surface, the equilibrium is determined by the adhesion between droplet and solid. And the adhesion can be expressed as follows: $W_a = \gamma_{sv} + \gamma_{lv} - \gamma_{sl}$, where γ_{sv}, γ_{lv} and γ_{sl} represent the surface tensions of solid-vapor, liquid-vapor, and solid-liquid interfaces, respectively. The cohesion between droplet and solid surface can be expressed as $W_c = 2\gamma_{lv}$. Therefore, the spreading parameter can be expressed as

$$S = W_a - W_c = \gamma_{sv} - (\gamma_{sl} + \gamma_{lv}). \tag{5.1}$$

When the spreading parameter is larger than 0, namely, the adhesion is stronger than cohesion, the droplet wets the solid surface.

5.2.1 Bridging Length and Time Scales

Physical, mechanical, chemical, and biological processes span length and time scales of many orders of magnitude [34–39]. As a dynamic process, moving contact line (MCL) problems or droplet spreading span 7–8 length and time scales from atomistic to continuum (Figure 5.2) [35]. A grand challenge in MCL is linking these vastly different length and time scales.

All these length and time scales originate from the competition among various forces [41,42], which indicates that they can be bridged using dimensionless numbers (Table 5.1). Dimensionless numbers provide clear physical interpretation of MCL problems under study and also produce valuable scale estimates. In wetting, gravity, inertia force, viscous force, van der Waals force, and electric field force affect droplet spreading at characteristic lengths of 10^{-3} m, 10^{-4} m, 10^{-8} m, 10^{-10}-10^{-9} m, and 10^{-10}-10^{-8} m, respectively.

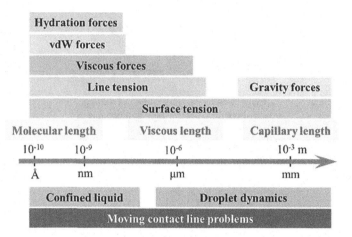

FIGURE 5.2 Schematic diagrams of various length scales in moving contact line problems. Ranges of several common forces at play in MCL problems are also illustrated. (Reprinted with permission from Zhao, Y.P., Bridging length and time scales in moving contact line problems. *Sci. China Phys. Mech.*, 59, 114631. Copyright 2016, Springer Nature.)

TABLE 5.1
Dimensionless Numbers and Characteristic Length Related to the Wetting

Force	Nondimensional Numbers	Characteristic Length (Water)
Gravity	Bond number: $\mathrm{Bo} = \rho g l_c^2 / \gamma_{lv}$	$l_c = \sqrt{\gamma_{lv}/\rho g}$ (10^{-3} m)
Inertia force	Weber number: $\mathrm{We} = \rho v^2 l_c / \gamma_{lv}$	$l_c = \gamma_{lv}/\rho v^2$ (10^{-4} m)
Viscous force	Ohnesorge number: $\mathrm{On} = \eta / \sqrt{\rho l_c \gamma_{lv}}$	$l_c = \eta^2/\rho \gamma_{lv}$ (10^{-8} m)
	Laplace number: $\mathrm{La} = \mathrm{On}^{-2}$	
Van der Waals force	Hamaker number: $\mathrm{Ha} = \dfrac{A}{\pi \gamma_{lv} l_c^2}$	$l_c = \sqrt{A/\gamma_{lv}}$ ($10^{-10} \sim 10^{-9}$ m)
Electric field force	EW number: $\eta_{EW} = \dfrac{\varepsilon V^2}{2 l_c \gamma_{lv}}$	$l_c = \dfrac{\varepsilon V^2}{\gamma_{lv}}$ ($10^{-10} \sim 10^{-8}$ m)

Abbreviations: ρ: density; g: acceleration of gravity; l_c: characteristic length; η: viscosity; A: Hamaker constant.

5.2.2 WETTING ON ROUGH SURFACES

In wetting, Young's equation is built on the assumption that a solid surface is smooth and homogeneous, and it does not take into account the surface roughness. However, the natural solid surfaces are generally rough in microscale and macroscale. Droplet wetting on rough surfaces exhibits different wetting and spreading behaviors than on smooth surfaces. Contact angles measured on rough surfaces are usually not consistent with that calculated by Young's equation. This phenomenon was found in many experiments and theoretical analysis in the twentieth century. Wenzel introduced a factor called "degree of roughness," which considers the surface roughness, and this factor is defined as the ratio of solid actual surface area to the projected surface area:

$$r_o = \frac{A_{sl(true)}}{A_{sl(projected)}} \tag{5.2}$$

where $A_{sl(true)}$ and $A_{sl(projected)}$ are the actual area and the projected area of solid surface, respectively. Obviously, $r_o \geq 1$, and increases with increasing roughness. The modified relation considering the surface roughness will be changed as follows:

$$\cos\theta_r = \frac{\mathrm{d}A_{lv}}{\mathrm{d}A_{sl(projected)}} = \frac{\mathrm{d}A_{sl(true)}}{\mathrm{d}A_{sl}}\frac{\mathrm{d}A_{lv}}{\mathrm{d}A_{sl(projected)}} = r_o\cos\theta_s \tag{5.3}$$

where θ_r and θ_s are the droplet contact angles on rough surface and smooth surface, respectively. The above equation is called the Wenzel equation, according to which, the rough surfaces will be more hydrophilic if the surface is originally hydrophilic, and more hydrophobic if the surface is originally hydrophobic. However, the Wenzel model does not consider the roughness morphology of the surface. Different roughness morphologies with the same roughness factor may cause different wetting behaviors. The wetting behavior due to different roughness morphologies was not been investigated until the discovery of fractal geometry.

In 1946, Cassie and Baxter further extended the Wenzel model, considering the rough surface as a composite surface that is composed of different materials, and the modified the wetting equation as follows:

$$\gamma_{lv}\cos\theta_r = f_1\left(\gamma_{sv1} - \gamma_{sl1}\right) + f_2\left(\gamma_{sv2} - \gamma_{sl2}\right) \tag{5.4}$$

or

$$\cos\theta_r = f_1\cos\theta_{f1} + f_2\cos\theta_{f2} \tag{5.5}$$

where f_1 and f_2 are the fractions of the surface area occupied by water and air, respectively. θ_{f1} and θ_{f2} are the contact angles of droplet on two materials that constitute the composite surface, respectively.

5.2.3 RIGID WETTING AND ELASTOWETTING

On rigid substrates, the static contact angle characterizes the wetting state of a droplet, and can be obtained from Young's equation. Young's contact angle can be obtained from the balance of surface tension vectors at the contact line of the droplet:

$$\vec{\gamma}_{sv} + \vec{\gamma}_{sl} + \vec{\gamma}_{lv} = 0 \tag{5.6}$$

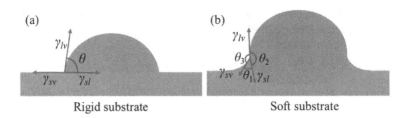

FIGURE 5.3 Droplet wetting states on (a) rigid and (b) soft substrates.

On ideally rigid substrates, the vertical forces of surface tensions can be ignored since they cause insignificant deformation of the solid surface. And Young's contact angle is built on the balance of the horizontal forces of surface tensions. The relation of forces in the horizontal direction is expressed as

$$\gamma_{sl} + \gamma_{lv} \cos\theta - \gamma_{sv} = 0 \tag{5.7}$$

As the solid elastic modulus decreases, the vertical forces increasingly deform the solid surface. For most soft materials, the vertical forces often cause significant deformation on the solid surfaces. Then, the relation of force balance at the contact line is described by Neumann's triangle [43]. Equation 5.6 leads to the following relation:

$$\frac{\gamma_{sl}}{\sin\theta_1} = \frac{\gamma_{sv}}{\sin\theta_2} = \frac{\gamma_{lv}}{\sin\theta_3} \tag{5.8}$$

θ_1, θ_2, and θ_3 are the angles depicted in Figure 5.3b. For droplets on immiscible liquid surface, the force balance at the contact line is also consistent with Neumann's triangle, for example, a silicone oil droplet on the surface of water.

5.2.4 WETTING ON ANISOTROPIC SURFACES

Anisotropic wetting happens generally on designed micro-channel surfaces, animal feather, leaves, and so on. The physical origin of anisotropic wetting is attributed to liquid contact line encountering physical discontinuity and chemical heterogeneity on the solid surfaces [44–46]. Micro- and nanometer scale topographical surface structures, such as parallel lines, grooves, wrinkles, and pillars, may translate to observable macroscopic changes in droplet wetting behavior. Similarly, surfaces with chemical anisotropies, such as parallel bands of chemical functionality with different surface energies, may influence droplet wetting behavior. In some cases, combining structural and chemical anisotropy produces unique wetting phenomena [44].

The droplet spreading on anisotropic surfaces is different from that on isotropic surfaces, and the wetting behavior and the static contact angles in different directions differ greatly. Based on these special phenomena, previous works have attempted to design surfaces that support movement of liquids in a single direction for guiding liquid flow, and the results show promise for enhancing microfluidic

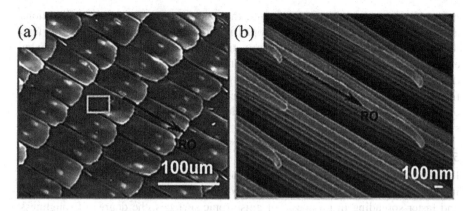

FIGURE 5.4 (a)–(b) SEM images of anisotropic wetting on a butterfly wing. This wing surface with anisotropic structure directs water to remove easily from wing. (Adapted from Zheng, Y. et al., Directional adhesion of superhydrophobic butterfly wings. *Soft Matter*, 3, 178–182. Copyright 2007, with permission from Royal Society of Chemistry.)

applications [47]. Gradients in chemical or structural surface features may also generate novel, complex anisotropic wetting behaviors. In nature, anisotropic wetting is observed on biological surfaces including certain plant leaves [48,49], flower petals [50,51], and bird feathers [52–54]. Super-hydrophobicity and anisotropic wetting of leaf surfaces encourage water droplet movement along the direction parallel to the leaf edge, while hindering movement in the direction perpendicular to the leaf edge.

In addition to plants, animals also exhibit and benefit from anisotropic wetting. Zheng et al. investigated directional liquid adhesion on super-hydrophobic butterfly wings [15]. In this instance, organized micro- and nanostructures facilitated water droplet removal from wings by promoting droplet movement along the radical outward direction of the body's central axis (Figure 5.4). The microscales and lamella-stacked nano-stripes result in anisotropic wetting characteristics [15].

5.3 EXPERIMENTS AND MOLECULAR DYNAMICS SIMULATIONS

5.3.1 WETTING DYNAMICS ON GROOVED SURFACES

When a droplet is deposited on a grooved surface, it is elongated in the direction parallel to the grooves, and pressed in the direction vertical to grooves. This leads to an elliptic shape. This is because of the absence of contact line pinning in the parallel direction and obvious contact line pinning in the vertical direction. When the surface structure changes, the shape of the droplet changes accordingly. Droplets on such grooved surfaces always have the special ability of directional transport, and this provides a useful way to control their movement. Usually, the contact angle of a droplet on solid surface is described by Young's equation, and the Wenzel and Cassie equations describe the effect of surface roughness on contact angle. However, a droplet placed on a grooved surface is also affected by the difference of topography in two different directions. As a result, the contact angles in different directions are different and the shape of the droplet is no longer spherical. There is obvious deficiency in the contact angle determination using the Wenzel or Cassie–Baxter equation. The droplet spreading in different directions on grooved substrates is also different. The velocity along the grooves is generally faster than in the direction vertical to the grooves. The spreading is accelerated under the effect of grooves. In this chapter, the apparent contact angle and velocity of droplet spreading on a grooved substrate are summarized.

5.3.1.1 Anisotropic Factor

Droplets on grooved surfaces often show some anisotropic wetting behavior, including varied apparent contact angle, elliptic contact line, stick-slip movement of MCL, and faster spreading in grooves. For anisotropic surfaces, the degree of roughness cannot totally reflect the effect of surface structure on the droplet wetting state. A parameter that accurately reflects the effect of surface structure anisotropy will be helpful to study wetting behavior on anisotropic surfaces. For a rough surface, the degree of roughness is defined as the ratio of the actual to projected surface areas. For the anisotropic surfaces, there are different degrees of roughness in vertical and parallel directions. A proposed anisotropic factor S_c was defined as follows [55]:

$$S_c = \frac{f_y}{f_x} \tag{5.9}$$

where f_x and f_y denote the roughnesses in the directions vertical and parallel to grooves, respectively. For grooved surfaces with a rectangular cross section, the roughness in the direction vertical to the grooves can be described as follows:

$$f_x = \frac{l_1 + l_2 + 2h_i}{l_1 + l_2} \tag{5.10}$$

where l_1 and l_2 are the width of grooves and ridge, respectively. h_i is the depth of the grooves. And in the direction parallel to the grooves, the roughness equals 1, and thus the anisotropic factor can be expressed as follows:

$$S_c = \frac{f_y}{f_x} = \frac{l_1 + l_2}{l_1 + l_2 + 2h} \tag{5.11}$$

For these corrugated surfaces the morphology of surface can be described as

$$w = A_m \cos \frac{2\pi x}{L} \tag{5.12}$$

where A_m and L are the amplitude and wavelength of the corrugated surface, respectively.

And the degree of roughness in the directions parallel and vertical to the grooves are defined as follows [55]:

$$f_x = (1/L) \int_0^L \sqrt{1 + (4\pi^2 A_m^2/L^2) \cos(2\pi x/L)^2} \, dx \qquad (5.13)$$

$$f_y = 1 \qquad (5.14)$$

And the anisotropic factor is defined as follows:

$$S_c = \frac{L}{\int_0^L \sqrt{1 + (4\pi^2 A_m^2/L^2) \cos(2\pi x/L)^2} \, dx} \qquad (5.15)$$

5.3.1.2 Static Contact Angle

A droplet on a grooved substrate has different apparent contact angles in different directions. The contact angle changes as the surface morphology varies. Previous works indicated that the contact angle in the direction parallel to the grooves was close to contact angle on the flat surface in microscale [30,33,56–58]. Kusumaatmaja et al. [59] showed that the contact angle in the parallel direction approximately obeys the Wenzel equation in many experiments. Despite numerous reports on contact angle in parallel direction, studies on contact angle in perpendicular direction are relatively few. Moreover, until now, the existing theories are not generally recognized. A systematic study of contact angle on a grooved surface from microscale to macroscale is needed.

The discrepancy in contact angles in different directions on a grooved surface brings inconvenience to the characterization of wetting properties. A uniform and effective method will be helpful. In the past, two contact angles of θ_x and θ_y where θ_x and θ_y are contact angles observed from perpendicular and parallel directions, respectively, were used to describe a droplet on a grooved surface [60,61]. However, this method is inconvenient in many aspects, for example, the definitions of hydrophobicity or hydrophilicity of grooved surface, the related wetting experiments in medicine, electronics, and so on. Fractal theory was adopted to describe the relationship between contact angle and surface morphology [49]. However, this theory does not consider the anisotropy of surface structure.

5.3.1.2.1 Static Contact Angle on a Corrugated Substrate

Droplets on grooved surfaces are usually elongated along the direction parallel to the groove, and contracted in the direction perpendicular to the groove. Thus, the droplet shows the shape of an elliptical cap for liquids with high surface tension, and a long strip for liquids with low surface tension (Figure 5.5). As the surface tension of liquid decreases, the droplet shape becomes more prolate [62].

The apparent contact angles of a droplet on a grooved surface are not equal in different viewing directions. In microscale, the contact angles in different viewing directions are the same. The variation in apparent contact angles from different viewing directions shows a monotonic decrease. For different grooved surfaces, the contact angles in the same viewing direction are different (Figure 5.6). When the anisotropic factor increases, the contact angle of droplet in the direction vertical to the grooves decreases, and the contact angle in the direction along the grooves increases.

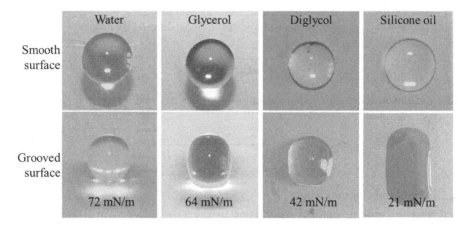

FIGURE 5.5 Droplet shapes of different surface tension liquids on smooth surfaces and grooved surfaces. From left to right, the liquids are water (72 mN/m), glycerol (64 mN/m), diglycol (42 mN/m), and silicone oil (21 mN/m), respectively. These droplets exhibit spherical caps on smooth surfaces and elongated shapes on grooved surfaces. Water, glycerol, and diglycol droplets show ellipsoidal cap shapes, and the silicone oil droplet shows a long strip shape on grooved surfaces. (From Wang, Z. et al., The effect of surface anisotropy on contact angles and the characterization of elliptical cap droplets. *Sci. China Technol. Sc.*, 61, 309–316. Copyright 2018, with permission from Springer Nature.)

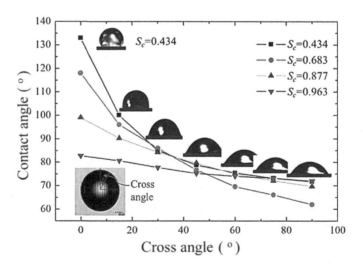

FIGURE 5.6 The change of contact angle with respect to cross angle on a corrugated substrate. Inserts: snapshots of a droplet from side view with $S_c = 0.434$. (From Wang, Z. and Zhao, Y.P., Wetting and electrowetting on corrugated substrates. *Phys. Fluids*, 29, 067101. Copyright 2017, with permission from American Institute of Physics.)

5.3.1.2.2 The Relation Between Contact Angle and Anisotropic Roughness

The anisotropic roughness affects contact angle on a grooved surface from microscale to macroscale [62]. The vertical coordinate represents the ratio of θ_c to θ_s, where θ_c is contact angle on grooved surface, and θ_s is contact angle on smooth surface. The horizontal coordinate represents the roughness ratio of parallel to vertical directions. Contact angles of a droplet on a grooved surface are larger than those on a flat surface with anisotropic rough morphology. Contact angles in vertical direction are larger than those in parallel direction. In microscale, contact angles change with surface morphology in a monotonous relationship. However, contact angles follow a nonlinear relation when the surface structure changes from microscale to macroscale.

On grooved substrates, contact angle first increases and then decreases with S_c in both vertical and parallel directions (Figure 5.7a). The trend of contact angle changing with S_c can be divided into two regions. contact angle increases with increasing S_c. In the first region, contact angle increases with increasing S_c. As S_c increases, the roughness plays an increasing role on contact angle. The reason is that a surface with infinitesimal roughness approaches a smooth surface.

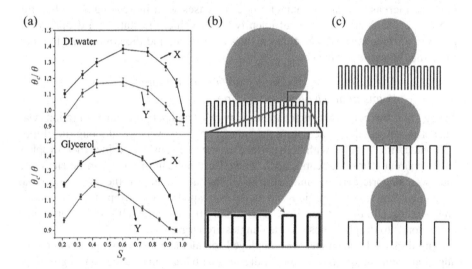

FIGURE 5.7 (a) The variation of contact angles on grooved substrate with rectangular cross section. Contact angles of DI water droplets (upper). Contact angles of glycerol droplets (lower). (b) Illustration of the propagation of a droplet on grooved surfaces. The droplet contact line will be driven to step forward under the attraction between droplet and the ridge. (c) A schematic diagram that illustrates the variation of contact angles of droplets on grooved surfaces with different groove width. The groove widths are (upper) 5 μm, (middle) 8 μm, (lower) 15 μm, and thus the semi-axes of the elliptic contact line in vertical direction are (upper) 27.5 μm, (middle) 20 μm, and (lower) 22 μm. The pinning force and contact angle increase with the decreasing semi-axis. The droplet on the grooved surface with groove width being 8 μm has largest pinning force compared to the other surfaces. (Adapted from Wang, Z. et al., The effect of surface anisotropy on contact angles and the characterization of elliptical cap droplets. *Sci. China Technol. Sc.*, 61, 309–316. Copyright 2018, with permission from Springer Nature.)

This phenomenon is interpreted with an example for illustration (Figure 5.7b). In the second region, contact angle decreases with increasing S_c, since the roughness effect was weakened with the increasing groove period. When the period of surface structure becomes infinite, the contact angle on a grooved surface is close to that on a smooth surface. Thus, there exists a critical value that distinguishes the two stages, and this value differs in vertical and parallel directions. The critical values are around 0.6 and 0.45 in vertical and parallel directions, respectively. The variation of contact angle of glycerol droplets is similar to that of deionized (DI) water droplets on a grooved surface [62].

For a grooved surface, S_c increases with increasing groove period (Figure 5.7). When a droplet is deposited on a grooved surface where the groove period is quite small, the contact line will be driven to step forward under the attraction between droplet and the next ridge in the vertical direction. The droplet reaches a stable state when the attractive force is balanced by the Laplace pressure. Driven by this attraction, the droplet propagates in vertical direction until it reaches equilibrium state. As the groove period decreases, the droplet propagates further in the vertical direction. For a droplet with a fixed volume, contact angle decreases as wetted area increases, that is, contact angle increases with increasing S_c. When the groove period decreases to a sufficiently small value, the state of a droplet on a grooved surface is similar to that on smooth surface, and the contact angle tends to be that on smooth surface.

5.3.1.2.3   The Calculation of Contact Angles at Arbitrary
Points of an Elliptic Contact Line

The relationship between contact angles in parallel and perpendicular directions was further investigated by calculating the contact angle at every point along the elliptical contact line. Liquids with high surface tension are considered since their droplets have apparent ellipsoidal cap shapes. The ellipsoidal cap droplets can be classified into two categories: (i) the small ellipsoidal cap model, and (ii) the large ellipsoidal cap model. The small ellipsoidal cap model corresponds to the droplet whose contact angles along elliptic contact line are less than 90°, and large ellipsoidal cap model corresponds to droplet whose contact angles are larger than 90°.

An ellipsoidal cap model was employed to calculate arbitrary apparent contact angles at points in the contact line for droplets with high surface tension (Figure 5.8). Ellipsoidal surfaces can be expressed with the function $x^2/a^2 + y^2/b^2 + z^2/c^2 = 1$ ($a > b > c > 0$ are the semi-major axes in X, Y, and Z directions, respectively). The cosine value of the angle between tangent plane and X-Y plane at any point (when X-Y plane is taken as droplet bottom, the angle will be contact angle) can be expressed as follows:

$$\cos\theta = \frac{2z_0/c^2}{\sqrt{\left(2x/a^2\right)^2 + \left(2y/b^2\right)^2 + \left(2z/c^2\right)^2}} \tag{5.16}$$

where $a, b, c,$ and z_0 are constants once the surface structure is fixed. The calculation of contact angles is based on the condition that $z = z_0$, while z_0 is the lowest level of

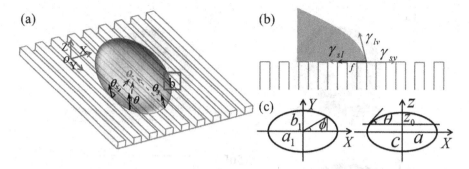

FIGURE 5.8 Schematic of ellipsoidal cap model. (a) θ_x, θ_y, and θ are shown. A mathematical model of ellipsoidal cap is used to calculate arbitrary contact angles along the elliptical contact line. X, Y, and Z directions are defined as shown above. (b) Droplet stays stable under the balance of surface tensions and pinning force on a grooved surface. (c) The geometrical models of the cross section and longitudinal section of ellipsoidal cap droplet. (From Wang, Z. et al., The effect of surface anisotropy on contact angles and the characterization of elliptical cap droplets. *Sci. China Technol. Sc.*, 61, 309–316. Copyright 2018, with permission from Springer Nature.)

the droplet in the z axis. In experiments, semi-major axes and contact angles in both directions can be measured easily. The cosine value of contact angle can be calculated when the above four parameters are known. Thus we obtain

$$\cos\theta_x = \frac{a\sqrt{a^2 - a_1^2}}{\sqrt{a^2c^2 + \left(a^2 - a_1^2\right)\left(b^2 - c^2\right)}} \tag{5.17}$$

$$\cos\theta_y = \frac{b\sqrt{b^2 - a_1^2}}{\sqrt{b^2c^2 + \left(b^2 - b_1^2\right)\left(a^2 - c^2\right)}} \tag{5.18}$$

$$a_1^2 c^2 = a^2(c^2 - z_0^2) \tag{5.19}$$

$$b_1^2 c^2 = b^2(c^2 - z_0^2) \tag{5.20}$$

where a_1 and b_1 are the semi-major axes of the ellipse at $z = z_0$. For small ellipsoidal cap droplets, z_0 will have positive value. For large ellipsoidal cap droplets, z_0 will have negative value. Combining Equations 5.17 through 5.20, we can determine the four parameters a, b, c, and z_0, and then the contact angle along the contact line can be obtained.

An example of a typical calculation indicates the validity (Figure 5.9). Droplet is made of glycerol, and S_c is 0.57. Four parameters were measured experimentally, and

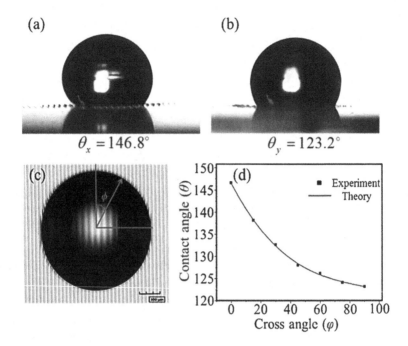

FIGURE 5.9 Images of a droplet on a grooved surface observed from different views. (a) Front view ($\phi = 0°$), and $\theta_x = 146.8°$. (b) Side view ($\phi = 90°$), and $\theta_y = 123.2°$. (c) Top view, with the outline of the droplet being elliptic. The cross angle ϕ is defined as the angle from the groove direction to the viewing direction. (d) The experimental data and predicted theoretical values of contact angles with respect to the cross angle. The predicted theoretical values agree well with experimental data. (From Wang, Z. et al., The effect of surface anisotropy on contact angles and the characterization of elliptical cap droplets. *Sci. China Technol. Sc.*, 61, 309–316. Copyright 2018, with permission from Springer Nature.)

θ_x and θ_y are 146.8° and 123.2°, respectively. The droplet image from top view shows the cross angle ϕ (acute angle, Figure 5.9c), which is defined as the angle from the groove direction to the measurement direction. From Figure 5.9d, we show that the theoretical values agree well with the experimental data. This result demonstrates that the ellipsoidal model is effective for the calculation of contact angles on grooved surfaces.

5.3.1.2.4 Composite Contact Angle for Ellipsoidal Cap Droplet

Contact angles of droplets on grooved surfaces vary at different positions on the elliptical contact line. This variation of contact angle creates many difficulties in the characterization of anisotropic wetting. Recently, Wang et al. [62] proposed a definition of hydrophobicity of a grooved surface, specifically for the non-uniform contact angles of droplets on grooved surface, where $\theta_x > 90°$ and $\theta_y < 90°$. As indicated before the contact angles around the contact line vary as a function of anisotropic factor (Figure 5.7a). This angle is called composite contact angle, and can be predicted from θ_x and θ_y, using a quadratic relationship:

$$\left(\frac{\theta_{co}}{\theta_x}\right)^2 + \left(\frac{\theta_{co}}{\theta_y}\right)^2 = 2 \tag{5.21}$$

where θ_{co} is composite contact angle, which can be rearranged to give composite contact angle as follows:

$$\theta_{co} = \sqrt{\frac{2\theta_x^2\theta_y^2}{\theta_x^2 + \theta_y^2}} \tag{5.22}$$

A cubical and other relationships are also available for calculating composite contact angle. However, the cubical relation shows excessive fluctuations compared to the quadratic relation. The form of composite contact angle above is simple yet unique compared to others. Equation 5.22 can address this composite contact angle problem, if four conditions are met:

1. Composite contact angle should be able to reflect the properties of droplet and material surface. Moreover, one composite contact angle corresponds to a given surface and liquid.
2. Composite contact angle should change with surface properties monotonously, and the relation between composite contact angle and materials surface properties should be regular and easy to measure.
3. The parameters in Equation 5.22 must be easy to obtain experimentally.
4. The expression must degenerate to that for homogeneous smooth surfaces.

θ_x, θ_y, and θ_{co} will be fixed when surface structure and liquid are given. The square of θ_x and θ_y in the expression ensures the uniqueness for the characterization of a given pair of surface and liquid. Furthermore, in experiments, the values of θ_x and θ_y are easy to obtain. When the droplet is on a smooth surface, θ_{co} equals θ_x and θ_y. The equation above can be degenerated to that on a smooth surface to satisfy the fourth condition. Once the experimental conditions are fixed, the composite contact angle can describe the wettability of the grooved surfaces.

5.3.1.3 Scaling Law of Droplet Spreading on Corrugated Surfaces

Liquid spreading is usually controlled by the competition between driving forces and energy dissipation. The driving force originates from the unbalance of surface tensions, expressed as $F_d = \gamma_{lv}(\cos\theta - \cos\theta_0)$, and energy dissipation is caused by viscous resistance at mesoscale and molecular friction at microscale of moving contact line. The derived scaling law is $l \sim t^n$, where l is the spreading length, and it has been corroborated by experiments and simulations, where $1/10 \le n \le 1/7$ is the scaling exponent generally. In recent years, some evidence showed that the spreading velocity is often not so slow. Yuan and Zhao [63] showed the spreading velocity will be much faster when the surface morphology is taken into account, and the spreading velocity on micropillar array surfaces obeys the scaling law of $l \sim t^{2/3}$ due to the acceleration effect by the pillars. When a droplet spreads on a grooved surface, the liquid generally exhibits continuous spreading behavior along the grooves (Y direction), while intermittent spreading behavior occurs vertical to the grooves [55]. The movement of contact line in the vertical direction (X direction) to the grooves shows a slip-stick movement. Wang and Zhao [55] indicated that the spreading of a droplet on a grooved surface obeys a scaling law of $l - t^{4/5}$ (Figure 5.10).

Under the action of hemi-wicking, not only the surface tension but also the Laplace pressure contribute to the spreading driving force. As a result, the scaling law of spreading has a larger scaling exponent. On a corrugated substrate, the structures may guide the liquid in the grooves, in a manner which is similar to wicking. For the droplet wetting on solid surfaces, Biance et al. [64] pointed out that the scaling law of droplets at low viscosities obeys $r/r_i \propto (t/\tau)^{2/(\varpi+2)}$, where r and r_i are the real-time radius and the initial radius of droplet, respectively. t and τ are the time and characteristic time, respectively. ϖ is a parameter related to spreading velocity [64]. The scaling law is $r/r_i \propto (t/\tau)^{1/2}$ for $\varpi = 2$, or $r/r_i \propto (t/\tau)^{2/3}$ for $\varpi = 1$ [65]. When $\varpi = 1/2$, the scaling law will be $r/r_i \propto (t/\tau)^{4/5}$.

The spreading of a droplet in the directions parallel and vertical to the grooves obeys the same scaling law. The difference between the two is that one is faster with a larger intercept in logarithmic coordinates, while the other is slower (Figure 5.10). In parallel direction, the spreading of liquid is continuous and in vertical direction it is intermittent. Figure 5.11 shows the snapshots of liquid spreading on a corrugated substrate. From 52 to 64 ms, the liquid continuously spreads forward along

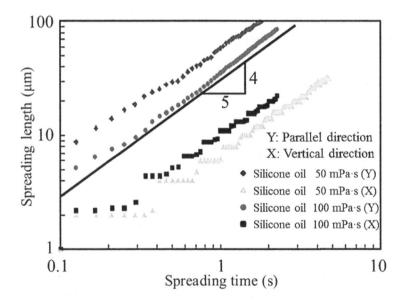

FIGURE 5.10 Dynamic wetting of silicone oil droplets of different viscosities on grooved surfaces. The spreading obeys the scaling law $l - t^{4/5}$. The spreading of droplet along the grooves shows a continuous spreading behavior, while in the vertical direction to the grooves, the droplet exhibits intermittent spreading behavior. (Adapted from Wang, Z. and Zhao, Y.P., Wetting and electrowetting on corrugated substrates. *Phys. Fluids*, 29, 067101. Copyright 2017, with permission from American Institute of Physics.)

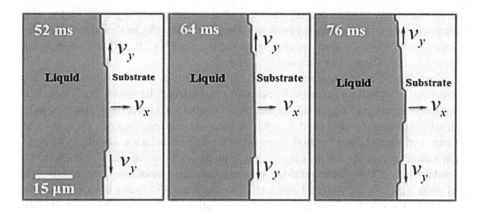

FIGURE 5.11 Microscale view of contact line movement. The spreading is continuous in the parallel direction (v_y) and intermittent in the vertical direction (v_x). (Adapted from Wang, Z. and Zhao, Y.P., Wetting and electrowetting on corrugated substrates. *Phys. Fluids*, 29, 067101. Copyright 2017, with permission from American Institute of Physics.)

the grooves for the whole period, while spreading of liquid is temporarily stopped by the periodic structure in the vertical direction. From 64 to 76 ms, the liquid along the grooves is still spreading forward, and the liquid in the vertical direction spreads forward for a certain distance. Viscosity also has an effect on the velocity of droplet spreading. With increasing viscosity, the velocity decreases. On the contrary, the velocity becomes larger with the viscosity of silicone oil being 50 mPa·s than 100 mPa·s (Figure 5.10).

The droplet spreads in the two directions and obeys the same scaling law, indicating that the mechanism is at work in both X and Y directions. During the wetting process, the liquid always spreads along the groove first for a certain distance; then because of inertia and extrusion of liquid in grooves, the liquid quickly enters into the next groove, leading to the spreading of droplet in X direction. This process is controlled by surface tensions. At the initial stage, the liquid stops at the front of the ridge under the balance of surface tensions, and the contact angle at microscale is the equilibrium contact angle. As the droplet spreads forward and the liquid expands in grooves, the contact angle increases, while the contact line stays still owing to contact angle hysteresis at the intermediate stage. As the liquid spreading continues, the contact angle becomes large enough for the liquid to quickly cross the ridge into the next groove. The spreading along the grooves is a result of the spreading and expanding of liquid in the parallel direction.

5.3.2 ELECTROWETTING (EW) ON STRUCTURED SURFACES

EW is the modification of wetting properties of a surface (which is typically hydrophobic) with an applied electric field. The EW behavior of droplets on variably charged surfaces was first explained by Lippmann in 1875 [66]. A. N. Frumkin used surface charge to change the shape of water drops in 1936. The term "EW" was introduced in 1981 by Beni and Hackwood [67] to describe an effect proposed for designing a new type of display device. EW using an insulating layer on top of a bare electrode was later studied by Bruno Berge in 1993 [68]. EW on this dielectric-coated surface is called electrowetting-on-dielectric (EWOD) [69] to distinguish it from the conventional EW on the bare electrode.

Nowadays, EW has become one of the most widely used tools for manipulating tiny drops of liquids on surfaces. Applications range from "lab-on-a-chip" devices to adjustable lenses and new kinds of electronic displays. Classical EW theories were initially based on an ideal model, in which the system is macroscopic, the surface is planar and smooth, and the liquid droplet is a perfect conductor. Owing to its great application potential, EW has drawn much attention from various fields. With the development of EW, more intensive studies are being conducted, concentrating on different aspects, such as real surface configurations. EW under high voltage and below micrometer level are taken into consideration. Contact angle hysteresis is an accompanying phenomenon that always affects EW as well as wetting. It can be caused by many factors, and will affect droplet actuation by influencing the mini-mum actuation voltage. In this review, we summarize the latest developments of EW on pillar-arrayed and corrugated substrates.

5.3.2.1   Extended EW Equations

EW on micropatterned layers of SU-8 photoresist with an amorphous Teflon coating was observed by Herbertson and coworkers in 2006 [70]. From the experimental results, they found that the cosine of the contact angle was proportional to the square of the voltage applied for increasing bias. However, this did not apply below 40 V and they suggested that this may be explained in terms of penetration of fluid into the pattern of the surface. They summarized the extended EW equation of droplets in Wenzel state by combining the Wenzel equation and Lippmann-Young equation, and gave the following expression [70]:

$$\cos\theta_r = r_o\left(\cos\theta_s + \eta_{EW}\right) \tag{5.23}$$

where r_o is the degree of roughness, θ_r and θ_s are the contact angles on rough surface and smooth surface, respectively, and η_{EW} is the electrowetting number. When the degree of roughness is close to 1, the extended equation degenerates into the classic EW equation, $\cos\theta = \cos\theta_0 + \varepsilon V^2/2\gamma_{lv}d$, where ε is the permittivity, V is the voltage, and d is the thickness of dielectric layer. In 2008, Dai and Zhao [71] developed the EW equation on rough surfaces, and gave the extended formation in Cassie–Baxter state, as follows:

$$\cos\theta_r = r_o f_1\left(\cos\theta_0 + \eta_{EW}\right) - f_2 \tag{5.24}$$

The extended equation will degenerate into the Cassie–Baxter equation when the electric filed is removed.

The extended EW equation on curved surfaces was given in 2012 by Wang and Zhao [72]. A curvature-modified EW number was introduced to describe the surface curvature effect. They found that the variation of the contact angle is curvature dependent; and is more significant with decreasing size of the system.

Compared to the planar case, the contact angle decreases on convex surfaces but increases on concave surfaces. And the extended EW equation is expressed as

$$\cos\theta = \cos\theta_0 + \frac{1}{2} \cdot \frac{cV^2}{\gamma_{lv}} \tag{5.25}$$

where $c = \varepsilon / \left[d\left(1 \pm \ell\right) \right]$, and $\ell = d/r_s$. d is the thickness of the dielectric film, and r_s is the radius of curvature substrate.

5.3.2.2 EW on a Pillar-Arrayed Surface

For a droplet on a pillar-arrayed surface, the EW will be complicated compared to that on smooth surfaces [70,73]. Under a weak voltage, the droplet stays in the Cassie state, and sits on a composite surface of vapor and solid [74]. As the voltage increases, the electromechanical force, that is, the Maxwell force, pulls the liquid and spreads it when the contact angle decreases. A Cassie-to-Wenzel wetting transition does not occur until the applied voltage exceeds a critical value, V_c. At the moment, the confined vapor cannot support the droplet on it, and the liquid-vapor interface suddenly breaks down [75]. Meanwhile, the droplet impales into the pillars. Under a high voltage $V > V_c$, the droplet is in the Wenzel state, in close contact with the solid.

With increasing voltage, the wettability of the solid increases. A precursor film, that is, a thin molecular film, is forward ahead of the nominal contact line [76]. Driven by the electric energy, the bulk droplet in one study [77] spread on the top of the precursor film. However, water did not completely wet the precursor film. In addition, the precursor film showed a hydrophobic feature to the bulk droplet above it under a weak voltage (6.2 V, Figure 5.12a). This is because of the ice-like and two-dimensional hydrogen-bond network in the precursor film. Further increasing the voltage resulted in stronger polarization of the water at the interface, reducing the contact angle below 90°. For $V \geq 18.6$ V, the liquid completely wet the solid surface and became a liquid film. The variation of energy with the square of the applied voltage V^2, whose slope represents the capacitance of the system, is plotted in black squares in Figure 5.12b. Typically, the droplet passes through two stages on pillar-arrayed surfaces. In the first stage, the droplet was deposited on the pillars with a confined vapor layer filling the gap among the pillars. The apparent contact angle of the water droplet on the pillars with $r_o = 1.710$ was about 140° without external voltage.

With increasing voltage, the Cassie-to-Wenzel transition occurs: the air pocket collapses and the droplet implodes into the pillars to change into the Wenzel state.

5.3.2.3 EW on a Corrugated Substrate

Surface morphology has an important influence on the EW of droplets. The experimental results of contact angle versus applied voltage are shown in Figure 5.13. The contact angle in vertical and parallel directions on corrugated substrate and on smooth surface were all measured. Contact angles in vertical and parallel directions maintain a fixed difference under applied voltage. The change of contact angle with applied voltage shows an obviously linear relationship. The contact angle on the smooth surface decreases slower than that on the corrugated surface under a small

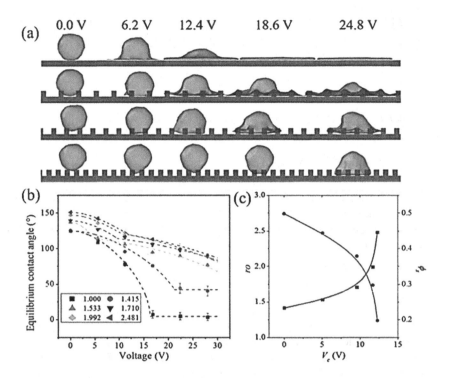

FIGURE 5.12 (a) Equilibrium configurations of a water droplet EW on a smooth surface, and pillar-arrayed surfaces with $r_o = 1.415$, $r_o = 1.710$, and $r_o = 2.482$, under different applied voltages. (b) Variation of apparent equilibrium contact angle with applied voltage. The symbols represent samples with different surface roughnesses. (c) Variation of the critical voltage, V_c, as a function of surface roughness r_o (black squares) and density of roughness ϕ_s (grey circles). The critical voltage V_c is defined as the voltage at which the Cassie-to-Wenzel wetting transition occurs. (Adapted from Yuan, Q. and Zhao, Y.P., Statics and dynamics of electrowetting on pillar-arrayed surfaces at the nanoscale, *Nanoscale*, 7, 2561–2567. Copyright 2015, The Royal Society of Chemistry.)

applied voltage. As a result, the contact angle on the smooth plane is close to the contact angle in vertical direction. When the applied voltage is beyond the critical value, the decrease of contact angle with applied voltage accelerates. As a result, the contact angle on the smooth plane is close to that in parallel direction on corrugated surfaces.

The process of contact angle changing with applied voltage can be thus divided into two distinct processes, which correspond to the Cassie state and the Wenzel state, respectively. When a droplet is deposited on a rough surface under voltage, at the initial stage, the droplet stays in the Cassie state. In the Cassie state, the spreading of droplets is controlled by the surface tensions and the electric field forces under applied voltage. The contact angle is determined by the equilibrium of these forces, which can be described by the Lippmann-Young equation. Given that the contact areas between the solid and the liquid on corrugated surfaces are less than that on smooth surfaces, by adopting the free energy minimization approach and considering

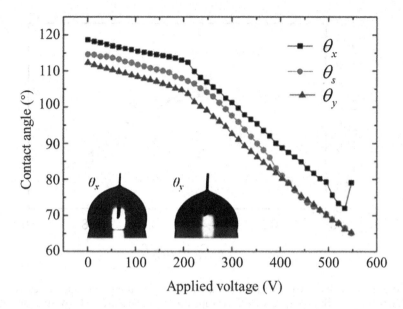

FIGURE 5.13 The EW curves on corrugated and smooth substrates. Squares and triangles represent CAs in vertical and Y directions on corrugated substrates and dots represent CAs on smooth surfaces. The changing of CA with applied voltage can be divided into two regions, and both show a linear relation of CA versus applied voltage. Insert: the image of droplet from side view under the applied voltage of 200 V. (From Wang, Z. and Zhao, Y.P., Wetting and electrowetting on corrugated substrates. *Phys. Fluids*, 29, 067101. Copyright 2017, with permission from American Institute of Physics.)

the actual contact areas κdA (κ is the ratio of contact areas to the projected areas, dA is the projected areas change), $dF = \left(\gamma_{sl} - \gamma_{sv} + \gamma_{lv}\cos\theta\right)rdA + dU - dW = 0$, the Lippmann-Young equation of $\cos\theta = \cos\theta_0 + (1/\kappa)\cdot(\varepsilon V^2/2\gamma_{lv}d)$ can be obtained, where $\kappa < 1$. Thus, the contact angle decreases more rapidly in the Cassie state. This results in that θ_s approaches θ_x and repulses θ_y. As the applied voltage increases, the state of droplets transitions from the Cassie state to the Wenzel state. When a droplet stays in the Wenzel state, θ_x and θ_y decrease slower than θ_s due to $\kappa > 1$. This makes θ_s close to θ_y and far from θ_x.

During the EW on a corrugated substrate, there exists a critical voltage, V_c, where the Cassie-to-Wenzel transition occurs. For a fixed size of surface pattern, V_c remains constant. When the size of surface structure changes, V_c also changes. Experiments for determining the critical voltages on samples with different S_c were conducted by Wang and Zhao [55] (Figure 5.14). The results showed that the critical voltage deceases as S_c increases. This shows that the Cassie-to-Wenzel transition is more likely to occur with a small anisotropic characteristic ratio.

Based on the above experimental results, a droplet spreads in the Cassie state at $V < V_c$ (Figure 5.15c), and the transition from the Cassie state to the Wenzel state happens when V reaches V_c. When V is above V_c, the droplet transforms into in Wenzel state (Figure 5.15d). The forces at liquid-solid interface include van der Waals forces, Laplace pressure, air pressure, hydrostatic pressure, and

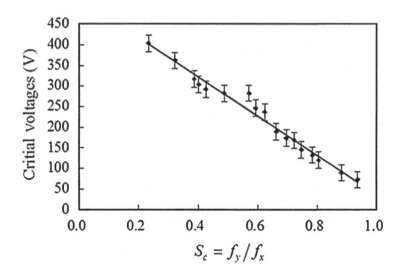

FIGURE 5.14 Critical voltages for the Cassie-to-Wenzel transition. The critical voltage decreases linearly with roughness factor. (From Wang, Z. and Zhao, Y.P., Wetting and electrowetting on corrugated substrates. *Phys. Fluids*, 29, 067101. Copyright 2017, with permission from American Institute of Physics.)

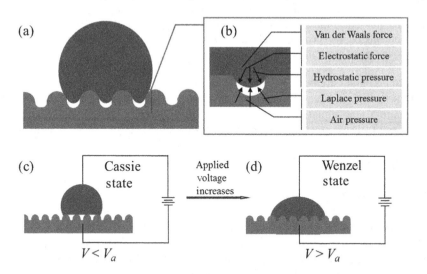

FIGURE 5.15 (a) Schematic diagram of droplet on a corrugated substrate. (b) The forces lead to the transition under applied voltage at the interface between solid and liquid. As the applied voltage increases, the electric field force increases. The contact of liquid and solid surface occurs when the applied voltage exceeds the critical voltage. (c) Droplet stays in the Cassie state at voltage below the critical voltage. (d) Droplet changes into the Wenzel state when voltage is above the critical value. (From Wang, Z., and Zhao, Y.P., Wetting and electrowetting on corrugated substrates. *Phys. Fluids*, 29, 067101. Copyright 2017, with permission from American Institute of Physics.)

Maxwell stress (Figure 5.15b). The Cassie-to-Wenzel transition was mainly controlled by the Laplace pressure and Maxwell stresses.

5.3.2.4 Extended EW Equation on Corrugated Substrate

Free energy minimization approach was used to derive the extended EW equation. The change of free energy can be expressed as follows:

$$dF = \gamma_{sl} dA - \gamma_{sv} dA + \gamma_{lv} dA \cos\theta + dU - dW \tag{5.26}$$

where γ_{sl}, γ_{sv}, and γ_{lv} represent the solid-liquid, solid-vapor, and liquid-vapor interfacial tension, respectively. dU and dW represent the changes in electric field energy in the insulating layer and the work done by electric power, respectively. For the corrugated substrates, assuming the ratio of the increased solid-liquid interface to total increased interface is α, and the ratio of increased actual area to increased apparent area is β. Then, Equation 5.26 can be expressed as follows:

$$dF = \gamma_{sl} \alpha\beta dA - \gamma_{sv} \alpha\beta dA + \gamma_{lv} (1-\alpha)\beta dA + \gamma_{lv} dA \cos\theta + dU - dW \tag{5.27}$$

A droplet stays stable on a surface when its free energy reaches a minimum, namely, $dF = 0$. Divide dF by dA, and make it equal to 0, then Eq. 5.27 becomes

$$\gamma_{sl} \alpha\beta - \gamma_{sv} \alpha\beta + \gamma_{lv} (1-\alpha)\beta + \gamma_{lv} \cos\theta + \frac{dU}{dA} - \frac{dW}{dA} = 0 \tag{5.28}$$

According to electromagnetics, $dU = \varepsilon E^2 dv$, where E is the electric field intensity, and v is the volume saturated with an electric field, $v = \int_0^d A(h)dh$, and h is the thickness of the insulating layer. The increased electric field energy can be approximately calculated as the situation of a droplet on smooth substrate, then the electric field energy can be written as follows:

$$\frac{dU}{dA} = \frac{1}{2}\varepsilon \int_0^d E^2 dh = \frac{1}{2}\varepsilon \int_0^d \frac{V^2}{d^2} dh = \frac{1}{2}\frac{\varepsilon V^2}{d} \tag{5.29}$$

The work done by electric power per unit area can be calculated as follows:

$$\frac{dW}{dA} = V\varepsilon E(d) = \frac{\varepsilon V^2}{d} \tag{5.30}$$

Assuming that the spreading velocities in the directions vertical and parallel to grooves are different, the equation of minimizing energy method should include the increased areas in the two directions. If the increased area in vertical direction is dA, the increased area in parallel direction is λdA (where λ is the ratio of velocities in parallel and vertical directions). Substituting θ_x and θ_y into Equation 5.27 to derive EW equation, the experimental results indicate that the contact angles in vertical and parallel directions maintain a fixed ratio. Then the equilibrium equation should be changed to

$$\gamma_{sl}(1+\lambda)\alpha\beta - \gamma_{sv}(1+\lambda)\alpha\beta + \gamma_{lv}(1+\lambda)(1-\alpha)\beta + \gamma_{lv}(\cos\theta_x + \lambda\cos\theta_y) + \frac{dU}{dA} - \frac{dW}{dA} = 0$$

(5.31)

When a droplet is in the Cassie state, namely, $\beta = 1$, Equation 5.31 can be written as follows:

$$\gamma_{sl}(1+\lambda)\alpha - \gamma_{sv}(1+\lambda)\alpha + \gamma_{lv}(1+\lambda)(1-\alpha) + \gamma_{lv}(\cos\theta_x + \lambda\cos\theta_y) + \frac{dU}{dA} - \frac{dW}{dA} = 0$$

(5.32)

Then simplifying Equation 5.32 and introducing the expression for electric filed energy and the work produced by electric power, we can get the following equation:

$$\gamma_{sl}\alpha - \gamma_{sv}\alpha + \gamma_{lv}(1-\alpha) + \frac{\gamma_{lv}}{1+\lambda}(\cos\theta_x + \lambda\cos\theta_y) - \frac{1}{2}\frac{\varepsilon V^2}{d} = 0$$

(5.33)

Then the classical form is given as

$$\frac{1}{1+\lambda}(\cos\theta_x + \lambda\cos\theta_y) = \alpha\frac{\gamma_{sv} - \gamma_{sl}}{\gamma_{lv}} + \frac{1}{2}\frac{\varepsilon V^2}{d\gamma_{lv}} - (1-\alpha)$$

(5.34)

or

$$\frac{1}{1+\lambda}(\cos\theta_x + \lambda\cos\theta_y) = \alpha\cos\theta_0 + \frac{1}{2}\frac{\varepsilon V^2}{d\gamma_{lv}} - (1-\alpha)$$

(5.35)

For an isotropic surface $\lambda = 1$, and the equation above can degenerate to [69]

$$\cos\theta = \alpha\cos\theta_0 + \frac{1}{2}\frac{\varepsilon V^2}{d\gamma_{lv}} - (1-\alpha)$$

(5.36)

The equations in vertical and parallel directions for anisotropic surface can be expressed as follows:

$$\frac{1}{1+\lambda}\cos(\varphi_x\theta_x + \psi_x) = \alpha\cos\theta_0 + \frac{1}{2}\frac{\varepsilon V^2}{d\gamma_{lv}} - (1-\alpha), \quad (X \text{ direction}) \quad (5.37)$$

$$\frac{1}{1+\lambda}\cos(\varphi_y\theta_y + \psi_y) = \alpha\cos\theta_0 + \frac{1}{2}\frac{\varepsilon V^2}{d\gamma_{lv}} - (1-\alpha), \quad (Y \text{ direction}) \quad (5.38)$$

where φ and ψ represent the combination of $\cos\theta_x$ and $\cos\theta_y$.

When a droplet stays in Wenzel state, namely, $\alpha = 1$, then equations can be written as follows:

$$\frac{1}{1+\lambda}\cos\left(\varphi_x\theta_x +\psi_x\right)= \beta\cos\theta_0 + \frac{1}{2}\frac{\varepsilon V^2}{d\gamma_{lv}}-\left(1-\beta\right), \quad \left(\text{X direction}\right) \quad (5.39)$$

$$\frac{1}{1+\lambda}\cos\left(\varphi_y\theta_y +\psi_y\right)= \beta\cos\theta_0 + \frac{1}{2}\frac{\varepsilon V^2}{d\gamma_{lv}}-\left(1-\beta\right). \quad \left(\text{Y direction}\right) \quad (5.40)$$

For the droplets with relatively small CA, the second-order approximation can be employed to analyze the EW equation. The Taylor series for $\cos\theta$ is expressed as

$$\cos\theta = 1-\frac{\theta^2}{2}+\frac{\theta^4}{4!}+o(\theta^4) \quad (5.41)$$

If the Taylor series is truncated to second order, and $\cos\theta$ is substituted into the EW equation, then, the EW equation can be expressed as

$$1-\frac{\theta^2}{2}= \cos\theta_0 + \frac{1}{2}\frac{\varepsilon V^2}{\gamma_{lv}d} \quad (5.42)$$

By comparing the functions $y = \cos\theta$ and $y =1-\theta^2/2$, we can see that when $\pi\theta/180$ is small, the two functions approximate each other. As the value of $\pi\theta/180$ becomes larger, the two functions separate from each other. When $\pi\theta/180$ is small enough, $\cos\theta$ and $1-\theta^2/2$ will approach 1. We can then introduce the parameter S_c to reconcile the difference. So, the extended EW equation can be expressed as follows:

$$\cos\left(S_c\theta\right)= f_1\left(\cos\theta_0 + \frac{1}{2}\frac{\varepsilon V^2}{d\gamma_{lv}}\right)- f_2 \quad (5.43)$$

where $f_1 = \alpha\beta\left(1+\lambda\right)$ and $f_2 =1-\alpha\beta$. When a droplet is on an isotropic surface with $S_c =1$, Equation 5.43 can degenerate into

$$\cos\theta = f_1\left(\cos\theta_0 + \frac{1}{2}\frac{\varepsilon V^2}{d\gamma_{lv}}\right)- f_2 \quad (5.44)$$

When $\beta =1$, Equation 5.44 can degenerate into the form deduced by Dai and Zhao (Eq. 5.24) [71]. For a droplet on smooth surfaces, namely, at $\alpha =1$, $\beta =1$, $f_1 =1$, and $f_2 = 0$. The extended EW equation degenerates to the classical form $\cos\theta = \cos\theta_0 + \varepsilon V^2/2d\gamma_{lv}$.

5.3.3 DROPLET WETTING ON PILLAR-ARRAYED SURFACES

When a droplet is released on a lyophilic pillar-arrayed surface, the base of the liquid penetrates into the space among the pillars, forming a fringe film. The upper part of the droplet collapses on the base of the fringe film, named the bulk. Compared to a smooth solid surface, the topography induces extensive specific effects to the spreading behavior [79–81]. MCL becomes a complex curved line and propagates in a special stepwise mode [65,82]. Wetting of a droplet on a microstructured surface has been investigated intensely for decades. Extrand and coworkers [83] elegantly reported that appropriate lyophilic pillar arrays could effectively drive partial wetting liquids to complete wetting. Later, by varying the geometry of the surface array and the liquid, Courbin et al. [82] found a diversity of final wetted shapes, including polygons and circles, of droplets on pillar-arrayed surfaces. Raj et al. [84] then showed a complete control on polygonal wetted shapes via the design of topographic or chemical heterogeneity on the surface. Jokinen et al. [85] developed a method of fabricating irregular pillars on a square array surface, and found directional wetting property of droplets on this surface, where droplets spread to irregular square-like shapes. Vrancken et al. [86] reported that droplet shapes depend on the array geometry, pillar shape, and array space.

5.3.3.1 The Wetting Shapes of Droplets

When a droplet was brought into contact with a micropillar-arrayed surface, the droplet spread into two parts (Figure 5.16). The lower part, namely the fringe, penetrated into the gaps among the pillars. The upper part, namely the bulk, spread on the base of the fringe and contributes to the expansion of the fringe. Usually, the fringe is in the front of the bulk, and spreads faster than the bulk in some specially structured surfaces.

For droplets with different viscosities, the liquid flows at a viscosity-dependent capillary velocity, so a droplet with a lower viscosity spreads faster. The bulk expanded on the base of the fringe, whose top was chemically and topologically smooth. Thus, the flow velocity of the bulk was isotropic, and its projection was always a circle. Observed under the microscope, the dark area means there is an interface intersecting with the substrate, while the bright area means the interface is parallel to the substrate. In the fringe region, the liquid surface is nearly parallel to polydimethylsiloxane (PDMS). Most of the incident energy transmits through the liquid and PDMS, and only small part of energy is reflected. In the contact line region, the liquid surface intersects with the substrate at a large angle. Therefore, little incident energy transmits, resulting in a dark outline in the contact line region. The fringe shape was influenced by its viscosity. The fringe propagated ahead of the bulk in the forest of micropillars. The pillars accelerated the approaching MCL, and pinned the leaving MCL. The distance between pillars varied with flow direction, causing the corresponding energy dissipation to be anisotropic. Therefore, the fringe spread in a direction-dependent velocity. Furthermore, the velocity was symmetric about the orthogonal and diagonal directions because of the arrangement of the pillars. In this way, the fringe evolved into an octagonal shape. For a liquid with a lower viscosity, its velocity and corresponding inertial force were larger. The fringe easily

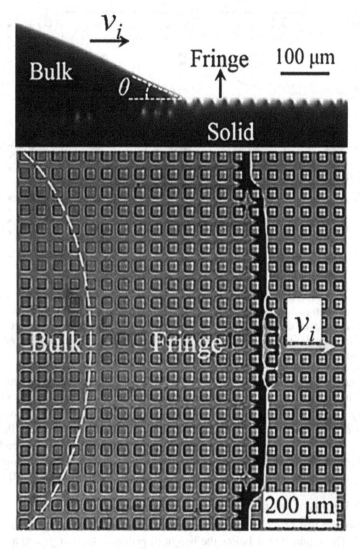

FIGURE 5.16 Side view and the top view of a droplet on a pillar-arrayed surface. Droplet spreading is divided into the bulk and the fringe, and the fringe is in the front of the bulk. (From Yuan, Q. et al., Dynamic spreading on pillar-arrayed surfaces: Viscous resistance versus molecular friction. *Phys. Fluids*, 26, 092104. Copyright 2014, with permission from American Institute of Physics.)

overcame the energy barrier along the diagonal direction of the pillars, becoming approximately a square. For a liquid with a higher viscosity, its velocity and corresponding inertial force were smaller. The MCL tried to achieve a kinetic balance in all directions, making the fringe roughly a circle (Figure 5.17).

Droplets adopt dynamic evolutions of projected shapes from initial circles to final bilayer polygons when deposited on hydrophilic rough surfaces. These dynamic processes are distinguished in two regimes on substrates. The bilayer structure of a

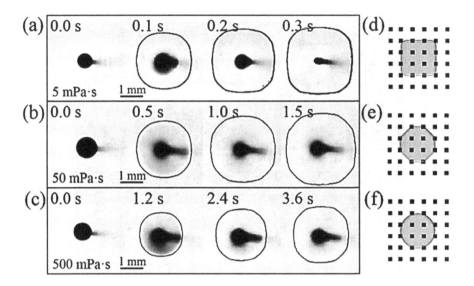

FIGURE 5.17 Experimental snapshots of the spreading of a droplet with viscosities of (a) 5 mPa·s, (b) 50 mPa·s, (c) 500 mPa·s on lyophilic pillars, and (d–f) illustrations of the square, octagonal, and circular shapes of the fringe. (From Yuan, Q. et al., Dynamic spreading on pillar-arrayed surfaces: Viscous resistance versus molecular friction. *Phys. Fluids*, 26, 092104. Copyright 2014, with permission from American Institute of Physics.)

droplet, induced by micropillars on the surface, was explained by the interaction between the fringe and the bulk. The evolution of polygonal shapes, following the symmetry of the pillar-arrayed surface, was analyzed by the competition effects of excess driving energy and resistance, which were induced by micropillars with increasing solid surface area fraction. Though the anisotropic droplets spread in different regimes, they obey the same scaling law $l - t^{2/3}$ (S is the wetted area and t is the spreading time), which is derived from the molecular kinetic theory [87,88].

5.3.3.2 Details of Droplet Wetting on Pillar-Arrayed Surfaces

When a droplet was brought into contact with the substrate, the liquid wedge began to spread. The liquid that is below the height of pillars is the fringe and above that is bulk. The fringe penetrated into the gaps among the pillars with a velocity faster than the bulk. With the increase of the distance between the fringe and the bulk, a transition region is gradually formed to link the liquid among the pillars and the liquid wedge. When the fringe flowed over the pillars, the liquid also spread faster along the height of the pillars and resulted in a ridge of liquid above the pillar top and a meniscus in the transition region. In this way, the fringe separated from the droplet and penetrated among the pillars. The snapshots sequence and three-dimensional spreading profiles of liquid on pillar-arrayed surfaces are shown in Figure 5.18.

Driven by the surface tension, the contact line moved along as normal at first. Once it touched the pillars at 0.2 ms, the liquid began to spread along the pillars (Figure 5.19). The radius of curvature of the liquid in the inset is about 2 μm. Hence, the local Laplace pressure (P_{cl}, about 103 Pa) provides additional driving energy to

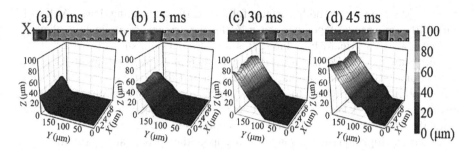

FIGURE 5.18 Snapshots sequence and three-dimensional spreading profiles of liquid ($\eta = 5$ mPa·s) on pillar-arrayed surfaces. (From Yuan, Q. et al., Dynamic spreading on pillar-arrayed surfaces: Viscous resistance versus molecular friction. *Phys. Fluids*, 26, 092104. Copyright 2014, with permission from American Institute of Physics.)

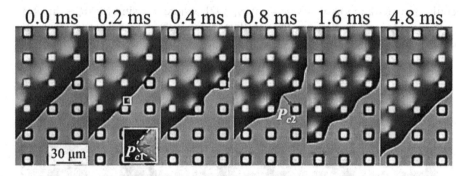

FIGURE 5.19 Snapshots sequence of liquid ($\eta = 5$ mPa·s) spreading on pillar-arrayed surfaces. The inset shows a close-up image of radius of curvature of the liquid labeled by a white dashed line. P_{c1} and P_{c2} represent the local capillary pressures. (From Yuan, Q. et al., Dynamic spreading on pillar-arrayed surfaces: Viscous resistance versus molecular friction. *Phys. Fluids*, 26, 092104. Copyright 2014, with permission from American Institute of Physics.)

spread faster velocity along the corner between the pillars and the substrate. The liquid near the pillars quickly forms the square shape and is pinned by the pillars at 0.4 ms. Meanwhile, the rest part of the MCL keeps spreading in a line with a slow velocity. As a result, a local Laplace pressure (P_{c2}, about 102 Pa) is formed and makes the MCL a wavy line at 0.8 ms. The local radius of curvature tends to become larger at 1.6 ms. The fast and slow part of the MCL tries to reach a balance state and finally return to a line. This process repeats in all the periods. The Laplace pressures at the interior corner and at the wavy MCL are the answers to the super-wetting of the pillar-arrayed surface.

For the droplet deposited on the pillar-arrayed lyophilic surface, the MCL advances with a characteristic capillary velocity of $v_{ca} \sim \gamma_{lv}/\eta \sim 0.1$ m·s^{-1}. The fringe propagates much faster than the bulk due to the driving force induced by the excess lyophilic surface. Obviously, the propagation of the droplet on a rough surface and a smooth surface follows different scaling laws. Although with different speeds depending on r_o (>1), the fringe approximately obeys a scaling law $l - t^{1/3}$. Meanwhile, the radius of a droplet on a smooth surface obeys a scaling law $l - t^{1/7}$, which was in agreement with previous experiments [89] and simulations [90]. The micropillar array accelerated the wetting process, causing the lyophilic solid surface to become super-lyophilic. When observed under microscope, the dark area means there is an interface intersecting with the substrate, while the bright area means the interface is parallel to the substrate.

Under a high-magnification microscope, the propagation of different parts of the fringe is quite different. In Figure 5.20, the fringe advances in the x direction. The MCL is initially linear on the substrate at a velocity of $v_i \sim 0.01$ mm$\cdot$s^{-1}. Once the MCL reaches the pillars, the excess driving force forces the MCL to accelerate, making the liquid propagate much faster at the interior corner between the pillar and the substrate, known as the Concus-Finn effect [91]. In an interior corner with opening angle 2α, the equilibrium velocity could be calculated as $v_e = f v_{ca} = f \gamma_{lv}/\eta$ [92], where $f = \sin\alpha \left(\cos\theta_e - \sin\alpha\right)/S_i$ is the topological coefficient for capillary flow at the interior corner. θ_e is the equilibrium contact angle. S_i is the slenderness ratio of the interior corner, that

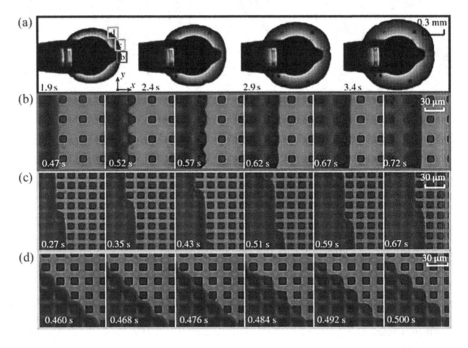

FIGURE 5.20 (a) Snapshots of the whole droplet under an optical microscope; (b–d) snapshots of different parts of the droplet labeled in (a), under a high magnification microscope and high-speed camera, (b) $\Delta t = 50$ ms, $r_o = 1.42$; (c) $\Delta t = 80$ ms, $r_o = 2.49$; (d) $\Delta t = 8$ ms, $r_o = 1.94$. (From Yuan, Q. and Zhao, Y.P., Multiscale dynamic wetting of a droplet on a lyophilic pillar-arrayed surface. *J. Fluid Mech.*, 716, 171–188. Copyright 2013, with permission from Cambridge University Press.)

is, the ratio of the length to the height of the liquid in the interior corner. So, the initial excess velocity could be estimated as

$$v_{ex} = \frac{\gamma_{lv}}{S_i\eta}\left[\sin\alpha_1\left(\cos\theta_0 - \sin\alpha_1\right) - \sin\alpha_2\left(\cos\theta_0 - \sin\alpha_2\right)\right] \qquad (5.45)$$

Because the slow flow on the flat substrate restrains the fast flow at the interior corner, the average velocity is about 0.1 mm·s^{-1}. The fast and slow parts of the fringe reach a dynamic balance. Then, the liquid is pinned around the pillars ($\Delta t = 0.62$ s) and does not advance until the liquid between the two pillars catches up. After the MCL passes the pillars ($\Delta t = 0.67$ s), the top of the pillars become blurry, which implied that the fringe climbed up the pillars and covered the top. The front of the fringe is dark owing to the inclined liquid–vapor interface. The MCL gradually decelerates to 0.01 mm s^{-1} and spreads again in a straight line ($\Delta t = 0.72$ s) on the substrate driven only by the interface energy. In this process, the area between the first and second rows of micro-pillars gradually become brighter, implying that the liquid had filled the space between the pillars and gradually make the liquid–vapor interface parallel to the substrate. This process repeats for every period of the pillars. And the average velocity of the fringe gradually decreases because of the energy dissipations in the flow and in the three-phase zone. So unlike Figure 5.20b, the liquid encounters the pillars in sequence. The fringe would also be accelerated [93] and pinned around the pillar one after another, making the MCL a complex "zipping" line. The progress is also periodically repeated. The zigzag lines at microscopic length scales forms a circle when observed in low magnification. In the dynamic spreading process of a droplet on the pillar-arrayed surface, the surface tension, the viscosity, and the inertia govern the liquid motions.

5.3.3.3 Scaling Law of Droplet Wetting on Pillar-Arrayed Surfaces

Wetting on these surfaces is usually controlled by the combined effects of surface topology, intrinsic wettability, and elasticity of the solid on the wetting process. The underlying mechanics was revealed by Yuan and Zhao [94] using molecular dynamics simulations. The direction-dependent dynamics of both liquid and pillars, especially at MCL, is revealed at atomic level. The flexible pillars accelerate the liquid when the liquid approaches, and pin the liquid when the liquid passes. The liquid deforms the pillars, resulting in energy dissipation at the MCL (Figure 5.21). Scaling analysis is performed based on molecular kinetic theory and is validated by the simulations.

The simulation domain is composed by flexible hydrophilic pillar-arrayed substrate and a water droplet. The droplet was modeled using the simple point charge/extended water model with viscosity $\eta = 0.729$ mPa·s, density $\rho = 994$ kg/m^3, and surface tension $\gamma_{lv} = 0.0636$ N/m, which are close to real water. In the simulations, the evaporated water molecules would quickly saturate the simulation box, so the evaporation effect is ignored. When brought into contact with hydrophilic pillars, a droplet collapses into two parts: the fringe penetrates into the spaces between pillars, and the bulk droplet spreads on the fringe. Meanwhile, the flexible pillars bend due to the spreading liquid. The tip deflections labeled by the arrows in Figure 5.21b all point to the center of the droplet, causing an increase in the local roughness, and

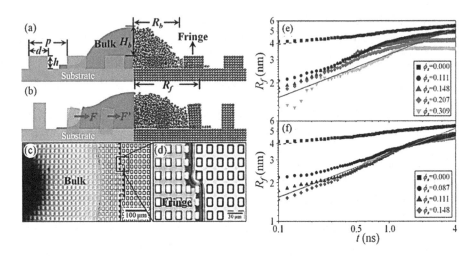

FIGURE 5.21 Illustrations of a droplet wetting and evolution of the fringe radius with respect to time: (a) on rigid pillar-arrays, and (b) on flexible pillar-arrays. (c–d) Droplet wetting on flexible PDMS pillar-arrays (Young's modulus is 2.2 MPa), (e) $h = 2.04$ nm, and (f) $h = 3.06$ nm. The black squares represent the radius of a droplet spreading on a smooth substrate. The dashed and solid lines represent the scaling laws $l - t^{1/7}$ and $l - t^{1/3}$, respectively. (Adapted from Yuan, Q. and Zhao, Y.P., Wetting on flexible hydrophilic pillar-arrays. *Sci. Rep.*, 3, 1944. Copyright 2013, Springer Nature.)

accelerating the liquid spreading. The interior corners formed by the pillars and the substrate provide additional driving forces to the liquid, making the velocity at the corner faster than the rest of the liquid. The initially circular droplet accommodates the arrangement of pillars and spreads in the shape of a square. Once the liquid reaches the pillars, the pillars are dragged toward the liquid. The fringe of the droplet advances in the space between the pillars and climbs up the pillars. The bulk droplet spreads on the base of the fringe, while the height of the bulk droplet decreases. Some of the pillars are dragged closer and interact with one another. The tip deflection near the MCL is approximately one order of magnitude larger than that away from the MCL. Especially when the liquid reaches flexible pillars, the deformation of these pillars is larger than those submerged in the liquid, which implies energy dissipation at the MCL. When the liquid advances to pass these flexible pillars, these pillars also deform to pin the liquid, preventing its further spreading.

5.4 SUMMARY AND CONCLUSIONS

Wetting and EW on structured surfaces was investigated. Given that the rapid development of applications of wetting and EW in self-assembly, droplet movement control, water-harvesting, self-cleaning surfaces, and so on, the underlying mechanism needs to be clearly revealed. In this review, some recent developments are summarized in detail; and static contact angle, EW, spreading shape, and the spreading scaling law are discussed in detail.

Anisotropic surfaces widely exist in nature and in industrial products, such as bird feathers, leaves, and microfluidics. The study of wetting on these surfaces is important and urgent. Furthermore, due to the anisotropy of corrugated substrates, there exist some difficulties in carrying out studies of wetting. In this review, the fundamental aspects of wetting on these surfaces are explained. On a corrugated substrate, the anisotropy of surface morphology has an important influence on the droplet contact angle. The contact angle first increases and later decreases as the anisotropic factor increases. This indicates that there exists a range of anisotropic value where the droplet has the maximum contact angle. A method to calculate the apparent contact angle and the definition of a composite contact angle will be helpful to the study of wetting on anisotropic surfaces. The EW on both pillar-arrayed and corrugated substrates exhibits the difference from that on smooth surfaces.

Currently, EW has become one of the most widely used tools for manipulating tiny droplets on surfaces. Applications range from "lab-on-a-chip" devices to adjustable lenses and new electronic displays. The EW on pillar-arrayed and anisotropic surfaces shows tremendous application potential in biology, medicine, materials science, and so on. In this review, the special phenomena of EW on these surfaces are summarized. EW on these surfaces can be divided into two stages, droplet in the Cassie and Wenzel states, respectively. The critical voltage changes linearly with anisotropic factor on corrugated substrates. The scaling law of droplets on these surfaces is also listed. Due to the capillary and semi-wicking, the spreading of droplets on these surfaces has a higher velocity. This phenomenon may be used for rapid water transport in future works.

ACKNOWLEDGMENTS

This work was jointly supported by the National Natural Science Foundation of China (NSFC, Grant No. U1562105), the Chinese Academy of Sciences (CAS) through CAS Interdisciplinary Innovation Team Project, the CAS Key Research Program of Frontier Sciences (Grant No. QYZDJ-SSW-JSC019), and the CAS Strategic Priority Research Program (Grant No. XDB22040401).

REFERENCES

1. V. Flauraud, M. Mastrangeli, G. D. Bernasconi, J. Butet, D. T. L. Alexander, E. Shahrabi, O. J. F. Martin and J. Brugger, "Nanoscale topographical control of capillary assembly of nanoparticles," *Nat. Nanotechnol.*, 12, 73–80 (2017).
2. U. Thiele, "Patterned deposition at moving contact lines," *Adv. Colloid Interface Sci.*, 206, 399–413 (2014).
3. T. A. Duncombe, E. Y. Erdem, A. Shastry, R. Baskaran and K. F. Böhringer, "Microfluidics: Controlling liquid drops with texture ratchets," *Adv. Mater.*, 24, 1497 (2012).
4. D. Mannetje, S. Ghosh, R. Lagraauw, S. Otten, A. Pit, C. Berendsen, J. Zeegers, D. van den Ende and F. Mugele, "Trapping of drops by wetting defects," *Nat. Commun.*, 5, 3559 (2014).
5. N. A. Malvadkar, M. J. Hancock, K. Sekeroglu, W. J. Dressick and M. C. Demirel, "An engineered anisotropic nanofilm with unidirectional wetting properties" *Nat. Mater.*, 9, 1023 (2010).

6. H. Chen, P. Zhang, L. Zhang, H. Liu, Y. Jiang, D. Zhang, Z. Han and L. Jiang, "Continuous directional water transport on the peristome surface of *Nepenthes alata*," *Nature*, 532, 85 (2016).

7. K. C. Park, P. Kim, A. Grinthal, N. He, D. Fox, J. C. Weaver and J. Aizenberg, "Condensation on slippery asymmetric bumps," *Nature*, 531, 78 (2016).

8. A. R. Parker and C. R. Lawrence, "Water capture by a desert beetle," *Nature*, 414, 33 (2001).

9. S. Pan, A. K. Kota, J. M. Mabry and A. Tuteja, "Superomniphobic surfaces for effective chemical shielding," *J. Am. Chem. Soc.*, 135, 578 (2012).

10. T. S. Wong, S. H. Kang, S. K. Y. Tang, E. J. Smythe, B. D. Hatton, A. Grinthal and J. Aizenberg, "Bioinspired self-repairing slippery surfaces with pressure-stable omniphobicity," *Nature*, 477, 443 (2011).

11. S. Anand, A. T. Paxson, R. Dhiman, J. D. Smith and K. K. Varanasi, "Enhanced condensation on lubricant-impregnated nanotextured surfaces," *ACS Nano*, 6, 10122–10129 (2012).

12. X. Chen, J. Wu, R. Ma, M. Hua, N. Koratkar, S. Yao and Z Wang, "Nanograssed micropyramidal architectures for continuous dropwise condensation," *Adv. Funct. Mater.*, 21, 4617–4623 (2011).

13. E. Bormashenko, "Progress in understanding wetting transitions on rough surfaces," *Adv. Colloid Interface Sci.*, 222, 92–103 (2015).

14. W. Barthlott and C. Neinhuis, "Purity of the sacred lotus, or escape from contamination in biological surfaces," *Planta*, 202, 1–8 (1997).

15. Y. Zheng, X. Gao and L. Jiang, "Directional adhesion of superhydrophobic butterfly wings," *Soft Matter*, 3, 178–182 (2007).

16. M. Gleiche, L. Chi, E. Gedig and H. Fuchs, "Anisotropic contact angle hysteresis of chemically nanostructured surfaces," *ChemPhysChem*, 2, 187–191 (2001).

17. M. Morita, T. Koga, H. Otsuka and A. Takahara, "Macroscopic wetting anisotropy on the line patterned surface of fluoroalkylsilane monolayers," *Langmuir*, 21, 911–918 (2005).

18. A. Dupuis, J. Léopoldès, D. G. Bucknall and J. M. Yeomans, "Control of drop positioning using chemical patterning," *Appl. Phys. Lett.*, 87, 024103 (2005).

19. D. Öner and T. J. McCarthy, "Ultrahydrophobic surfaces: Effects of topography length scales on wettability," *Langmuir*, 16, 7777–7782 (2000).

20. J. Bico, C. Marzolin and D. Quéré, "Pearl drops," *EPL (Europhysics Letters)*, 47, 220–226 (1999).

21. S. Brandon, N. Haimovich, E. Yeger and A. Marmur, "Partial wetting of chemically patterned surfaces: The effect of drop size," *J. Colloid Interface Sci.*, 263, 237–243 (2003).

22. Z. Yoshimitsu, A. Nakajima, T. Watanabe and K. Hashimoto, "Effects of surface structure on the hydrophobicity and sliding behavior of water droplets," *Langmuir*, 18, 5818–5822 (2002).

23. L. Zhu, Y. Xiu, J. Xu, P. A. Tamirisa, D. W. Hess and C. Wong, "Superhydrophobicity on two tier rough surfaces fabricated by controlled growth of aligned carbon nanotube arrays coated with fluorocarbon," *Langmuir*, 21, 11208–11212 (2005).

24. X. Gao, X. Yao and L. Jiang, "Effects of rugged nanoprotrusions on the surface hydrophobicity and water adhesion of anisotropic micropatterns," *Langmuir*, 23, 4886–4891 (2007).

25. W. Ming, D. Wu, R. van Benthem and G. de With, "Superhydrophobic films from raspberry like particles," *Nano Lett.*, 5, 2298–2301 (2005).

26. H. Kusumaatmaja and J. M. Yeomans, "Controlling drop size and polydispersity using chemically patterned surfaces," *Langmuir*, 23, 956–959 (2007).

27. L. M. Fidalgo, C. Abell and W. T. S. Huck, "Surface induced droplet fusion in microfluidic devices," *Lab Chip*, 7, 984–986 (2007).

28. J. Ou and J. P. Rothstein, "Direct velocity measurements of the flow past drag-reducing ultrahydrophobic surfaces," *Phys. Fluids*, 17, 103606 (2005).

29. P. Joseph, C. Cottin-Bizonne, J. M. Benoit, C. Ybert, C. Journet, P. Tabeling and L. Bocquet, "Slippage of water past superhydrophobic carbon nanotube forests in microchannels," *Phys. Rev. Lett.*, 97, 156104 (2006).

30. Y. Chen, B. He, J. Lee and N. A. Patankar, "Anisotropy in the wetting of rough surfaces," *J. Colloid Interface Sci.*, 281, 458–464 (2005).

31. J. Y. Chung, J. P. Youngblood and C. M. Stafford, "Anisotropic wetting on tunable micro wrinkled surfaces," *Soft Matter*, 3, 1163–1169 (2007).

32. R. D. Narhe and D. A. Beysens, "Nucleation and growth on a superhydrophobic grooved surface," *Phys. Rev. Lett.*, 93, 076103 (2004).

33. J. T. Hirvi and T. A. Pakkanen, "Wetting of nanogrooved polymer surfaces," *Langmuir*, 23, 7724–7729 (2007).

34. Y. P. Zhao, *Physical Mechanics of Surfaces and Interfaces*, Science Press, Beijing (2012).

35. Y. P. Zhao, "Moving contact line problem: Advances and perspectives," *Theoretical and Applied Mechanics Letters*, 4, 034002 (2014).

36. B. F. Zhang, K. Li and J. Zhao, "Regularity analysis of wrinkles under the action of capillary force in an annular thin film," *Sci. China Phys. Mech.*, 57, 1574–1580 (2014).

37. L. Y. Wang, F. C. Wang, F. Q. Yang and H. A. Wu, "Molecular kinetic theory of boundary slip on textured surfaces by molecular dynamics simulations," *Sci. China Phys. Mech.*, 57, 2152–2160 (2014).

38. P. F. Hao, C. J. Lv, Z. H. Yao and F. Niu, "Wetting property of smooth and textured hydrophobic surfaces under condensation condition," *Sci. China Phys. Mech.*, 57, 2127–2132 (2014).

39. P. F. Hao, C. J. Lv, F. L. Niu and Y. Yu, "Water droplet impact on superhydrophobic surfaces with microstructures and hierarchical roughness," *Sci. China Phys. Mech.*, 57, 1376–1381 (2014).

40. Y. P. Zhao, "Bridging length and time scales in moving contact line problems," *Sci. China Phys. Mech.*, 59, 114631 (2016).

41. G. W. Wang, "Radial deformation and adhesion of carbon nanotube," *Sci. China Phys. Mech.*, 57, 1569–1574 (2014).

42. F. C. Wang, F. Yang and Y. P. Zhao, "Size effect on the coalescence-induced self-propelled droplet," *Appl. Phys. Lett.*, 98, 053112 (2011).

43. L. Chen, E. Bonaccurso, T. Gambaryan-Roisman, V. Starov, N. Koursari and Y. Zhao, "Static and dynamic wetting of soft substrates," *Curr. Opin. Colloid Interface Sci.*, 36, 46–57 (2018).

44. D. Xia, L. M. Johnson and G. P. López, "Anisotropic wetting surfaces with one dimensional and directional structures: Fabrication approaches, wetting properties and potential applications," *Adv. Mater.*, 24, 1287–1302 (2012).

45. D. Chatain, "Anisotropy of wetting," *Annu. Rev. Mater. Res.*, 38, 45–70 (2008).

46. J. F. Oliver, C. Huh and S. G. Mason, "Resistance to spreading of liquids by sharp edges," *J. Colloid Interface Sci.*, 59, 568–581 (1977).

47. O. Bliznyuk, H. P. Jansen, E. S. Kooij, H. J. Zandvliet and B. Poelsema, "Smart design of stripe-patterned gradient surfaces to control droplet motion," *Langmuir*, 27, 11238–11245 (2011).

48. K. Liu, X. Yao and L. Jiang, "Recent developments in bio-inspired special wettability," *Chem. Soc. Rev.*, 39, 3240–3255 (2010).

49. L. Feng, S. Li, Y. Li, H. Li, L. Zhang, J. Zhai, Y. Song, B. Liu, L. Jiang and D. Zhu, "Superhydrophobic surfaces: From natural to artificial," *Adv. Mater.*, 14, 1857–1860 (2002).

50. H. M. Whitney, M. Kolle, P. Andrew, L. Chittka, U. Steiner and B. J. Glover, "Floral iridescence, produced by diffractive optics, acts as a cue for animal pollinators," *Science*, 323, 130–133 (2009).

51. L. Feng, Y. Zhang, J. Xi, Y. Zhu, N. Wang, F. Xia and L. Jiang, "Petal effect: A super-hydrophobic state with high adhesive force," *Langmuir*, 24, 4114–4119 (2008).

52. Y. Liu, X. Chen and J. H. Xin, "Hydrophobic duck feathers and their simulation on textile substrates for water repellent treatment," *Bioinspir. Biomim.*, 3, 046007 (2008).

53. E. Bormashenko, Y. Bormashenko, T. Stein, G. Whyman and E. Bormashenko, "Why do pigeon feathers repel water? Hydrophobicity of pennae, Cassie-Baxter wetting hypothesis and Cassie-Wenzel capillarity-induced wetting transition," *J. Colloid Interface Sci.*, 311, 212–216 (2007).

54. P. G. de Gennes, "Wetting: Statics and dynamics," *Rev. Mod. Phys.*, 57, 827–863 (1985).

55. Z. Wang and Y. P. Zhao, "Wetting and electrowetting on corrugated substrates," *Phys. Fluids*, 29, 067101 (2017).

56. J. Yang, F. R. A. J. Rose, N. Gadegaard and M. R. Alexander, "Effect of sessile drop volume on the wetting anisotropy observed on grooved surfaces," *Langmuir*, 25, 2567–2571 (2009).

57. Y. Zhao, Q. Lu, M. Li and X. Li, "Anisotropic wetting characteristics on submicrometer-scale periodic grooved surface," *Langmuir*, 23, 6212–6217 (2007).

58. W. Li, G. Fang, Y. Li and G. Qiao, "Anisotropic wetting behavior arising from super-hydrophobic surfaces: Parallel grooved structure," *J. Phys. Chem. B*, 112, 7234–7243 (2008).

59. H. Kusumaatmaja, R. J. Vrancken, C. W. M. Bastiaansen and J. M. Yeomans, "Anisotropic drop morphologies on corrugated surfaces," *Langmuir*, 24, 7299–7308 (2008).

60. M. S. Bell, A. Shahraz, K. A. Fichthorn and A. Borhan, "Effects of hierarchical surface roughness on droplet contact angle," *Langmuir*, 31, 6752–6762 (2015).

61. F. Chen, D. Zhang, Q. Yang, X. Wang, B. Dai, X. Li, X. Hao, Y. Ding, J. Si and X. Hou, "Anisotropic wetting on microstrips surface fabricated by femtosecond laser," *Langmuir*, 27, 359–365 (2011).

62. Z. Wang, E. Chen and Y. P. Zhao, "The effect of surface anisotropy on contact angles and the characterization of elliptical cap droplets," *Sci China Technol. Sci.*, 61, 309–316 (2018).

63. Q. Yuan and Y. P. Zhao, "Multiscale dynamic wetting of a droplet on a lyophilic pillar-arrayed surface," *J. Fluid Mech.*, 716, 171–188 (2013).

64. A. L. Biance, C. Clanet and D. Quéré, "First steps of the spreading of a liquid droplet," *Phys. Rev. E*, 69, 016301 (2004).

65. H. P. Kavehpour, B. Ovryn and G. H. McKinley, "Evaporatively-driven Marangoni instabilities of volatile liquid films spreading on thermally active substrates," *Colloids Surfaces A*, 206, 409 (2002).

66. G. Lippmann. Relations entre les phénomènes électriques et capillaires. Gauthier-Villars (1875).

67. G. Beni and S. Hackwood, "Electrowetting displays," *Appl. Phys. Lett.*, 38, 207–209 (1981).

68. B. Berge, "Électrocapillarité et mouillage de films isolants par l'eau," *C. R. Acad. Sci. Paris*, 317, Série II, 157–163 (1993).

69. J. Lee, "Microactuation by continuous electrowetting and electrowetting: Theory, fabrication, and demonstration," PhD Thesis, University of California, Los Angeles, CA (2000).

70. D. L. Herbertson, C. R. Evans, N. J. Shirtcliffe, G. McHale and M. I. Newton, "Electrowetting on superhydrophobic SU-8 patterned surfaces," *Sensor Actuat. A Phys.*, 130, 189–193 (2006).

71. W. Dai and Y. P. Zhao, "The nonlinear phenomena of thin polydimethylsiloxane (PDMS) films in electrowetting," *Int. J. Nonlin. Sci. Num.*, 8, 519–526 (2007).

72. Y. Wang and Y. P. Zhao, "Electrowetting on curved surfaces," *Soft Matter*, 8, 2599–2606 (2012).
73. V. Bahadur and S. V. Garimella, "Electrowetting based control of static droplet states on rough surfaces," *Langmuir*, 23, 4918–4924 (2007).
74. T. Koishi, K. Yasuoka, S. Fujikawa, T. Ebisuzaki and X. C. Zeng, "Coexistence and transition between Cassie and Wenzel state on pillared hydrophobic surface," *Proc. Natl Acad. Sci. USA*, 106, 8435–8440 (2009).
75. P. Papadopoulos, L. Mammen, X. Deng, D. Vollmer and H. J. Butt, "How superhydrophobicity breaks down," *Proc. Natl Acad. Sci. USA*, 110, 3254–3258 (2013).
76. Q. Yuan and Y. P. Zhao, "Precursor film in dynamic wetting, electrowetting, and electro-elasto-capillarity," *Phys. Rev. Lett.*, 104, 246101 (2010).
77. C. Wang, H. Lu, Z. Wang, P. Xiu, B. Zhou, G. Zuo, R. Wan, J. Hu and H. Fang, "Stable liquid water droplet on a water monolayer formed at room temperature on ionic model substrates," *Phys. Rev. Lett.*, 103, 137801 (2009).
78. Q. Yuan and Y. P. Zhao, "Statics and dynamics of electrowetting on pillar-arrayed surfaces at the nanoscale," *Nanoscale*, 7, 2561–2567 (2015).
79. G. McHale, S. M. Rowan, M. I. Newton and N. A. Käb, "Estimation of contact angles on fibers," *J. Adhesion Sci. Technol.*, 13, 1457–1469 (1999).
80. F. C. Wang and H. A. Wu, "Pinning and depinning mechanism of the contact line during evaporation of nano droplets sessile on textured surfaces," *Soft Matter*, 9, 5703–5709 (2013).
81. Y. S. Yu, "Circular plate deformed by a sessile droplet," *J. Adhesion Sci. Technol.*, 28, 1970–1979 (2014).
82. L. Courbin, E. Denieul, E. Dressaire, M. Roper, A. Ajdari and H. A. Stone, "Imbibition by polygonal spreading on microdecorated surfaces," *Nat. Mater.*, 6, 661 (2007).
83. C. W. Extrand, S. I. Moon, P. Hall and D. Schmidt, "Superwetting of structured surfaces," *Langmuir*, 23, 8882–8890 (2007).
84. R. Raj, S. Adera, R. Enright and E. N. Wang, "High-resolution liquid patterns via three-dimensional droplet shape control," *Nat. Commun.*, 5, 4975 (2014).
85. V. Jokinen, M. Leinikka and S. Franssila, "Microstructured surfaces for directional wetting," *Adv. Mater.*, 21, 4835–4838 (2009).
86. R. J. Vrancken, M. L. Blow, H. Kusumaatmaja, K. Hermans, A. M. Prenen, C. W. M. Bastiaansen, D. J. Broera and J. M. Yeomansb, "Anisotropic wetting and de-wetting of drops on substrates patterned with polygonal posts," *Soft Matter*, 9, 674–683 (2013).
87. Q. Yuan, X. Huang and Y. P. Zhao, "Dynamic spreading on pillar-arrayed surfaces: Viscous resistance versus molecular friction," *Phys. Fluids*, 26, 092104 (2014).
88. E. Chen, Q. Yuan, X. Huang and Y. P. Zhao, "Dynamic polygonal spreading of a droplet on a lyophilic pillar-arrayed surface," *J. Adhesion Sci. Technol.*, 30, 2265–2276 (2016).
89. R. Fetzer, M. Ramiasa and J. Ralston, "Dynamics of liquid displacement," *Langmuir*, 25, 8069–8074 (2009).
90. J. D. Coninck and T. D. Blake, "Wetting and molecular dynamics simulations of simple liquids," *Annu. Rev. Mater. Res.*, 38, 1–22 (2008).
91. P. Concus and R. Finn, "On the behaviour of a capillary surface in a wedge," *Proc. Natl Acad. Sci. USA*, 63, 292–299 (1969).
92. M. M. Weislogel and S. Lichter, "Capillary flow in an interior corner," *J. Fluid Mech.*, 373, 349–378 (1998).
93. R. Seemann, M. Brinkmann, E. J. Kramer, F. F. Lange and R. Lipowsky, "Wetting morphologies at microstructured surfaces," *Proc. Natl Acad. Sci. USA*, 102, 1848–1852 (2005).
94. Q. Yuan and Y. P. Zhao, "Wetting on flexible hydrophilic pillar-arrays," *Sci. Rep.*, 3, 1944 (2013).

SYMBOLS

a	semi-axis of ellipsoid (m)
A	Hamaker constant
A_m	amplitude of the corrugated surface (m)
$A_{sl(\text{true})}$	actual area of solid surface (m^2)
$A_{sl(\text{projected})}$	projected area of solid surface (m^2)
b	semi-axis of ellipsoid (m)
c	semi-axis of ellipsoid (m)
d	thickness of dielectric layer (m)
dA	the projected areas change (m^2)
E	electric field intensity (V/m)
f_1	fraction of surface area occupied by water
f_2	fraction of the surface area occupied by air
f_x	roughness in vertical direction to grooves
f_y	roughness in parallel direction to grooves
F	free energy (J)
F_d	driving force originating from the unbalance of surface tensions (N)
g	acceleration of gravity (m/s^2)
h	depth of the grooves (m)
l	spreading length (m)
l_1	width of grooves (m)
l_2	width of ridge (m)
l_c	characteristic length (m)
L	wavelength of the corrugated surface (m)
n	scaling exponent
r	real-time radius of droplet (m)
r_i	initial radius of droplet (m)
r_o	roughness factor
r_s	radius of curved substrate (m)
S	spreading parameter (N/m)
S_c	anisotropic factor
S_i	slenderness ratio of the interior corner
t	time (s)
U	electric field energy in insulating layer (J)
v	volume saturated with an electric field
v_{ca}	characteristic capillary velocity (m/s)
v_e	equilibrium velocity (m/s)
v_{ex}	initial excess velocity (m/s)
v_x	spreading velocity in X direction (m/s)
v_y	spreading velocity in Y direction (m/s)
v_i	velocity of MCL (m/s)
V	voltage (V)

V_c	critical voltage (V)
W	electric power (J)
W_a	adhesion between droplet and solid (N/m)
W_c	cohesion between droplet and solid (N/m)
z_0	the lowest level of the droplet in the z axis (m)
α	ratio of the increased solid-liquid interface to total increased interface
β	ratio of increased actual area to increased apparent area
γ_{lv}	surface tension of liquid-vapor (N/m)
γ_{sl}	surface tension of solid-liquid (N/m)
γ_{sv}	surface tension of solid-vapor (N/m)
ε	permittivity (F/m)
η	viscosity (Pa·s)
η_{EW}	electrowetting number
θ	real-time contact angle (°)
θ_0	initial contact angle (°)
θ_1	the angle between solid-vapor and solid-liquid interfaces (°)
θ_2	the angle between solid-liquid and liquid-vapor interfaces (°)
θ_3	the angle between liquid-vapor and solid-vapor interfaces (°)
θ_c	contact angle on grooved surface (°)
θ_{co}	composite contact angle (°)
θ_{f1}	contact angles of droplet on material I (°)
θ_{f2}	contact angles of droplet on material II (°)
θ_e	equilibrium contact angle (°)
θ_r	droplet contact angles on rough surface (°)
θ_s	droplet contact angle on smooth surface (°)
θ_x	contact angle observed in vertical direction to grooves (°)
θ_y	contact angle observed in parallel direction to grooves (°)
κ	ratio of contact areas to the projected areas
λ	ratio of velocities in parallel and vertical directions
ρ	density (kg/m³)
τ	characteristic time (s)
ϕ	cross angle (°)
ϖ	parameter related to spreading velocity

6 A Quantitative Approach to Measure Work of Adhesion

Ratul Das, Jie Liu, Sakshi B. Yadav, Sirui Tang, Semih Gulec, and Rafael Tadmor

CONTENTS

6.1 INTRODUCTION

Work of adhesion, unlike surface tension, has not achieved optimum measurability even today. Methods to measure surface tension, such as the Wilhelmy Plate method (1863) [1], Tate's falling drop method (1864) [2], and Dupré's concept to determine relationship between the solid–liquid work of adhesion and the surface tension via contact angle measurements (1869) [3], are still widely used to measure surface tension but work of adhesion measurements are restricted to estimation via the Young-Dupré equation [3–23].

The contact angle values required in the Young-Dupré equation can further complicate the measurement since these values are affected by hysteresis and unknown differences between macroscopic and nanoscopic contact angle values. Thus, this equation has been used qualitatively and its shortcomings have been reported [11,24].

Therefore, "In the absence of any data on the energy of adhesion," [4] and with the shortcomings of the existing methods [11,24], there is a need for developing a method that can help quantify work of adhesion and not restrict it to a mere estimation.

Intuitively, work of adhesion is, literally, work for separating the liquid from the solid without any dependence on contact angle measurements. Such an experimental setup was built by Boreyko and Chen [25] wherein solid–liquid separation was induced by adding vibrational energy to the system. Our setup is also based on the same premise, except that it is better controlled and allows for a gradual increase in the pull-off force acting on the drop. Such a separation via controlled pull-off force agrees with Tate's analogy (described in the following two paragraphs) to measure the liquid surface tension using the *falling drop weight* method [2].

In this method, liquid drops from the end of a tube are allowed to fall into a container until enough are collected to accurately determine the weight per drop. Liquid surface tension, $\gamma_{LV}\left(\frac{N}{m}\right)$, can be determined by Tate's law as follows:

$$\gamma_{LV} = \frac{mg}{\pi D_{tube}} \tag{6.1}$$

where m (Kg) is the mass of the liquid at the moment of separation of the drop from the tube tip, $g\left(\frac{m}{s^2}\right)$ is the gravitational acceleration, and the $D_{tube}(m)$ is the inner diameter of the tube tip if the liquid does not wet the tip; otherwise, the D_{tube} is the outer diameter of the tube tip. Equation 6.1 describes ideal Tate's law which assumes that no drop volume is left on the tube tip when the drop falls.

In practice, the drop weight obtained is smaller than the "ideal" value mg because, as demonstrated in Adamson and Gast's book [6], the drop that falls is smaller than the entire drop under the tube (part of the liquid is left on the tube). Yet, the *necking* is initiated by the entire weight under the tube, and therefore the entire drop weight (and not just the weight that falls) needs to be considered. The later *necking* at a lower position along the drop is not related to the initiation of the process because the force that induced the fall is the entire weight under the tube. Similarly, based on this analogy, the force that induces the spontaneous solid–liquid area reduction in our experiment is associated with the entire drop (and not what is torn away from it at a later stage).

In this chapter, like Tate's setup, we try to increase the drop weight; however, the former method does so by inflating the drop, whereas we do so by increasing the normal force acting on the drop [26–30] (described later in detail). In the drop weight method, the weight of the falling drop in Equation 6.1 is larger than the real value of the weight of drop due to the drop formation. In the calculation with known value of the diameter of the tube tip, the unknown mass of the falling drop is the main reason causing the error in that system. In our system, the mass of the drop is prescribed; the measurement of drop diameter is the main source of error.

In the drop weight method, capillary forces influence the solid–liquid separation, whereas in our method, we look for a *critical* depinning force beyond which no further pinning force can prevent the separation (further discussion about determining critical depinning force is given in later sections of this chapter).

In addition to the drop weight method, the Wilhelmy plate method [1] of surface tension measurement is applied commonly. A clean thin plate connected to a balance is oriented perpendicular to the interface between liquid–liquid or air–liquid, and the force acting on the plate results from the wetting. Once wetting occurs, a wetted perimeter, the triple line, forms which would move during the measurement process. This technique is used to measure the force (F) via the balance and calculate surface tension based on Wilhelmy equation:

$$\gamma_{LV} = \frac{F}{l\cos(\theta)} \tag{6.2}$$

where l (m) is the length of triple line and θ (°) is the contact angle. However, the triple line on the plate is sensitive to the defects (surface irregularities and chemical and topographical inhomogeneities) on the solid surface [31,32] and prevents accurate measurement of force.

6.2 THEORY

6.2.1 CENTRIFUGAL ADHESION BALANCE (CAB)

6.2.1.1 Design of CAB

CAB (Figure 6.1a and b), designed at Surface Science Lab, Lamar University, is a desktop device used to conduct studies pertaining to tribology and surface characterization. The schematic diagram in Figure 6.2a is the initial prototype (first generation) of CAB and shows its working parts. It consists of a rotating arm which moves perpendicular to the gravitational field. One end of the arm has a closed chamber which houses a goniometer (sample holder, light source, and a camera) and the other end has a counterbalance. The sample holder can tilt perpendicular to the rotation. In the past, this tilt was introduced manually but the new model is equipped with a motorized and automated tilt mechanism that is synchronized with the centrifugal rotation.

FIGURE 6.1 (a) The outside of centrifugal adhesion balance (CAB), shown with all the doors closed. (b) CAB goniometer.

(a)

(b)

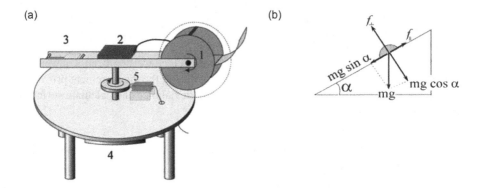

FIGURE 6.2 (a) Schematic of the first generation of CAB showing it's working parts: (1) closed chamber with sample holder, camera, and light source; (2) control box; (3) counterbalance; (4) DC motor; and (5) encoder to monitor angular velocity. (b) Free body diagram of the forces acting on a drop in a tilt plate method. (Reprinted with permission from Tadmor, R., et al., *Phys. Rev. Lett.*, 103, 26, 1–4, 2009. Copyright 2009 by the American Physical Society.)

Figure 6.2b is the free-body diagram of a drop and shows the forces acting on it in a conventional tilt plate experiment. However, as the drop slides, both lateral and normal forces change simultaneously which violates the basic research principle—change only one parameter at a time. Thus, when the drop motion is being investigated it is essential to change *only one parameter at a time*—either lateral force or normal force. This control of parameters makes analyzing of data easier and offers a better understanding of results. CAB is novel, one of a kind device that can control individual forces or *manipulate forces*.

Consider Figure 6.3a, it is the free body diagram of a drop in CAB's sample holder. It experiences two kinds of forces—gravitational (mg), due to its own weight, and centrifugal ($\omega^2 R$), due to rotation of CAB's arm. Both these forces are controlled

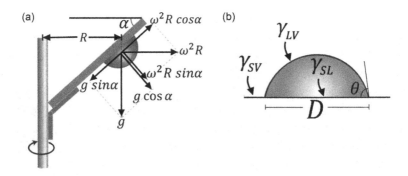

FIGURE 6.3 (a) Alignment for total zero lateral force ($\omega^2 R \cos\alpha = g \sin\alpha$). At this alignment, the drop can only move (fly) normal to the surface. (b) Some drop parameters used in this study. (Reprinted with permission from Tadmor, R. et al., *Langmuir*, 33, 15, 3594–3600, 2017. Copyright 2017 American Chemical Society.)

and changed by the tilt angle (α) of goniometer and increasing angular velocity (ω) of the arm, respectively (see Equations 6.3 and 6.4). The angular velocity changes at the rate of 1 RPM/s and the goniometer tilt changes accordingly, to either maintain $f_{\parallel} = 0$ or $f_{\perp} = 0$, as per the experimental requirement. For the work of adhesion study, the drop alignment in Figure 6.3a is such that $f_{\parallel} = 0$ is maintained throughout the experimental run.

CAB combines centrifugal and gravitational forces to manipulate normal and lateral forces, acting on the drop as follows:

$$f_{\parallel} = m\left(\omega^2 R \cos\alpha - g \sin\alpha\right) \tag{6.3}$$

$$f_{\perp} = m(\omega^2 R \sin\alpha + g \cos\alpha) \tag{6.4}$$

where $f_{\perp}$ and $f_{\parallel}$ (both measured in N) are the normal and lateral forces acting on the drop, $\omega\left(\frac{rad}{s}\right)$ is the CAB angular velocity, R (m) is the drop's distance from the CAB's center of rotation, α (rad) is the tilt angle with respect to the horizontal, and m (gm) is the drop's mass. Figure 6.3b shows some drop parameters used in this study.

6.2.1.2 Working of CAB

Figure 6.4a shows the assembly of goniometer—a light source, sample holder, and a camera—which is controlled by a tablet-style computer (it is part of counterweight as shown in Figure 6.4b). Live feed of the drop on the sample holder is captured by the camera and is sent to a computer, nearby in the lab. Rotation of arm and tilt of goniometer are controlled by a PID controller via feedback mechanism.

Via CAB, a user can manipulate both normal and lateral forces acting on a drop and, also, obtain other parameters such as drop diameter, drop height, contact angle, and so on. Contact angle measurement (obtained by using conventional goniometer) can give inaccurate value of work of adhesion because of hysteresis. We can overcome this flaw with CAB since it measures work of adhesion directly.

FIGURE 6.4 (a) Picture of the goniometer assembly of the CAB: (1) CCD camera; (2) syringe holder; (3) sample holder and optic dome; and (4) light source. (b) Picture of the rotating arm of the CAB with the counterbalance: (1) goniometer; (2) DC motor (inside the cylinder); and (3) tablet computer (counterbalance).

Similarly, the Wilhelmy plate method has certain drawbacks as well. This method can measure advancing and receding contact angles for a known liquid surface tension and measures the force required to move a contact line of constant length along a surface. CAB, on the other hand, can measure the force required to *shrink* the triple line to a critical point beyond which the triple line spontaneously *shrinks* without applying any additional force.

Wilhelmy plate method is also affected by surface irregularities and heterogeneities [31,32]. In CAB, we seek a critical depinning point (which occurs toward the end of the experiment) for our calculations, and since surface deformities do not affect that stage of experiment (or beyond) it has no influence on the work of adhesion measurement.

6.2.2 Measuring the Work of Adhesion Using CAB

To measure the work of adhesion by CAB, one major task is to maintain zero $f_{\parallel}$ during the whole operation. In accordance with Equation 6.3, $f_{\parallel} = 0$ gives

$$\tan \alpha = \frac{\omega^2 R}{g} \tag{6.5}$$

Two parameters, α (rad) and ω $\left(\frac{rad}{s}\right)$, which are tilt angle and CAB angular velocity, respectively, can be adjusted to achieve the equality in Equation 6.5. As CAB starts rotating, ω increases from 0, and α changes simultaneously according to Equation 6.5. Thus, the pair of forces, centrifugal and gravitational components in lateral direction, acting on the drop always have opposite direction but equal magnitude.

It is essential to ensure that no lateral movement of the drop occurs while the normal force gradually increases. In brief, only the force in the normal direction is increasing, but in the lateral direction, it is 0.

The experimental setup follows the Dupré gedanken experiment [3]. The solid–liquid work of adhesion, $W_{SL}\left(\frac{J}{m^2}\right)$ is given by the following equation:

$$W_{SL} = \gamma_S + \gamma_L - \gamma_{SL} \tag{6.6}$$

where $\gamma_S\left(\frac{N}{m}\right)$ is the solid-vapor, $\gamma_L\left(\frac{N}{m}\right)$ is the liquid-vapor, and $\gamma_{SL}\left(\frac{N}{m}\right)$ is the solid–liquid surface tensions, respectively. The Dupré equation (Equation 6.6) combined with Young–Laplace equation below:

$$\cos\theta = \frac{\gamma_s - \gamma_{SL}}{\gamma_L} \tag{6.7}$$

results in Equation 6.8, where θ is the contact angle that the liquid drop makes with the solid surface:

$$W_{SL} = \gamma_L\left(1 + \cos\theta\right) \tag{6.8}$$

Equation 6.8 is called the Young-Dupré equation [3] and it relates work of adhesion to the equilibrium contact angle.

In accordance with the Tate's law experiment to measure liquid surface tension, the work of separation per area equals the pull-off (separation) force per triple line circumference. For the separation force to equal the adhesion force, it is required that the analogy to Tate's law (as discussed earlier) will be insensitive to the direction of the separation process, since the liquid–liquid separation is vertical to the solid–liquid separation. Such a requirement is justified if we consider adhesion studies done on adhesive tapes [33] where the peeling force is insensitive to the peeling direction for tapes with low elastic modulus. Yet, more studies are needed to address it theoretically. Considering the adhesive tape understanding and Tate's analogy, the work of separation per area equals the pull off (adhesion) force per triple line circumference [26], that is,

$$W_{SL} = \frac{F_D}{\pi D_P} \qquad (6.9)$$

where F_D (N) is the pull-off force, which is perpendicular to the interface, D_P (m) is the diameter of the triple line when the drop is pulled off, and πD_P is the drop's pull-off circumference. Subscript P in D_P stands for pull-off.

W_{SL} is related to contact area and its unit is $\frac{J}{m^2}$ which is the same as surface energy. Also, surface energy has the dimension of force per unit length that is commonly used for surface tension. Similarly, work of adhesion can also be described as force per unit length.

The definition of adhesion force is contained in the definition of separation force. Adhesion force is equal to separation force when the direction of liquid–liquid separation is vertical to solid–liquid separation. In our study, liquid–liquid separation is perpendicular to the surface, while solid–liquid separation is along the surface with the retraction of the drop. It is an analogy of Tate's law, namely, separation force is insensitive to the direction of separation [26].

Since W_{SL} is an intensive property, it is independent of the circumference. The circumference of the drop, πD, varies during the separation process, but W_{SL} can be considered as a constant for any specific drop and it can be calculated when $D = D_P$.

$$\frac{\partial \left(\dfrac{F_D}{\pi D} \right)}{\partial (\pi D)} \Bigg|_{D=D_P} = 0 \qquad (6.10)$$

Theoretically, the force versus circumference graphs are depicted in Figure 6.5. There is a linear relationship between pull-off force F_D and pull-off circumference πD_P (Figure 6.5a). Normal force $F_\perp$ (adhesion force) from Equation 6.4 is increased with the reduced drop's circumference. Regarding the shrinkage that occurs during the CAB experimental run, the normal force $F_\perp$ would cause shrinkage of the circumference up to a point until $F_\perp$ intersects F_D and the corresponding πD is πD_P. Beyond this point, $F_\perp$ is no more required to shrink the triple line as represented by the horizontal line.

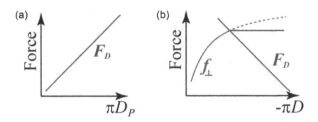

FIGURE 6.5 Schematics plotting of the Dupré force variation with (a) the drop's pull-off circumference. (b) The drop's circumference as a negative abscissa (i.e., negative x-axis), superimposed on the drop retention force, $f_\perp$. Following $f_\perp$ intersection with F_D (the negative abscissa represents a positive time axis), the circumference will need to continue decreasing spontaneously without further force investment, as represented by the line horizontal to the x-axis. (Reprinted with permission from Tadmor, R. et al., *Langmuir*, 33, 15, 3594–3600, 2017. Copyright 2017 American Chemical Society.)

6.3   EXPERIMENTAL

6.3.1   Materials

Silicon wafers, obtained from Virginia Semiconductor, Fredericksburg, VA (diameter: 76.2 mm ± 0.3 mm, center thickness: 381 μm ± 25 μm, orientation: ±0.9°, dopant: Boron, resistivity: 0.0034–0.0046 Ω-cm), were cut into 2 cm × 2 cm squares. Ethanol (99.5%, 200 proof absolute), distilled water (0.1 μm filtered Molecular Biology Reagent), hydrochloric acid (37%), toluene (99.5%), and octadecyltrimethoxysilane (90% technical grade, CAS No. 3069-42-9), and hydrofluoric (HF) acid (48%) were obtained from SigmaAldrich and used for the self-assembly process [34].

6.3.2   Preparation of Silanized Silica

Step 1: Surface hydrophilization
The silicon wafer substrate was rinsed with ethanol and distilled water and then dried in an oven at 100°C for 30 minutes. The substrate was then transferred to a UV/Ozone Cleaner for 45 minutes to remove any organic contaminants on it.

Step 2: Silanization process
Silanization took place in a closed beaker. A solution of octadecyltrimethoxysilane (OTS) and toluene with volume ratio of 99:1 (toluene:OTS) was prepared and heated to 70°C. The surface from Step 1 was immersed in this solution and kept in it for 3 hours [35]. Then the wafer was washed with water several times to remove any traces of toluene and dried in the oven at 80°C for 45 minutes. The wafer was then ready to use.

To obtain HF-etched silanized surface, we immersed the silicon surface into the HF solution for 1 hour before Step 1, then followed the same procedures for the surface hydrophilization and silanization process.

6.3.3 MEASUREMENT OF WORK OF ADHESION

The experiments were conducted in the environment of $23 \pm 1°C$ and 96% of relative humidity lab environment. The silanized silicon surface was placed in the sample holder and covered with an optic dome. *Above saturation* method [36–39] was used to maintain high humidity inside the optic dome to prevent drop evaporation.

For this study, as mentioned earlier, CAB is run at $f_\parallel = 0$ condition. The change in the CAB's angular velocity and the goniometer tilt angle are recorded in real time in the nearby computer, along with the time stamp (HH:MM:SS). The drop images are recorded (along with the time stamp) at the rate of 15 frames per second and stored in the tablet computer. The drop images are analyzed using CAB's Drop Image Analyzer software, to obtain drop parameters such as contact angle, position, circumference, normal force, and so on. The data obtained can then be represented in the plots as shown in Figures 6.8 and 6.9 and analyzed.

6.4 RESULTS AND DISCUSSION

To measure the work of adhesion, the normal force acting on the drop or the pull-off force is one datum (a specific data point shown in Figures 6.8 and 6.10) in an increasing force curve, $f_\perp$. The gradually increasing nature of $f_\perp$, required to detach a liquid drop from solid surface, is not due to work of adhesion but rather due to contact angle hysteresis [9,35], which is not the focus of this chapter. Irrespective of this fact, we see that $f_\perp$ value increases as D decreases. Thus, if we place the trend of the pull-off force shown in Figure 6.5a on the curve of $f_\perp$ versus D, we obtain what we see in Figure 6.5b. To visualize the procedure, we consider a decreasing circumference $(-\pi D)$ abscissa as shown in Figure 6.5b. At $f_\perp = F_D$ and $D = D_P$, the drop diameter, D, will decrease spontaneously, the applied normal force will not have enough time to increase significantly before the drop snaps from the solid surface. The point beyond which the drop shrinks spontaneously without any further force investment is identified by the line parallel to the x-axis in Figure 6.5b. To identify this value experimentally, we look for the first moment in the experimental data where

$$\frac{\partial f_\perp}{\partial (\pi D)} = 0 \tag{6.11}$$

At this point the force corresponds to work of adhesion, shown in Equation 6.9:

$$f_\perp = F_D = \pi D_P W_{SL} \tag{6.12}$$

After which the triple line circumference will decrease without any additional force, and the circumference reduction will be spontaneous, as can be seen from Figure 6.6.

Figure 6.6 shows selected drop images from an experimental run. Two processes occurred one after another: first, there was a reduction in the solid–liquid interfacial area and then the drop elongated, which implies an increase in the liquid–air interfacial area. The reason for this order is that the reduction in solid–liquid interfacial area initiates drop elongation due to mass conservation while the opposite

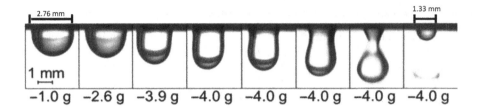

2.76 mm   1.33 mm

1 mm

−1.0 g   −2.6 g   −3.9 g   −4.0 g   −4.0 g   −4.0 g   −4.0 g   −4.0 g

FIGURE 6.6   Selected images of a sessile, water drop detachment from a silanized (C18) silicon surface. For this experiment, the CAB is run at an increasing normal force (effective gravity) which pulls on the drop (−1.0 g to −4.0 g is the change in effective gravity, acting on the drop, during the CAB run). (Reprinted with permission from Tadmor, R. et al., *Langmuir*, 33, 15, 3594–3600, 2017. Copyright 2017 American Chemical Society.)

is not true: drop elongation does not force a reduction in solid–liquid surface area (unless the solid–liquid surface area is weaker, which means that it retracts first). As the solid–liquid area is reduced, the drop becomes narrow mainly at the vicinity of the solid, and less at the drop apex (this is due to hydrostatic considerations: the lower hydrostatic pressure at higher locations equals the Laplace pressure at that point, and therefore the curvature is smaller or even negative, ending up in a bell shape). This results in a new drop shape that has a narrower neck. At some point the force pulling on the drop above the neck will equal the neck capillary pulling force, and a liquid–liquid snap will occur. This happens long after the time at which a spontaneous reduction in the solid–liquid interfacial area is initiated (several frames). *Therefore, the point considered for the pull-off is when the solid–liquid area starts to reduce spontaneously,* namely, without an increase in the normal force and not when the liquid–liquid separation occurs a few frames later as a result of the change in the drop shape that the solid–liquid retraction induces.

Additionally—in analogy to two springs in series in which treating one of the springs can be done irrespective of the other—we are concerned with only the solid–liquid area change, this interfacial change is essential to determine work of adhesion, irrespective of the drop elongation (which is a separate study) (Figure 6.7). The analogy here is the two interfaces whose area is being changed due to the increasing normal force: the solid–liquid area reduction and the liquid–air area elongation. Unlike

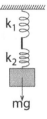

k_1

k_2

mg

FIGURE 6.7   Schematic of a system of two springs in a series, with spring constants k_1 and k_2, and a weight of mass m. (Reprinted with permission from Tadmor, R. et al., *Langmuir*, 33, 15, 3594–3600, 2017. Copyright 2009 by the American Physical Society.)

the system of two springs, the solid–liquid interfacial area reduction, reduces the liquid–air circumference (at the triple contact line), making further elongation easier. This, together with *necking* above the triple line (and possible liquid–air undulations), causes the thinnest part of the liquid–air interface (*neck*) to snap and, maybe, a little drop to be left behind. It appears that in all systems encountered, the spontaneous shrinkage of the solid–liquid contact area precedes the necking. This spontaneous drop shrinkage is an essential condition to measure and represent solid–liquid interaction, or else there is no solid–liquid area change to follow. In other words, we can treat independently each of the two entities (solid–liquid interface and liquid–air interface), provided that the solid–liquid contact area reduces before the necking.

As expressed in Equation 6.12, the important parameter is not the effective gravity but rather the force pulling the drop and the resulting circumference. Figure 6.8a shows the overall trend, while Figure 6.8b magnifies the final stage. The first point that obeys Equation 6.12 is marked with an arrow in Figure 6.8b.

Taking the values from the plot and substituting in Equation 6.12, we obtain the following:

$$W_{SL} = \frac{f_\perp}{\pi D_P} = \frac{361\,\mu N}{7.19\,mm} = 50.2\,\frac{mJ}{m^2} \tag{6.13}$$

From Equation 6.8, the contact angle that corresponds to the work of separation above is 107.6°, while the measured apparent advancing angle was 107°. This value is in line with the repeated data in the references [8, 22, 40], which shows that the measured apparent contact angles are lower than the angles back-calculated from the work of adhesion.

Contact angle hysteresis is affected by surface roughness and heterogeneity, and these parameters also affect the process of solid–liquid separation. An apparent contact angle that is lower than the theoretically predicted one was also observed by

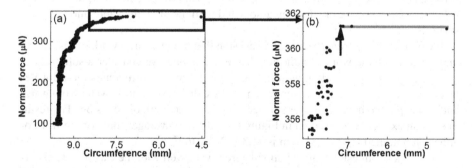

FIGURE 6.8 Water drop triple line circumference on silanized (C18) silicon surface versus the effective gravitational force (normal force) pulling on the drop. Drop size is 9.2 μL. (a) Data points for the entire duration of the experiment. (b) Data points for the last 3 seconds of the experiments that show spontaneous drop shrinkage. (Reprinted with permission from Tadmor, R. et al., *Langmuir*, 33, 15, 3594–3600, 2017. Copyright 2017 American Chemical Society.)

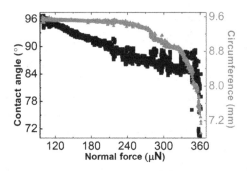

FIGURE 6.9 The apparent contact angle (■) and circumference (▲) of 9.2 µL water drop on a silanized (C18) silicon versus pull-off force (normal force) on the drop. (Reprinted with permission from Tadmor, R. et al., *Langmuir*, 33, 15, 3594–3600, 2017. Copyright 2017 American Chemical Society.)

Fan et al. [41]. They showed that the apparent macroscopic experimental cutoff that distinguishes between hydrophobic and hydrophilic surfaces is lower than the theoretical 90° value (which may correspond to a nanoscopic angle that differs from the macroscopic observed value). In turn, the surface irregularity is related to somewhat irregular contact angles (see Figure 6.9). Nonetheless, the point at which the spontaneous separation occurs is the same, regardless of the way the drop reached that critical value.

The wafer used for the plot in Figure 6.9 is rather smooth and has a low contact angle hysteresis of 13.8° when measured under normal gravity conditions (with θ_A (1 g) = 103° and θ_R (1 g) = 89.2°). It is only when we modify (gradually increase) the effective gravity that we obtain a higher contact angle hysteresis.

The contact angle hysteresis associated with surface heterogeneities can vary from drop to drop due to the large area covered by a typical drop. However, if the surface has a defect-free area at the triple line, then once the normal pull-off force is higher than the work of adhesion associated with the area, the triple line will start shrinking spontaneously.

It can be implied from the above discussion that the measured work of separation may be close to the work of adhesion. To test this idea, we consider a solid–liquid system with a known work of adhesion. Glass is known to form a nanometric water layer on it [42], so in a glass–water system, we are in fact separating water from water and can expect to have the water surface tension as the work of adhesion. The result of such an experiment is shown in Figure 6.10, and the average value for the work of adhesion obtained for such system is 71.3 ± 2.3 mJ/m².

The work of adhesion value obtained agrees very well with the classic Tate experiment and literature values [42]. The macroscopic (observed) contact angle value of this system is roughly 20°, which is in agreement with measurements of Pashley and Kitchener [42] and is lower than the back-calculated contact angle based on the work of separation experiments (but again corresponds well with references [11,16,36]). Therefore, we conclude that our work of separation measurements correlate well with the thermodynamic work of adhesion.

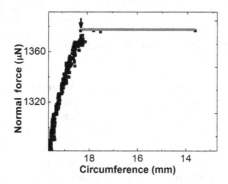

FIGURE 6.10　Water drop triple line circumference on a glass surface versus the effective gravitational force pulling on the drops (zoomed in at the end of the run). (Reprinted with permission from Tadmor, R. et al., *Langmuir*, 33, 15, 3594–3600, 2017. Copyright 2017 American Chemical Society.)

We also show below several images from the 13-minute long CAB experiment of separating water from glass (see Figure 6.11).

Surface roughness plays a critical role in wetting phenomenon, in the same way that it affects drop spreading [27,28]. It also affects the spontaneous retraction of triple line and hence the work of adhesion measurements. A Wenzel regime is expected to have a higher work of adhesion, while a Cassie regime will result in a lower value. The preliminary results that we have obtained show it is indeed the case, such an experiment was conducted using a HF-etched silanized (C18) silicon and water. The value of the work of adhesion obtained, 60.2 mJ/m², is significantly higher when compared to the unetched surface shown before and this change in the work of adhesion value results from the change in the surface roughness.

In Table 6.1 we compare our work of separation results with literature estimates for the work of adhesion. We consider these categories: (1) the work of adhesion based on the contact angle measurements (Young-Dupré equation) or theoretically predicted value (when it exists); (2) surface tension of the liquid; and (3) work of

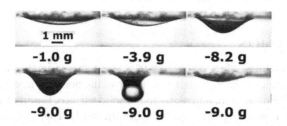

FIGURE 6.11　Selected images of a sessile, water drop on glass surface. During the experiment, the effective gravity (pull-off force) acting on the drop is gradually increased until the drop snaps (−1.0 g to −9.0 g is the change in effective gravity, acting on the drop, during the CAB run). (Reprinted with permission from Tadmor, R. et al., *Langmuir*, 33, 15, 3594–3600, 2017. Copyright 2017 American Chemical Society.)

TABLE 6.1

Comparing W_{SL} from CAB with Existing Work of Adhesion in Current Literature

Work of Adhesion (WoA) from CAB (W_{SL})	Theoretical Prediction for WoA (W_T)	Is W_T Different from Liquid Surface Tension[a]?	Agreement with Ref [11,16,36] That $W_{Thermo} < W_\theta$
Water on OTS-coated Si $W_{SL} = 51.9 \pm 1.5$ mJ/m²	50.3 mJ/m² $< W_T <$ 85.7 mJ/m² (corresponding to the Young-Dupré equation solved for $\theta_{A\_max} = 107.5°$, $\theta_{R\_min} = 79°$)[b]	Yes: 51.9 ± 1.5 mJ/m² $\neq 72.5$ mJ/m² for water at 96% relative humidity	Yes:[c] $51.9 < 66$ mJ/m², i.e., W_T that corresponds to 95°
Water drop separated from water nanolayer on glass $W_{SL} = 71.3 \pm 2.4$ mJ/m²	$W_T = 72.5$ mJ/m² if the water layer covering the glass can slip or $W_T = 145$ mJ/m² if the water layer covering the glass cannot slip	No: 71.3 ± 2.4 mJ/m² ≈ 72.5 mJ/m² for water at 96% relative humidity	Yes:[c] $71.3 < 141$ mJ/m², i.e., W_T that corresponds to 15.5°

Source: Tadmor, R. et al., *Langmuir*, 33, 15, 3594–3600, 2017.

a   The surface tension value is for the relative humidity conditions in the CAB chamber, i.e., 96%.

b   The maximal and minimal values are obtained at different effective gravities using the CAB. At normal gravity (1 g), the contact angle hysteresis is smaller ($\theta_{A\_max} = 103°$, $\theta_{R\_min} = 89.2°$) for this smooth surface.

c   The literature (refs [11,16,36]) shows that the thermodynamic work of adhesion, W_{Thermo}, is typically lower than that obtained from the "as placed" contact angle, W_θ.

adhesion comparisons that exists in the literature. We see from Table 6.1 that the work of separation values that we obtain correlate well with understanding of the work of adhesion in current literature [11,16,36].

6.5   CONCLUSION

We used the centrifugal adhesion balance to apply an increasing normal force, which is the net of gravitational and centrifugal forces in the normal direction. The normal force detaches the drop perpendicularly to the solid surface. Based on our technique, we demonstrate how to experimentally obtain the work of separation. The value of the work of separation correlates well with the expected work of adhesion and is independent of drop size or initial conditions, which is in agreement with the Dupré equation.

ACKNOWLEDGMENT

This study was supported by NSF grants CMMI-1405109, CBET-1428398, and CBET-0960229.

REFERENCES

1. L. Wilhelmy, "Ueber die Abhängigkeit der Capillaritäts-Constanten des Alkohols von Substanz und Gestalt des benetzten festen Körpers," *Ann. der Phys. und Chemie*, 195, 6, 177–217 (1863).
2. T. Tate, "On the magnitude of a drop of liquid formed under different circumstances," *London, Edinburgh, Dublin Philos. Mag. J. Sci.*, 27, 181, 176–180 (1864).
3. A. M. Dupré, *Théorie mécanique de la chaleur*, Paris, France: Gauthier-Villars (1869).
4. J. Israelachvili, *Intermolecular and Surface Forces*, Elsevier (2011).
5. I. You, T. G. Lee, Y. S. Nam, and H. Lee, "Fabrication of a micro-omnifluidic device by omniphilic/omniphobic patterning on nanostructured surfaces," *ACS Nano*, 8, 9, 9016–9024 (2014).
6. A. W. Adamson and A. P. Gast, *Physical Chemistry of Surfaces*, 6th ed., 124, 5. New York: John Wiley & Sons (1977).
7. D. A. Dikin, S. Stankovich, E. J. Zimney, R. D. Piner, G. H. Dommett, G. Evmenenko, S. T. Nguyen, and R. S. Ruoff, "Preparation and characterization of graphene oxide paper," *Nature*, 448, 7152, 457–460 (2007).
8. J. J. Kuna, K. Voïtchovsky, C. Singh, H. Jiang, S. Mwenifumbo, P. K. Ghorai, M. M. Stevens, S. C. Glotzer, and F. Stellacci, "The effect of nanometre-scale structure on interfacial energy," *Nat. Mater.*, 8, 10, 837–842 (2009).
9. S. Rose, A. Prevoteau, P. Elzière, D. Hourdet, A. Marcellan, and L. Leibler, "Nanoparticle solutions as adhesives for gels and biological tissues," *Nature*, 505, 7483, 382–385 (2014).
10. T. M. Schutzius, S. Jung, T. Maitra, G. Graeber, M. Köhme, and D. Poulikakos, "Spontaneous droplet trampolining on rigid superhydrophobic surfaces," *Nature*, 527, 7576, 82–85 (2015).
11. T. Salez, M. Benzaquen, and É. Raphaël, "From adhesion to wetting of a soft particle," *Soft Matter*, 9, 45, 10699 (2013).
12. G. P. Maier, M. V. Rapp, J. H. Waite, J. N. Israelachvili, and A. Butler, "Adaptive synergy between catechol and lysine promotes wet adhesion by surface salt displacement," *Science*, 349, 6248, 628–632 (2015).
13. G. P. Maier, M. V. Rapp, J. H. Waite, J. N. Israelachvili, and A. Butler, "Adaptive synergy between catechol and lysine promotes wet adhesion by surface salt displacement," *Science*, 349, 6248, 628–632 (2015).
14. S. Cho, S. Kim, J. H. Kim, J. Zhao, J. Seok, D. H. Keum, J. Baik, D. H. Choe, K. J. Chang, K. Suenaga, and S. W. Kim, "Phase patterning for ohmic homojunction contact in $MoTe_2$," *Science*, 349, 6248, 625–628 (2015).
15. P. G. de Gennes, F. Brochard-Wyart, and D. Quéré, *Capillarity and Wetting Phenomena*, New York: Springer New York (2004).
16. Y. Lu, S. Sathasivam, J. Song, C. R. Crick, C. J. Carmalt, and I. P. Parkin, "Robust self-cleaning surfaces that function when exposed to either air or oil," *Science*, 347, 6226, 1132–1135, (2015).
17. H. Y. Erbil, "The debate on the dependence of apparent contact angles on drop contact area or three-phase contact line: A review," *Surf. Sci. Rep.*, 69, 4, 325–365 (2014).

18. D. Seveno, T. D. Blake, and J. De Coninck, "Young's Equation at the Nanoscale," *Phys. Rev. Lett.*, 111, 9, 96101, (2013).

19. A. Y. Stark, D. M. Dryden, J. Olderman, K. A. Peterson, P. H. Niewiarowski, R. H. French, and A. Dhinojwala, "Adhesive interactions of geckos with wet and dry fluoropolymer substrates," *J. Royal Soc. Interface*, 12, 108, 20150464 (2015).

20. H. E. N'guessan, A. Leh, P. Cox, P. Bahadur, R. Tadmor, P. Patra, R. Vajtai, P. M. Ajayan, and P. Wasnik, "Water tribology on graphene," *Nat. Commun.*, 3, 1242 (2012).

21. D. H. Kim, M. C. Jung, S. H. Cho, S. H. Kim, H. Y. Kim, H. J. Lee, K. H. Oh, and M. W. Moon, "UV-responsive nano-sponge for oil absorption and desorption," *Sci. Rep.*, 5, 1, 12908 (2015).

22. K. Voitchovsky, J. J. Kuna, S. A. Contera, E. Tosatti, and F. Stellacci, "Direct mapping of the solid-liquid adhesion energy with subnanometre resolution," *Nat. Nanotechnol.*, 5, 6, 401–405 (2010).

23. E. Raphael and P. G. De Gennes, "Rubber-rubber adhesion with connector molecules," *J. Phys. Chem.*, 96, 10, 4002–4007 (1992).

24. L. Heepe, A. E. Kovalev, A. E. Filippov, and S. N. Gorb, "Adhesion failure at 180 000 frames per second: Direct observation of the detachment process of a mushroom-shaped adhesive," *Phys. Rev. Lett.*, 111, 10, 1–5 (2013).

25. J. B. Boreyko and C. H. Chen, "Restoring superhydrophobicity of lotus leaves with vibration-induced dewetting," *Phys. Rev. Lett.*, 103, 17, 1–4 (2009).

26. R. Tadmor, R. Das, S. Gulec, J. Liu, H. N'guessan, M. Shah, P. Wasnik, and S. B. Yadav, "Solid–Liquid Work of Adhesion," *Langmuir*, 33, 15, 3594–3600 (2017).

27. C. W. Extrand, "Comment on Solid–Liquid Work of Adhesion," *Langmuir*, 33, 36, 9241–9242 (2017).

28. S. Gulec, S. Yadav, R. Das, and R. Tadmor, "Reply to Comment on Solid–Liquid Work of Adhesion," *Langmuir*, 33, 48, 13899–13901 (2017).

29. S. B. Yadav, R. Das, S. Gulec, J. Liu, and R. Tadmor, "The Interfacial Modulus of a Solid Surface and the Young's Equilibrium Contact Angle Using Line Energy," In: *Advances in Contact Angle, Wettability and Adhesion*, K. L. Mittal, Ed., Hoboken, NJ: John Wiley & Sons, 131–143 (2018).

30. R. Tadmor, S. B. Yadav, S. Gulec, A. Leh, L. Dang, H. N'guessan, R. Das, M. Turmine, and M. Tadmor., "Why drops bounce on smooth surfaces," *Langmuir*, 34, 15, 4695–4700 (2018).

31. E. Chen, Q. Yuan, X. Huang, and Y. P. Zhao, "Dynamic polygonal spreading of a droplet on a lyophilic pillar-arrayed surface," *J. Adhes. Sci. Technol.*, 30, 20, 2265–2276 (2016).

32. Q. Yuan, X. Huang, and Y. P. Zhao, "Dynamic spreading on pillar-arrayed surfaces: Viscous resistance versus molecular friction," *Phys. Fluids*, 26, 9, 92104 (2014).

33. Y. Sugizaki, T. Shiina, Y. Tanaka, and A. Suzuki, "Effects of peel angle on peel force of adhesive tape from soft adherend," *J. Adhes. Sci. Technol.*, 30, 24, 2637–2654 (2016).

34. B. W. Ninham and P. Lo Nostro, *"Molecular Forces and Self Assembly,"* Cambridge, UK: Cambridge University Press (2010).

35. H. Einati, A. Mottel, A. Inberg, and Y. Shacham-Diamand, "Electrochemical studies of self-assembled monolayers using impedance spectroscopy," *Electrochim. Acta*, 54, 25, 6063–6069 (2009).

36. N. Belman, K. Jin, Y. Golan, J. N. Israelachvili, and N. S. Pesika, "Origin of the contact angle hysteresis of water on chemisorbed and physisorbed self-assembled monolayers," *Langmuir*, 28, 41, 14609–14617 (2012).

37. P. S. Wasnik, H. E. N'guessan, and R. Tadmor, "Controlling arbitrary humidity without convection," *J. Colloid Interface Sci.*, 455, 212–219 (2015).

38. W. Xu, R. Leeladhar, Y. T. Kang, and C. H. Choi, "Evaporation kinetics of sessile water droplets on micropillared superhydrophobic surfaces," *Langmuir*, 29, 20, 6032–6041 (2013).
39. R. Tadmor, P. Bahadur, A. Leh, H. E. N'Guessan, R. Jaini, and L. Dang, "Measurement of lateral adhesion forces at the interface between a liquid drop and a substrate," *Phys. Rev. Lett.*, 103, 26, 1–4 (2009).
40. A. P. Defante, T. N. Burai, M. L. Becker, and A. Dhinojwala, "Consequences of water between two hydrophobic surfaces on adhesion and wetting," *Langmuir*, 31, 8, 2398–2406 (2015).
41. J. G. Fan, X. J. Tang, and Y. P. Zhao, "Water contact angles of vertically aligned Si nanorod arrays," *Nanotechnology*, 15, 5, 501–504 (2004).
42. R. Pashley and J. Kitchener, "Surface forces in adsorbed multilayers of water on quartz," *J. Colloid Interface Sci.*, 71, 3, 491–500 (1979).

7 Adhesion of Bacteria to Solid Surfaces

Nabel A. Negm, Dina A. Ismail,
Sahar A. Moustafa, and Maram T.H. Abou Kana

CONTENTS

7.1 INTRODUCTION

Bacterial adhesion has been described as the balance of attractive and repulsive interactions between bacteria and solid surfaces. The type of interactions depends on several factors, including the nature of the bacterial strain, nature of the surface, nutrients concentration in the environment, and the interaction of bacterial cells with the other members in their community. The adhesive nature of bacteria is related to various outer membrane features such as pili, flagella, proteins, and lipopolysaccharides (LPSs) [1]. Adhesion of bacteria is governed not only by long-range forces such as steric and electrostatic interactions, but also by short-range forces such as van der Waal forces, acid–base, hydrogen bonding, and bio-specific interactions [2].

7.2 FORMS OF BACTERIAL COMMUNITIES

Bacteria generally exist in one of two types of population: (a) planktonic or freely existing in bulk solution, and (b) sessile, as a unit attached to a surface or within the confines of a biofilm. Plankton are collections of organisms that live in water. The name comes from the Greek word *planktos*, meaning "drifter." Plankton are carried by waves, tides, and currents. Plankton exists in many different shapes and sizes. Plankton are grouped into three categories (Figure 7.1): plants (phytoplankton), animals (zooplankton), and bacteria (bacterioplankton) [3].

Zooplankton range in size from tiny microscopic organisms, such as protozoans and rotifers, to larger plankton called macroplankton, such as jellyfish, shrimp, and fish

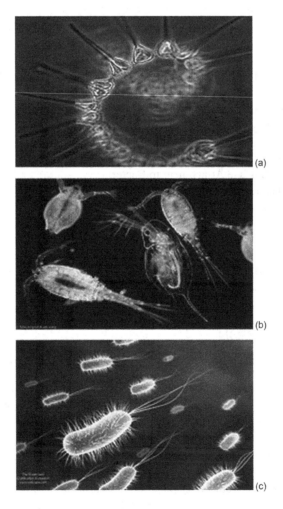

FIGURE 7.1 The three categories of planktons in the environment: (a) phytoplankton, (b) zooplankton, and (c) bacterioplankton. (From Agbeti, M.D., *Can. J. Fish. Aquat. Sci.*, 49, 1171–1175, 1992.)

TABLE 7.1

Classification of Planktons According to Their Size

Category	Size Range	Description
Femtoplankton	0.02–0.2 μm	Marine viruses
Picoplankton	20–200 μm	Small eukaryotic protists, bacteria
Nanoplankton	2–20 μm	Heterotrophic nanoflagellates feeding on bacteria
Microplankton	20–200 μm	Protozoans like ciliates
Mesoplankton	0.2–20 mm	Metazoans (copepod, medusa, etc.)
Macroplankton	20–200 mm	Membrane of hydromedusae, myside
Megaplankton	>200 mm	Metazoans (jellyfish)

Source: Fuhrman, A. J., *Nature*, 399, 541–548, 1999.

larvae. Unlike the phytoplankton, they are not dependent on light for food. They migrate to the surface of water at night to feed on phytoplankton or small zooplankton and sink deep below surface during the daylight to avoid being eaten by larger animals. Macroplankton range in size from 0.02 mm to 16 cm as represented in Table 7.1 [4].

Phytoplankton need light and nutrients to survive. The phytoplankton population may grow rapidly at high temperatures and in the presence of high nutrients. These dense "blooms" can change the color of the seawater so much that it can be seen from space. Phytoplankton can harm the environment since they block sunlight from reaching the bottom in shallow areas. During these blooms, the phytoplankton die and sink to the bottom where they decompose. This process consumes oxygen in the deep water where fish, lobsters, crabs, and other animals live.

A biofilm is defined as a microbial-derived sessile community characterized by cells that attach to a surface or to each other (Figure 7.2) [5]. Simply, it is the attachment of bacterial cells to the surface achieved with the help of gelatinous extracellular polymeric substances mainly composed of polysaccharides, proteins, and nucleic acids [5]. The attachment exhibits an altered phenotype with respect to growth rate and gene transcription compared to their free-floating (so-called planktonic form) counterparts [6].

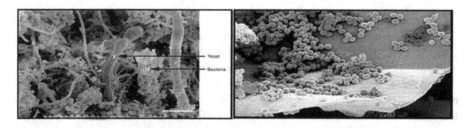

FIGURE 7.2 Electron micrograph of biofilm-forming microorganism. Left: distribution of bacteria and yeast in the biofilm; right: bacterial germs in the biofilm. (From Costerton, J.W., *Int. J. Antimicrob. Agents*, 11, 217–221, 1999. https://gimasterycourse.com/wp-content/uploads/Chronic-Candida-The-Role-of-Biofilm-in-Candida-Pathogenicity-and-Treatment-Options-9-21-15.pdf.)

7.3   CHARACTERISTICS OF A BIOFILM

A biofilm can be defined as an organized group of microorganisms living within a self-produced matrix of polymeric substances that get attached to diverse surfaces [7]. These microbial collectives are found to be ubiquitous in almost every environment [8]. Biofilms can be found on both biotic and abiotic surfaces [9]. Biofilms have been seen on liquid surfaces as a floating mat and as well as in submerged state [10]. Biofilms contain either homogeneous or heterogeneous communities of bacteria, embedded in a matrix of extracellular polymeric substance (EPS). An EPS mainly consists of polysaccharides, in addition to biomolecules such as proteins, lipids, and nucleic acids [9]. Polymers such as glycopeptides, lipids, and lipopolysaccharides form a scaffold that holds the biofilm together [11]. Investigation of the EPS coat in a biofilm has led to the discovery that biofilms are hydrogels with viscoelastic properties [12,13], which allows them to withstand mechanical stress. The nutrients that are present in the matrix of EPS are trapped for the use by the bacteria. The water present in the matrix is also efficiently trapped by hydrogen bonding with the hydrophilic polysaccharides in EPS [14].

7.4   PRINCIPLES OF BIOFILM FORMATION

The process of biofilm formation is complex, but generally involves five processes:

1. Growth of a surface conditioning film.
2. Close of microorganisms with the surface.
3. Reversible and irreversible adhesion of microbes to the conditioned surface.
4. Growth and division of the microorganisms with colonization of the surface, micro association, and biofilm formation; phenotype changes (changes in shape without change in the gene code); and genotype changes (changes in the gene codes).
5. Biofilm cell detachment/scattering.
   The five processes are explained in the next section.

7.4.1   THE CONDITIONING FILM

The conditioning layer is the substance where the biofilm grows, and it is generally composed of several organic and inorganic particles. These particles originate as contaminants in the bulk fluid and precipitate on the substrate where the biofilm is formed due to the gravitational force or flow movement. Consequently, the particles become part of the conditioning layer. The settled particles affect the properties of the conditioning film. Also, surface charge, surface potential, and surface tensions are positively altered by the interactions of the conditioning layer with substrate. The substrate surface provides anchorage and nutrients that supplement growth of the adhered bacterial community [15].

7.4.2   REVERSIBLE ADHESION

In the initial process of adhesion, planktonic and dispersed (not attached to the surface of substrate) bacterial cells are moved from bulk stream to the conditioned surface by physical forces or by bacterial supplements such as flagella. Some of these cells reach the surface and adhere. In addition, a number of reversibly adsorbed cells remain immobilized and become irreversibly adsorbed. It was reported that the physical attractive forces to bacteria such as flagella, fimbriae, and pili can overcome the physical repulsive forces of the surface due to the electrical double layer at the surface [16].

7.4.3   COLONY GROWTH

After the individual bacterial cells have adhered to the surface, the cells start to divide individually, in a process called binary division, and steeply spread outward and upward from the adhesion point to form bacterial groups called bacterial clusters [17]. This type of growth converts the adhered bacterial biofilm into a mushroom-like structure, which allows transporting of nutrients to deep bacterial colonies within the adhered biofilm.

7.4.4   BIOFILM DEVELOPMENT

The adhered phase of bacterial growth corresponds to a situation wherein the rate of cell division equals the bacterial cells death rate. At high cellular population in the colony, a series of cellular signaling mechanisms are generated from the crowded biofilm, called "quorum sensing" [18]. In quorum sensing, a number of inducers (chemical and peptide signals in high concentrations, e.g., homoserine lactones) are used to excite genetic expression of both mechanical and enzymatic processors of alginates, which form a fundamental part of the extracellular matrix. Enzymes are produced by the community itself which breaks down polysaccharides holding the biofilm together and aggressively releasing surface bacteria for colonization onto fresh substrates (Figure 7.3).

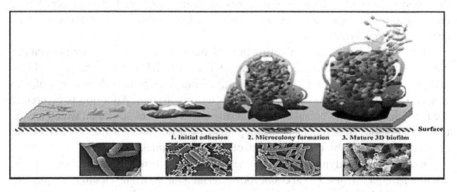

FIGURE 7.3   Steps for release of bacteria from biofilm to the environment: (1) initial adhesion of microorganisms to the surface; (2) formation of microcolony in the adsorbed biofilm; and (3) mature biofilm. (http://www.cresa.cat/blogs/sociedad/en/espanol-biofilms-bacterianos-por-que-deberia-importarnos/ Friday April 19th, 2013.)

7.5   MECHANISM OF BACTERIAL ADHESION

Individual bacterial cells are freely suspended in the bulk of the fluids before adhering to different surfaces. Motile bacteria are found in one of the three regions of the fluids: (1) in the bulk liquid, where the bacterial cells are not influenced by the surface; (2) in the near-surface of the bulk liquid, where the cells are influenced by the hydrodynamic movement effects of the surface; and (3) in the constrained near-surface, where the cells are influenced by both hydrodynamic movement effects of the surface and physicochemical effects of the surface such as van der Waals and electrostatic forces [19].

The role of physicochemical effects on bacterial adhesion to surfaces has been extensively reviewed. Two physicochemical approaches that were initially considered distinctly different are available to describe microbial adhesion interactions. In the thermodynamic approach [20], the interacting surfaces are assumed to physically contact each other under conditions of thermodynamic equilibrium, that is, reversible adhesion. The thermodynamic approach is based on surface-free energies of the interacting surfaces and does not include an explicit role for electrostatic interactions. Alternatively, the classical DLVO (Derjaguin, Landau, Verwey, Overbeek) approach describes the interaction energies between the interacting surfaces, based on Lifshitz-van der Waals and electrostatic interactions and their decay with separation distance [21]. Both approaches have shown merits for microbial adhesion, when certain collections of strains and species are considered. But both have failed so far to yield a generalized description of all aspects of microbial adhesion valid for each and every strain [22].

Van Oss et al. [23] introduced a so-called extended DLVO theory by including short-range Lewis acid base interactions in the classical DLVO approach. Inclusion of acid–base interactions in the classical DLVO approach probably [24] implies that hydrophobic attractive [25] and hydrophilic repulsive [26] forces can be accounted for in colloid and surface science in a more formal way.

The thermodynamic theory [27] is the first physicochemical approach that has been used to describe bacterial attachment to surfaces. It takes into account the various types of attractive and repulsive interactions, such as van der Waals, electrostatic, or dipole moment but expresses them collectively in terms of free energy, a thermodynamic term.

This approach requires an estimation of numerical values of thermodynamic parameters, that is, surface free energy of the bacterial and substratum surfaces and surface free energy (or surface tension) of the suspending solution, in order to calculate the Gibbs adhesion energy for bacterial adhesion. Adhesion is favored if the free energy per unit surface area is negative as a result of adhesion, which means that spontaneous attachment is accompanied by a decrease in free energy of the system, as predicted by the second law of thermodynamics.

From the thermodynamic point of view, there are three different theories that are more frequently used to account for bacterial adhesion. The Neumann equation, an approach based on polar-dispersion components, and the Lewis acid–base theory [28,29]. Each of these theories attributes a different role to the nature and the molecular details of surfaces and interfaces involved in the adhesion process. They are not generalizations or refinements of the same approach and do not depict different shades of the same subject, and they are incompatible.

Neumann's theory accepts that a single contact angle is sufficient to characterize the field of forces arising from the solid surface and that the molecular details do not affect the adhesion efficiency.

The polar-dispersion approach has succeeded in predicting the adhesion between phases where no specific interfacial interactions exist. However, the assumption that matter interacts through forces arising from permanent dipoles and that this kind of interaction, like the dispersion one, is symmetrical is in strong conflict with the present view of intermolecular interactions in condensed phases.

The electron-donor electron-acceptor approach is the most advanced theory that uses currently accepted physical knowledge to account for interfacial interactions [30]. It demonstrates that the permanent dipole contribution to intermolecular forces is negligibly small and that acid–base, and, in particular, hydrogen bonding is responsible for the interactions. However, the accuracy of the quantitative outcome of this theory is still debated.

In the DLVO theory, the interaction energy is distance dependent. The thermodynamic approach is an equilibrium model that does not allow for a kinetic interpretation. Generally, it is very difficult to obtain accurate values for bacterial surface free energies because these surfaces possess complex chemistry and hydration in vivo. Thus, calculations of free energy changes during adhesion may be incorrect. Furthermore, the thermodynamic theory applies to closed systems where no energy is put into the system from outside. Thus, the application of thermodynamic theory has not been entirely successful in explaining or predicting all the various attachment behaviors observed in bacterial systems. However, this approach has helped to explain an increasingly common observation: in numerous cases increased hydrophobicity of the solid surface or of the bacteria surface tends to result in increased numbers of attached cells. For two surfaces to come together resulting in molecular interactions, adsorbed water must be displaced. If the surface is highly hydrated, such water displacement is energetically unfavorable and may be impossible to overcome by the counteracting attractive interactions. It is clear from all the above that neither the DLVO nor the thermodynamic approach can fully explain bacterial adhesion.

For this reason an extended DLVO theory [31] has been suggested in which the hydrophobic/ hydrophilic interactions are included. So, the total adhesion energy can be expressed as in Equation 7.1:

$$\Delta G^{adh} = \Delta G^{vdW} + \Delta G^{dl} + \Delta G^{AB} \tag{7.1}$$

where ΔG^{vdW} and ΔG^{dl} are the "classic" van der Waals (vdW) interactions and double layer interactions and ΔG^{AB} relates to acid–base interactions.

Equation 7.1 introduces a component that, in principle, describes attractive hydrophobic interactions and repulsive hydration effects, which are 10–100 times stronger than the vdW interactions of surfaces in direct contact. The distance dependence, which is important in the calculation of the total adhesion energy, is given from the classical DLVO theory for the vdW and the double layer interactions and the distance dependence of the adhesion energy component ΔG^{AB} decays exponentially from its value at close contact. From these considerations it can be concluded that the application of physicochemical theory, although it has helped to explain some observations, has not been entirely successful in predicting the various attachment behaviors observed in bacterial systems. Thus, it may simply be a manifestation of the difficulty of applying a physical theory to biological systems. The complexity of bacterial surface polymer composition, as well as the change in polymer composition and synthesis with changing environmental conditions or time, can explain much of the variability in experimental observations of bacterial attachment.

The DLVO theory [32,33] has been used to describe the total interaction (V_{total}) between a cell and a surface as a balance between two additive factors: V_A represents the van der Waals interactions (generally attractive) and V_R represents the repulsive interactions due to the overlap between the electrical double layers of the cell and the substrate (Coulomb interactions, generally repulsive due to the negative charge of cells and substrate). Although the DLVO theory could account for the low levels of bacterial attachment to negatively charged surfaces observed experimentally, it cannot explain the variety of attachment behaviors observed with other types of surfaces or in solutions with appreciable electrolytes. It could be argued that the DLVO theory describes only one of several components of the attachment process. However, it does not describe the various molecular interactions that would come into play when polymers at the bacterial surface enter into contact with molecular groups on the substrate as well as any conditioning film. Moreover it does not account for structures and molecules on bacterial surfaces that affect cell-surface distance and the exact type of interaction, for the substrate roughness and the fact that correlation between surface charge and adhesion is not straightforward (the effect of charge is more important for adhesion of hydrophilic than hydrophobic cells) [25].

At low and moderate fluid velocities, nonmotile bacteria adhere to surfaces. At high fluid velocity, nonmotile bacteria are transported by the flow of fluids and do not adhere [34]. Motile bacteria attach to surfaces regardless of fluid velocity. Importantly, the difference between motile and nonmotile bacteria occurs only when motile cells actuate their flagella; bacteria with nonfunctional flagella adhere in the same way as cells lacking flagella [34] (Figure 7.4). The density of detached bacteria is commonly located in the range of 1.06–1.13 g mL^{-1} [35] and leads to the slow deposition of cells onto surfaces from suspension in bulk liquid. Sedimentation rates for different marine bacteria have been measured and range from 10–30 µmh^{-1} [36].

Interestingly, the density of *E. coli* cells increases when they change into their stationary phase (inactive state, spores), which facilitates *E. coli*'s rapid deposition on surfaces [37]. Populations of cells that make contact with surfaces may thus be increased in their stationary phase. In contrast, the density of *V. parahaemolyticus* cells decreases in their stationary phase [38].

After the bacterial cells contact the surfaces, their attachment occurs in two steps. The first step of bacterial adhesion is the initial attachment, which is reversible and occurs rapidly in about 1 minute [39,40]. This phase involves hydrodynamic and electrostatic interactions with the surface. During this period, the adhesion force between bacteria and surface increases rapidly. A similar phenomenon was reported about the attachment of polystyrene beads to a surface [41] and was attributed to some physicochemical interactions and not biological interactions. These interactions include the loss of interfacial water, structural changes in surface molecules, and repositioning of the cell body to maximize attachment to the surface [42].

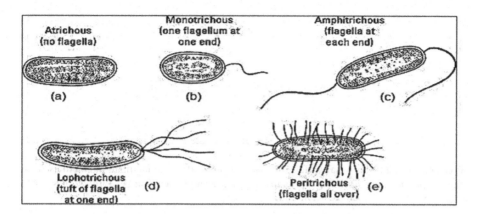

FIGURE 7.4 Bacterial cells: (a) without flagella; and (b–e) with different types of flagella. (From Jucker, B.A. et al., *Environ. Sci. Technol.*, 32, 2909–2915, 1998.)

Most bacteria possess predominantly net negative surface charge during the early stationary phase of cell growth [43], which enhances their interaction with the positively charged surfaces. This effect is diminished in media with highly ionic strength due to charge screening [44]. Quorum sensing in *E. coli* causes an increase in the negative charge on cell surfaces, which may facilitate the interaction of bacteria with surfaces during the initial stages of biofilm formation [45].

The second step in bacterial adhesion is an irreversible step and occurs on a time scale of several hours. This step involves van der Waals interactions between the hydrophobic region of the outer cell membrane of the bacteria and the surface. It also involves biological interaction involving several proteins that fix the initially adhered cells during their transition from reversible to irreversible attachment. The cytoplasmic *P. aeruginosa* protein SadB is required for transition from reversible to irreversible attachment [46]. In *E. coli*, both LPS and pili increase the initial rate of cell attachment and the rate of transition to irreversible attachment [47]. In *Pseudomonas fluorescens*, ABC transporter and secreted protein are required for irreversible attachment [48]. Irreversible attachment is also facilitated by the production of EPS secreted by cells in biofilms that are attached to surfaces, which anchors cells to surfaces [6].

The interaction of bacterial cells occurs preferentially to the surface and is dependent on their type and the forces affecting this interaction. This was observed from the fact that different bacteria occupying the same position do not necessarily interact with the same surfaces. For example, *Streptococcus mutans* binds to the surface of teeth but not to the tongue; while *Streptococcus salivarius* has opposite preference [49].

Bacteria have several different classes of extracellular organelles that mediate specific attachment to surfaces, including flagella, pili (also called fimbriae), and curli fibers (Figure 7.5) [50]. These organelles are frequently terminated with proteins which act as adhering promoters, which bind to molecules present on the surface. Two *E. coli* examples include type I pili, which binds to glycoproteins that present alpha-D-mannose, and type IV pili, which bind to phosphatidyl ethanolamine [51]. Several species of oral bacteria adhere specifically to proline-rich proteins in saliva [52,53].

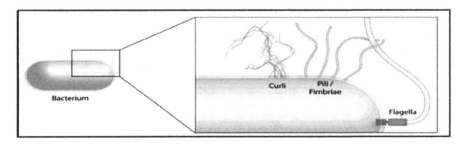

FIGURE 7.5 The bacterial surface has several organelles that facilitate interactions with substrates, including curli fibers, pili, and flagella. (From Beachey, E.H., *J. Infect. Dis.*, 143, 325–345, 1981.)

An interesting example of specific attachment involves *Flavobacterium johnsoniae*. SprB is an adhering promoter that is required for *F. johnsoniae* on agar (not on glass). SprB binds to agar surfaces, moves along the cell surface, and produces cell motion relative to the agar surface. Predicted homologs of SprB in the *F. johnsoniae* genome may facilitate cell attachment and motility on surfaces other than agar [54].

Reversible cell attachment does not necessarily lead to irreversible attachment during surface colonization. In *E. coli*, weak cell adhesion is facilitated by the pilus tip adhesion throughout FimH. FimH mediates binding to glycoproteins that have N-linked oligosaccharides presenting terminal mannose residues [55]. FimH binding to mono-mannose is dependent on shear stress, with a certain threshold level of force required to switch from weak to strong adhesion [56]. Anderson and coworkers observed that a wild type strain of *E. coli* that adhered weakly to mannosylated surfaces colonized the surface more rapidly than a mutant strain which attached strongly (irreversibly) to the surface [57].

7.6 BENEFITS OF SURFACE ATTACHMENT TO BACTERIA

Adhering to surfaces provides several advantages to bacteria. Horizontal attachment of bacterial cells to surfaces stimulates their growth, particularly in nutrient-poor environments. Due to the increase of contact surface between the cells and the sediment, organic material suspended in liquid during its deposition on the cellular surfaces increases the local concentration of nutrients [58]. Similarly, increasing the substrate surface area (e.g., by adding glass beads to a culture container) provides more area on which nutrients can adsorb, enabling the growth of the cells at nutrient concentrations that are too low to support bacterial growth [59].

Caulobacter crescentus is an attractive illustration of a bacterium species that takes advantage of surface attachment to optimize nutrient uptake. *C. crescentus* oscillates between stalked cells that adhere tightly to surfaces using a protein holdfast and motile cells that lack this organelle and instead have a polar flagellum. This phenotypic switch makes it possible for cells to adapt to both nutrient-rich (favoring motility) and nutrient-poor (favoring adhesion) environments [60].

In addition to facilitating nutrient capture, surface attachment allows some bacteria to obtain the necessary metabolites and co-factors directly from the surfaces to which they adhere. *Shewanella* and other genera of bacteria that grow on metal surfaces can use metals such as iron and magnesium as terminal electron acceptors in respiration [61,62].

Extracellular organelles facilitate the transport of ions between cells and surfaces. For example, *Geobacter sulfurreducens* uses pili to conduct charge transport between cells and surfaces. *Shewanella oneidensis* uses an outer membrane protein complex to form an electron bridge between the periplasm and the extracellular environment [63].

Bacteria attached to surfaces often exist as biofilms, which play several protective roles. The EPS secreted by cells in biofilms attached to surfaces provides protection from mechanical damage and shear caused by fluid flow [64,65]. Additionally, biofilms often exhibit resistance to antibiotic treatment [66]. Several different mechanisms contribute to this resistance, including the barrier function of the biofilm matrix; the presence of dormant persistent cells and highly resistant small colony variants; and up-regulation of several biofilm-specific antibiotic resistance genes [67,68].

Recently, it was demonstrated that cells adhered to surfaces but not associated with biofilms have resistance profiles that are similar to biofilm cells [69,70]. This does not require mutations in genomic DNA, as the process is reversible and cells become susceptible to antibiotics after detachment. John et al. [69] suggested that surface attachment facilitates antibiotic resistance by two primary mechanisms: (1) reducing the net negative charge on bacterial cells, and (2) enhancing the stability of the membrane.

However, attachment to surfaces is one way in which bacterial communities can attain such high densities. In addition to antibiotic resistance, cells in biofilms often gain protection from predators. When exposed to protozoa, the bacterium *Serratia marcescens* rapidly forms surface-associated microcolonies an early stage in biofilm formation that protects cells from attack by these predators [71].

Bacteria that are attached to surfaces, particularly those associated with biofilms, may become specialized in some functions in the bacterial community to cells in other regions [72]. In *Bacillus subtilis* biofilms, motile cells, EPS-producing cells, and spore-forming cells are localized at different regions within the biofilm. Strains of *B. subtilis* that are unable to form structured biofilms do not sporulate, which suggests that localization/specialization is required for the formation of bacterial spores [73]. Similarly, surface-associated microcolonies and biofilms of *Pseudomonas aeruginosa* contain groups of cells that display different susceptibility to antimicrobial agents [74,75].

Surface sensing is a precursor to swarming (Figure 7.6) which is considered an important adaptive behavior involving morphological changes in the cellular shape to ease the adaptive mechanism between the bacterial cells and the different morphological surfaces [76] that facilitates rapid community growth and migration of bacterial communities [77]. Swarming motility was reported in at least 15 different genera of bacteria from different natural habitats [76].

Several mechanisms of bacterial adaptation to different surfaces have been reported during bacterial swarming, including reduced susceptibility to antibiotics [78,79] and mutualistic interactions with fungal spores [80]. Mutualistic interaction is defined as an interaction between individuals of different species that results in positive (beneficial) effects on per capita reproduction and/or survival of the interacting populations. In *B. subtilis* swarming colonies, spatially distinct groups of cells express different

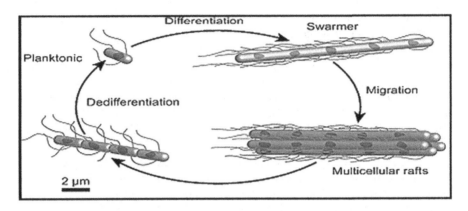

FIGURE 7.6 Swarming of bacterial cells starts from planktonic microorganism to multicellular rafts. (From Purevdorj-Gage, B. et al., *J. Gen. Microbiol.*, 151, 1569–1576, 2005.)

levels of flagellin (the protein that assembles into the flagellar filament) and have different cell morphologies [81,82]. Similarly, *Proteus mirabilis* cells with distinct cell morphologies are found in different regions of swarming communities [83,84].

Cells in bacterial communities such as swarms or biofilms interact with each other in several different ways. Bacteria are able to communicate through the use of small molecule chemical messengers in a process referred to as quorum sensing [85]. The dense packing of cells in bacterial communities facilitates an increase in the concentration of small molecules that transfer information between cells and trigger physiological changes [86]. The type of the small molecules and their concentration at the surfaces enhance the exchange of chemical information within biofilms and communities attached to surfaces [87–89]. Lateral gene transfer is also enhanced in biofilms compared to planktonic cells freely suspended in fluids [90]. Additionally, surface-associated growth induces phenotypes that promote natural immunity in *Vibrio cholerae* [91]. Myxobacteria cells associated with biofilms even exchange outer membrane proteins and lipids [92].

7.7 DRAWBACKS OF SURFACE ATTACHMENT TO BACTERIA

Adhering to surfaces also has several disadvantages to bacteria, including inhibition of motility, often due to a "switch" in the activation of genes involved in motility and adhesion. For example, gene coding for flagella may be turned off by the same transcriptional regulator that turns on genes for extracellular matrix production [93–95]. Inhibiting cell motility prevents cells from searching for optimal environments when nutrients become depleted. Bacterial cells may overcome this disadvantage (and others) in certain environments by sensing surfaces and triggering surface-associated phenotypes that activate motility and prevent adhesion. Some pathogenic bacterial strains also use surface sensing as a trigger to up-regulate virulence factors as a prelude to invasion of the host [96].

7.8 CELL ATTACHMENT TO DIFFERENT CLASSES OF MATERIALS

Bacterial attachment to surfaces is not desirable in a wide range of areas, including in lubrication [97], medical implants [98], water purification systems [99], and many more industrial processes [100]. The biofouling of metal ship hulls frequently begins with bacterial adhesion before progressing to larger marine organisms even in the use of antifouling coatings [101].

Thermodynamics plays a central role in regulating the binding of bacteria to surfaces. Cells attach preferentially to hydrophilic materials (i.e., materials with a high surface energy) when the surface energy of the bacterium is larger than the surface energy of the liquid medium containing the bacterial suspended cells. The surface energy of bacteria is typically smaller than the surface energy of liquids in which cells are suspended, and this mismatch causes cells to attach preferentially to the hydrophobic materials [20].

Bacteria are able to attach to the surfaces of a wide variety of materials, including glass, aluminum, stainless steel, organic polymers, and fluorinated materials such as poly(tetrafluoro ethylene) [102].

Bacteria can also attach to surfaces that initially resist the attachment of cells. This process occurs through the deposition of a layer of proteins including proteins found naturally in the environment as well as those secreted by the bacteria. The proteins condition the surface and mask functional groups that reduce bacterial cells adhesion [103,104]. The formation of conditioning layers presents a challenge for creating surfaces that are bacteria-resistant. An excellent case in point is surfaces that have quaternary ammonium salts, which are bactericidal, before conditioning layers are deposited [105,106]. Despite the challenges presented by biofouling, the development of surfaces that resist bacterial attachment is an active area of research. Several strategies have been described for reducing the attachment of bacteria, including controlling surface chemistry [107] and controlling structural properties of surfaces [108] (Figure 7.7).

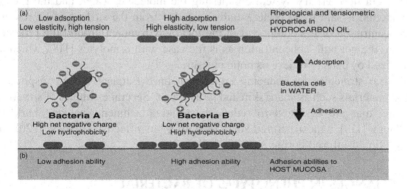

FIGURE 7.7 The capacity of bacteria to adhere is a function of: (a) physicochemical charges, and (b) hydrophobic properties of bacteria. These properties can be measured quantitatively using rheological and tensiometric methods as illustrated in the upper part of the figure. Through bacterial adsorption at a hydrophobic interface, the interfacial elasticity is increased, and depending on the bacterial characteristics, interfacial tension can be decreased. (From de Wouters, T. et al., *PLoS One*, 10, 1–17, 2015.)

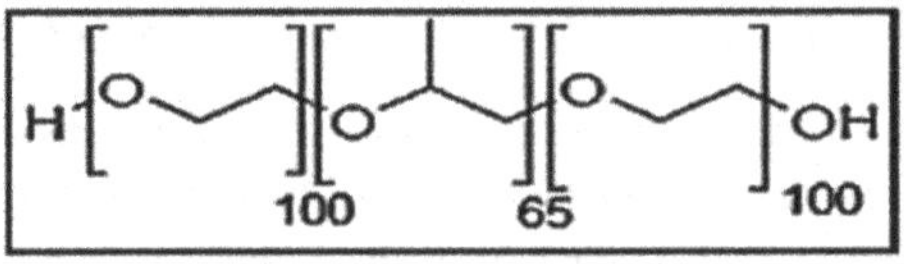

FIGURE 7.8 Chemical structure of Pluronic F-127 used as an anti-adhesive coating to prevent cells adhesion. (From Okano, T. et al., *J. Control. Release*, 36, 125–133, 1995.)

Zwitterionic surfaces such as betaines [109–111] and silver-impregnated surfaces that slowly release silver ions are among the most effective chemical strategies for inhibiting the attachment of bacteria to surfaces. Another strategy is the use of thermo-responsive hydrogels, such as poly(N-isopropyl acrylamide). Above a critical temperature, these polymers undergo a phase transition and present a hydrophobic surface that facilitates cell attachment and growth. When the temperature is reduced, the polymer swells and presents a hydrophilic surface, and adsorbed cells are released from the surface [112,113].

Polymer brushes, in which one end of a polymer is attached to a surface and the polymer chain is extended into solution, have been successfully used as anti-adhesion coatings [114]. Surfaces coated with Pluronic F-127 (Figure 7.8) reduce the initial attachment and growth rate of *Staphylococcus aureus* and *Staphylococcus epidermidis* and enable the removal of biofilms consisting of these organisms using fluid flow. However, Pluronic F-127 had no effect on *Pseudomonas aeruginosa* adhesion and biofilm removal [115]. Polymer brushes can also be functionalized with antimicrobial peptides, which enhance the overall antimicrobial activity of the surface [116].

Kane et al. [117] presented a hypothesis that relates the ability of a given molecule to make surfaces protein-resistant with its ability to be excluded from the surface of a protein in a ternary system comprising the molecule, water, and the protein. A compound is said to be preferentially excluded from the surface of a protein if the concentration of the compound in the local environment around the protein is less than the net bulk concentration as is the case with osmolytes [118], which are synthesized by cells to relieve osmotic pressure.

Physical strategies for reducing attachment have frequently been inspired by natural materials such as shark skin and lotus leaves. Several examples of structured materials that are designed to reduce bacterial attachment include Sharklet™ technology [119], nanostructured surfaces with a low effective stiffness [120], and slippery liquid-infused porous surfaces [121].

7.9 CHANGES IN PHENOTYPES OF BACTERIAL CELLS UPON ADHESION ON SURFACES

Many of the changes that happened to the bacterial cells are morphological changes (phenotype changes) to enhance and facilitate its adhesion to the surfaces. For example, cells of *S. aureus* preferentially localize fibronectin-binding protein when presented with surfaces that display fibronectin (Figure 7.9) [122].

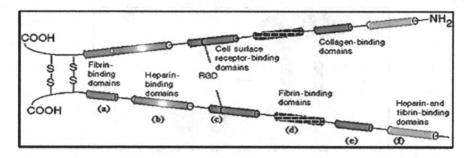

FIGURE 7.9 Structure of fibronectin-binding protein: (a) fibrin-binding domains, (b) heparin-binding domains, (c) cell surface receptor-binding domains, (d) fibrin binding domains, (e) collagen-binding domains, and (f) heparin- and fibrin-binding domains. (From Epstein, A.K. et al., *Proc. Nat. Acad. Sci. USA*, 109, 13182–13187, 2012.)

During *E. coli* cell adhesion, a decrease in the outer membrane protein (OmpX) (Figure 7.10) increases the production of EPS and increases antibiotic susceptibility [123].

During biofilm formation, quorum sensing alters the charge on *E. coli* cells [45] which influences binding to surfaces presenting electrostatically attractive charges. Some organisms, including *Helicobacter pylori* [124] and *S. aureus* [122], produce organelles for adhesion only upon cell contact with host tissues.

In response to surfaces, *P. aeruginosa* activates a protecting system that consists of a complex of protein (Wsp), which helps the chemotaxis system (chemotaxis is the movement of bacterial cells in response to a chemical stimulus) [125] and dynamically localize in cells. WspA (Figure 7.11) is a chemoreceptor-like protein that senses surfaces and produces a signal through the other Wsp protein that ultimately leads to phosphorylation of WspR [126].

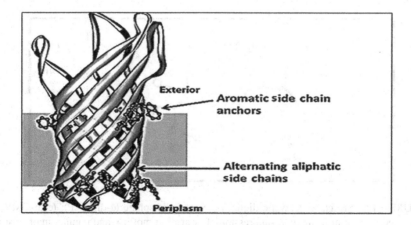

FIGURE 7.10 Structure of outer membrane protein. Regions defined by arrows show the aromatic side chains and alternative aliphatic side chains which are responsible for anchoring the cells to the surfaces. (From Lower, S.K. et al., *Biophys. J.*, 99, 2803–2811, 2010.)

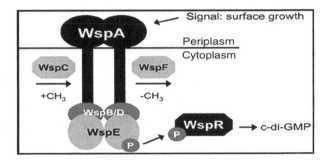

FIGURE 7.11 Structure of chemoreceptor-like protein WspA. WspA is predicted to be a membrane-bound methyl-accepting chemotaxis protein that detects an unknown signal when grown on a surface. (From Lower, S.K. et al., *Biophys. J.*, 99, 2803–2811, 2010.)

Phosphorylated WspR catalyzes the synthesis of cyclic diguanylate monophosphate (c-di-GMP) [127], which is implicated in biofilm formation; the suppression of swarming motility in *P. aeruginosa* [128,129]; and the regulation of several motility or attachment-related systems in bacteria [130]. In *P. aeruginosa*, c-di-GMP (Figure 7.12) affects the activity of the transcription factor FleQ. FleQ is the master regulator of flagella gene expression (Figure 7.13). FleQ also inhibits the expression of genes required for EPS synthesis. The surface-induced increase and subsequent role of c-di-GMP in *P. aeruginosa* are among the best-characterized mechanisms of bacterial surface sensing, and provide an example of the influence of physical properties of the interfaces on bacterial biochemistry and physiology [131].

Pathogenic bacteria frequently decouple their division from growth and form filaments in response to the presence of host surfaces [132]. The filamentation of

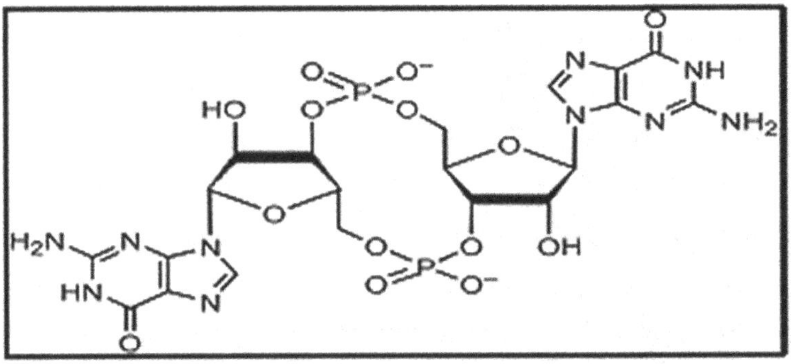

FIGURE 7.12 Structure of cyclic diguanylate monophosphate (c-di-GMP); (c-di-GMP is associated in biofilm formation, suppression of swarming motility and regulation of several motility- or attachment-related systems in bacteria). (From Güvener, Z.T. and Harwood, C.S., *Mol. Microbiol.*, 66, 1459–1473, 2007; Hickman, J.W. et al., *Proc. Natl. Acad. Sci. USA*, 102, 14422–14427, 2005; Kuchma, S.L. et al., *J. Bacteriol.*, 189, 8165–8178, 2007.)

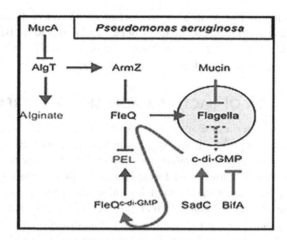

FIGURE 7.13 Role of FleQ transcription factor in flagella gene expression. (From Simm, R. et al., *Mol. Microbiol.*, 53, 1123–1134, 2004.)

pathogenic *E. coli* facilitates the escape of cells from the host immune response during urinary tract infections [133]. *Agrobacterium tumefaciens* also filaments upon contact with plant host tissues [134]. Filamentation also occurs during surface sensing-associated process of swarming (Figure 7.14).

During swarming, a wide variety of changes occur in the global transcription of bacterial genes [135–137]. These changes produce significant alterations in cell morphology, including an increase in the surface density of flagella and cell length [84]. Swarming cells also regulate several infection-related genes, including protease, urease, hemolysin, and proteins that facilitate host invasion [96,135].

Interestingly, the physical properties of surfaces may influence cell morphology and community structure. The marine bacterium SW5 adheres and grows on both hydrophobic and hydrophilic surfaces; however, its growth into communities is influenced by surface properties. Cells adhere uniformly to hydrophobic surfaces,

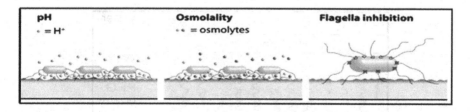

FIGURE 7.14 Mechanism of surface sensing of bacteria. Cells adhering to surfaces may trap ions and small molecules in the thin layer of fluid positioned between the cell body and the surface, which forms a microenvironment that has different properties than the bulk liquid. The decrease in the pH of fluids in close proximity to surface-associated cells enhances the proton motive force and directly affects cellular bioenergetics. (From Wang, Q. et al., *Mol. Microbiol.*, 52, 169–187, 2004.)

form microcolonies, and grow into tightly packed multilayer biofilms. Fewer cells attach to hydrophilic surfaces, and changes in cell division lead to the formation of chains of cells that are >100 μm long. These chains become loosely entangled to form relatively unstructured and less densely packed biofilms.

7.10   PREVENTION OF BACTERIAL ADHESION TO SURFACES

The attachment of bacteria to a surface leads to subsequent colonization resulting in the formation of a biofilm [138,139]. This adhesion of bacteria to a surface is mediated by different types of interactions which can be specific, for example through a protein film that might have formed on the surface, or through nonspecific interactions [140]. Biofilm formation on biological implants like catheters, prosthetic devices, and contact lenses leads to infection [141]. Biocontamination is of great concern when it comes to life-threatening infectious diseases caused by attached bacteria, especially when caused by antibiotic-resistant strains [13,138,142].

Several techniques and methods were reported to prevent the adhesion of bacteria to the surfaces. Immobilizing polyethylene glycol (PEG) is one of the most commonly used approaches to impart protein resistance to a surface. The antifouling properties of PEG-based coatings have been widely reported in the literature [143–145]. Physical adsorption, chemical adsorption, covalent attachment, and block or graft copolymerization are some of the techniques that have been used to attach PEG to surfaces [146]. Jeon et al. [147] studied the basis for the exceptional protein resistance of PEG-functionalized surfaces. They considered the PEG chains to be terminally attached to a hydrophobic substrate placed in water. The protein was modeled as a block of infinite length placed parallel to the substrate, separated by the polymer chains and water in between (Figure 7.15). According to their studies, the approach of the protein toward the substrate results in compression of the PEG chains leading to steric repulsive elastic forces.

Even though PEG is the most commonly used substance to impart protein resistance to a surface, it has a tendency to auto-oxidize and form aldehydes and ethers in the presence of oxygen [148]. This causes the PEG-coated surfaces to lose their protein resistance. Thus, it is important to design alternative molecules that resist protein adhesion. Whitesides and coworkers [149,150] have investigated the

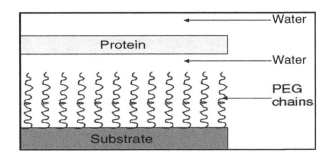

FIGURE 7.15   Schematics from the theoretical study of Jeon et al. [148] showing a protein of infinite size in water with a solid substrate having terminally attached polyethylene glycol (PEG) chains.

properties that confer such exceptional protein resistance ability to self-assembled monolayers or (SAMs) with oligo(ethylene glycol) groups and used these properties to design other protein-resistant surfaces.

Kane et al. [116] proposed a hypothesis that relates the ability of kosmotrope-based protein-resistant surfaces to render protein-resistance with an ability to be excluded from the surface of a protein in a ternary system comprising that molecule, water, and the protein. A compound is said to be preferentially excluded from the surface of a protein if the concentration of the compound in the local environment around the protein is less than that in the bulk concentration (Figure 7.16), as is the case with osmolytes [117], which are synthesized by cells to relieve osmotic pressure. Kosmotropes stabilize the structures of proteins by being preferentially excluded from the protein surface [151].

Trimethyl amine-N-oxide (TMAO) is a kosmotrope that has the ability to counteract the protein denaturant urea by imparting stability to the folded state of the protein. The unfolded state is unstable in the presence of TMAO due to the osmophobic protein backbone being exposed in the unfolded state as opposed to the folded state [152].

The most common approach for preventing the adhesion of proteins or microbes involves the functionalization of surfaces with PEG or oligo(ethylene glycol) groups. Major advances in recent years include the development of new approaches for the functionalization of surfaces with PEG, the development of alternatives to PEG for resisting protein and microbial adhesion, and approaches based on the control of surface modulus and topography. Even though most of these surfaces resist the adsorption of proteins and bacteria, they do not address the issue of deactivating the microbes, which is more vital for combating microbial contamination. As far as coatings for marine antifouling applications are concerned, surfaces with low surface energy or with an optimized surface topography show promising results.

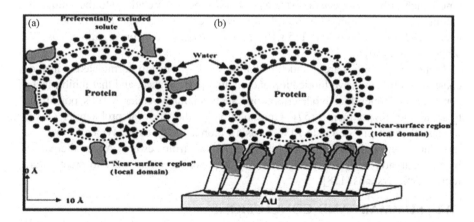

FIGURE 7.16 Schematic illustration of (a) a solute that is completely excluded from the surface of the protein (the local domain), and (b) a protein that does not adsorb onto a surface. (From Baltzer-Mattsby, I. et al., *Appl. Environ. Microbiol.*, 55, 2681–2689, 1989.)

7.11   MICROBIAL ADHESION IN LUBRICATION
AND TRIBOLOGICAL SYSTEMS

Metalworking fluids (MWFs) are commonly used in manufacturing and machining industries for lubricating. MWFs can be categorized into three classes: (i) oil-based or straight oils, (ii) oil-in-water emulsions with high oil concentration (soluble oils) or lower oil concentration (semisynthetic MWF), and (iii) synthetics without oils [153]. MWF formulations contain a variety of additives such as emulsifiers, biocides, function modifiers, extreme pressure additives, antifoam agents, and corrosion inhibitors. Straight oils are stable toward microbial growth because water is necessary for the microbial life cycle. The presence of water contaminations in straight oil MWF systems enhances the microbial growth. Microbial growth is a major issue with metalworking fluids that are soluble oils, synthetics, and semisynthetics.

Historically, microbial contamination of MWFs became a problem in the metalworking industries, due to the potential adverse health effects and negative impact on their tribological performance.

Degradation of MWFs by bacteria changes the fluids' viscosity and the produced acids, which lowered the pH of these fluids, and causes corrosion and leaks in the metalworking fluid handling system. Anaerobic bacteria, specifically the sulfate-reducing bacteria, produce hydrogen sulfide and other offensive and toxic gases which affect the health of the handlers of MWFs. Excessive microbial growth may result in plugging filters and ports and may interfere with the metalworking operation. More importantly, the microbial contaminants may result in adverse health effects to workers exposed to MWF aerosols.

Water-based MWFs are excellent nutritional sources for many kinds of bacteria and fungi. The predominant microbial species routinely recovered from MWFs are virtually identical to those routinely recovered from natural water systems. Attempts to manage microbial growth by the addition of biocides may result in the emergence of biocide-resistant strains. Even if biocides are added to the fluids, the soluble oils are favorable for microbial growth such as gram-negative bacteria, for example, *Pseudomonas* spp. [154,155], *Pseudomonas aeroginosa*, and *Klebsiella pneumoniae* [156].

Metalworking fluids are chemically complex and specifically formulated to include constituents that improve lubrication and cooling performance, and that inhibit metal corrosion and microbial biodeterioration. Such mixtures render MWFs potentially toxic to the environment [157]. One solution to the disposal problem is on-site biological treatment of waste MWFs, using bioreactor systems. Typically they are inoculated with unidentified microbial communities from sewage. As an alternative, bioaugmentation with selected strains may improve the opportunity to create more reproducible systems [158].

7.12   SUMMARY AND CONCLUSIONS

In this chapter, adhesion of microorganisms on different surfaces was described. The mechanism of the adhesion was illustrated from the first step of cellular microorganisms to the surface. Formation of microbial colonies was reviewed and described. The role of adhesion of microorganisms was correlated to the tribological systems including microbial corrosion, biocides, and the influence of the different microorganisms on the tribological properties of the metalworking fluids.

REFERENCES

1. Y. Chao and T. Zhang, "Probing roles of lipopolysaccharide," *Langmuir*, 27, 11545–11553 (2011).
2. S. M. Hinsa, M. Espinosa-Urgel, J. L. Ramos and G. A. O'Toole, "Transition from reversible to irreversible attachment during biofilm formation by *Pseudomonas fluorescens* WCS365 requires an ABC transporter and a large secreted protein," *Mol. Microbiol.*, 49, 905–918 (2003).
3. M. D. Agbeti, "Relationship between diatom assemblages and trophic variables: A comparison of old and new approaches," *Can. J. Fish. Aquat. Sci.*, 49, 1171–1175 (1992).
4. A. J. Fuhrman, "Marine viruses and their biogeochemical and ecological effects," *Nature*, 399, 541–548 (1999).
5. J. W. Costerton, "Introduction to biofilm," *Int. J. Antimicrob. Agents*, 11, 217–221 (1999).
6. K. C. Marshall, R. Stout and R. Mitchell, "Mechanism of the initial events in the sorption of marine bacteria to surfaces," *J. Genetic Microbiol.*, 68, 337–342 (1970).
7. J. Hurlow, K. Couch, K. Laforet, L. Bolton, D. Metcalf and P. Bowler, "Clinical biofilms: A challenging frontier in wound care," *Adv. Wound Care*, 4, 295–301 (2015).
8. M. R. Parsek and P. K. Singh, "Bacterial biofilms: An emerging link to disease pathogenesis," *Annu. Rev. Microbiol.*, 57, 677–701 (2003).
9. M. E. Cortes, J. Consuegra and R. D. Sinisterra, "Biofilm formation, control and novel strategies for eradication," *Curr. Res. Technol. Adv.*, 2, 896–905 (2011).
10. R. Vasudevan, "Biofilms: Microbial cities of scientific significance," *J. Microbiol. Exp.*, 1, 1–14 (2014).
11. H. C. Flemming and J. Wingender, "The biofilm matrix," *Nature Rev. Microbiol.*, 8, 623–633 (2010).
12. P. Stoodley, R. Cargo, C. J. Rupp, S. Wilson and I. Klapper, "Biofilm material properties as related to shear-induced deformation and detachment phenomena," *J. Ind. Microbiol. Biotechnol.*, 29, 361–367 (2002).
13. L. Hall-Stoodley, J. W. Costerton and P. Stoodley, "Bacterial biofilms: From the natural environment to infectious diseases," *Nature Rev. Microbiol.*, 2, 95–108 (2004).
14. M. Kostakioti, M. Hadjifrangiskou and S. J. Hultgren, "Bacterial biofilms: Development, dispersal, and therapeutic strategies in the dawn of the post antibiotic era," *Cold Spring Harb. Perspect. Med.*, 3, 1–23 (2013).
15. M. D. Rodney, "Biofilm formation: A clinically relevant microbiological process," *Clin. Infect. Dis.*, 33, 1387–1392 (2001).
16. L. A. De Weger, C. I. van der Vlugt, A. H. Wijfjes, P. A. Bakker, B. Schippers and B. Lugtenberg, "Flagella of a plant-growth stimulating *Pseudomonas fluorescens* strains are required for colonization of potato roots," *J. Bacteriol.*, 169, 2769–2773 (1987).
17. P. Hall-Stoodley, "Developmental regulation of microbial biofilms," *Curr. Opinion Biotechnol.*, 13, 28–33 (2002).
18. B. L. Bassler, "How bacteria talk to each other: Regulation of gene expression by quorum sensing," *Curr. Opinion Microbiol.*, 2, 582–587 (1999).
19. M. A. S. Vigeant, R. M. Ford, M. Wagner and L. K. Tamm, "Reversible and irreversible adhesion of motile *Escherichia coli* cells analyzed by total internal reflection aqueous fluorescence microscopy," *Appl. Environ. Microbiol.*, 68, 2794–2801 (2002).

20. D. R. Absolom, F. V. Lamberti, Z. Policova, W. Zingg, C. J. van Oss and A. W. Neumann, "Surface thermodynamics of bacterial adhesion," *Appl. Environ. Microbiol.*, 46, 90–97 (1983).

21. P. R. Rutter and B. Vincent, "The adhesion of microorganisms to surfaces, physicochemical aspects. In: *Microbial Adhesion to Surfaces*, R. C. W. Berkeley, J. M. Lynch, J. Melling, P. R. Rutter, and B. Vincent, Eds., Ellis Horwood, London, UK, 79–91 (1980).

22. M. C. M. Van Loosdrecht, J. Lyklema, W. Norde and A. J. B. Zehnder, "Bacterial adhesion: A physicochemical approach," *Microb. Ecol.*, 17, 1–15 (1989).

23. C. J. van Oss, R. J. Good and M. K. Chaudhury, "The role of van der Waals forces and hydrogen bonds in 'hydrophobic interactions' between biopolymers and low energy surfaces," *J. Colloid Interface Sci.*, 111, 378–390 (1986).

24. C. J. van Oss, R. J. Good and M. K. Chaudhury, "Additive and non-additive surface tension components and the interpretation of contact angles," *Langmuir*, 4, 884–891 (1988).

25. J. Wood and R. Sharma, "How long is the long-range hydrophobic attraction?" *Langmuir*, 11, 4797–4802 (1995).

26. M. Elimelech, "Indirect evidence for hydration forces in the deposition of polystyrene latex colloids on glass surfaces," *J. Chem. Soc. Faraday Trans.*, 86, 1623–1624 (1990).

27. M. Morra and C. Cassinelli, "Bacterial adhesion to polymer surfaces: A critical review of surface thermodynamic approaches," *J. Biomater. Sci. Polymer*, 9, 55–74 (1997).

28. F. M. Etzler, "Characterization of surface free energies and surface chemistry of solids," In: *Contact Angle, Wettability and Adhesion*, K. L. Mittal Ed., Vol. 3, pp. 219–264, CRC Press, Boca Raton, FL (2003).

29. F. M. Etzler, "Determination of the surface free energy of solids: A critical review," *Rev. Adhesion Adhesives*, 1, 3–45 (2013).

30. K. L. Mittal Ed., *Acid-Base Interactions: Relevance to Adhesion Science and Technology*, Vol. 2, CRC Press, Boca Raton, FL (2000).

31. M. Hermansson, "The DLVO theory in microbial adhesion," *Colloids Surfaces B*, 14, 105–119 (1999).

32. B. A. Jucker, A. J. B. Zehnder and H. Harms, "Quantification of polymer interactions in bacterial adhesion," *Environ. Sci. Technol.*, 32, 2909–2915 (1998).

33. P. A. Pethica, "Microbial and cell adhesion. In: *Microbial Adhesion to Surfaces*, R. C. W. Berkeley, J. M. Lynch, J. Melling, P. R. Rutter and B. Vincent Eds., pp. 19–45, Ellis Horwood, Chichester, UK (1980).

34. J. W. McClaine and R. M. Ford, "Reversal of flagella rotation is important in initial attachment of *Escherichia coli* to glass in a dynamic system with high and low ionic strength buffers," *Appl. Environ. Microbiol.*, 68, 1280–1289 (2002).

35. H. E. Kubitschek, "Buoyant density variation during the cell cycle in microorganisms," *Critical Rev. Microbiol.*, 14, 73–97 (1987).

36. K. Inoue, M. Nishimura, B. B. Nayak and K. Kogure, "Separation of marine bacteria according to buoyant density by use of the density-dependent cell sorting method," *Appl. Environ. Microbiol.*, 73, 1049–1053, 2007.

37. H. Makinoshima, A. Nishimura and A. Ishihama, "Fractionation of *Escherichia coli* cell populations at different stages during growth transition to stationary phase," *Mol. Microbiol.*, 43, 269–279 (2002).

38. T. Nishino, B. B. Nayak and K. Kogure, "Density-dependent sorting of physiologically different cells of *Vibrio parahaemolyticus*," *Appl. Environ. Microbiol.*, 69, 3569–3572 (2003).

39. N. P. Boks, H. J. Busscher, H. C. van der Mei and W. Norde, "Bond-strengthening in *staphylococcal* adhesion to hydrophilic and hydrophobic surfaces using atomic force microscopy," *Langmuir*, 24, 12990–12994 (2008).

40. N. P. Boks, H. J. Kaper, W. Norde, H. J. Busscher and H. C. van der Mei, "Residence time dependent desorption of *Staphylococcus epidermidis* from hydrophobic and hydrophilic substrata," *Colloids Surf. B*, 67, 276–278 (2008).

41. J. M. Meinders, H. C. Van der Mei and H. J. Busscher, "Deposition efficiency and reversibility of bacterial adhesion under flow," *J. Colloid Interface Sci.*, 176, 329–341 (1995).

42. H. J. Busscher, W. Norde, P. K. Sharma and H. C. van der Mei, "Interfacial rearrangement in initial microbial adhesion to surfaces," *Curr. Opinion Colloid Interface Sci.*, 15, 510–517 (2010).

43. H. Hayashi, H. Seiki, S. Tsuneda, A. Hirata and H. Sasaki, "Influence of growth phase on bacterial cell electro-kinetic characteristics examined by soft particle electrophoresis theory," *J. Colloid Interface Sci.*, 264, 565–568 (2003).

44. L. D. Renner and D. B. Weibel, "Physicochemical regulation of biofilm formation," *MRS Bull*, 36, 347–355 (2011).

45. K. E. Eboigbodin, J. R. Newton, A. F. Routh and C. A. Biggs, "Bacterial quorum sensing and cell surface electro-kinetic properties," *Appl. Microbiol. Biotechnol.*, 73, 669–675 (2006).

46. N. C. Caiazza and G. A. O'Toole, "SadB is required for the transition from reversible to irreversible attachment during biofilm formation by *Pseudomonas aeruginosa* PA14," *J. Bacteriol.*, 186, 4476–4485 (2004).

47. Y. Chao and T. Zhang, "Probing roles of lipopolysaccharide, type 1 fimbria, and colanic acid in the attachment of *Escherichia coli* strains on inert surfaces," *Langmuir*, 27, 11545–11553 (2011).

48. S. M. Hinsa, M. Espinosa-Urgel, J. L. Ramos and G. A. O'Toole, "Transition from reversible to irreversible attachment during biofilm formation by *Pseudomonas fluorescens* WCS365 requires an ABC transporter and a large secreted protein," *Mol. Microbiol.*, 49, 905–918 (2003).

49. E. H. Beachey, "Bacterial adherence: Adhesion-receptor interactions mediating the attachment of bacteria to mucosal surfaces," *J. Infect. Dis.*, 143, 325–345 (1981).

50. R. Van Houdt and C. W. Michiels, "Role of bacterial cell surface structures in *Escherichia coli* biofilm formation," *Res. Microbiol.*, 156, 626–633 (2005).

51. K. J. Spears, A. J. Roe and D. L. Gally, "A comparison of enteropathogenic and enterohaemorrhagic *Escherichia coli* pathogenesis," *FEMS Microbiol. Lett.*, 255, 187–202 (2006).

52. F. Newman, J. A. Beeley and T. W. MacFarlane, "Adherence of oral microorganisms to human parotid salivary proteins," *Electrophoresis*, 17, 266–270 (1996).

53. A. Amano, S. Shizukuishi, H. Horie, S. Kimura, I. Morisaki and S. Hamada, "Binding of *Porphyromonas gingivalis* fimbriae to proline-rich glycoproteins in parotid saliva via a domain shared by major salivary components," *Infect. Immun.*, 66, 2072–2077 (1998).

54. S. S. Nelson, S. Bollampalli and M. J. McBride, "SprB is a cell surface component of the *Flavobacterium johnsoniae* gliding motility machinery," *J. Bacteriol.*, 190, 2851–2857 (2008).

55. J. Bouckaert, J. Mackenzie, J. L. de Paz, B. Chipwaza, D. Choudhury, A. Zavialov, K. Mannerstedt et al., "The affinity of the FimH fimbrial adhesion is receptor-driven and quasi-independent of *Escherichia coli* pathotypes," *Mol. Microbiol.*, 61, 1556–1568 (2006).

56. L. M. Nilsson, W. E. Thomas, E. Trintchina, V. Vogel and E. V. Sokurenko, "Catch bond-mediated adhesion without a shear threshold: Trimannose versus monomannose interactions with the FimH adhesion of *Escherichia coli*," *J. Biol. Chem.*, 281, 16656–16663 (2006).

57. B. N. Anderson, A. M. Ding, L. M. Nilsson, K. Kusuma, V. Tchesnokova, V. Vogel, E. V. Sokurenko and W. E. Thomas, "Weak rolling adhesion enhances bacterial surface colonization," *J. Bacteriol.*, 189, 1794–1802 (2007).

58. C. E. Zobell, "The effect of solid surfaces upon bacterial activity," *J. Bacteriol.*, 46, 39–56 (1943).

59. H. Heukelekian and A. Heller, "Relation between food concentration and surface for bacterial growth," *J. Bacteriol.*, 40, 547–558 (1940).

60. J. S. Poindexter, "The *Caulobacters*: Ubiquitous unusual bacteria," *Microbiol. Rev.*, 45, 123–179 (1981).

61. M. C. Grantham and P. M. Dove, "Investigation of bacterial-mineral interactions using fluid tapping mode atomic force microscopy," *Geochim. Cosmochim. Acta*, 60, 2473–2480 (1996).

62. K. H. Nealson and S. E. Finkel, "Electron flow and biofilms," *MRS Bull.*, 36, 380–384 (2011).

63. K. P. Nevin, B. C. Kim, R. H. Glaven, J. P. Johnson, T. L. Woodard, B. A. Methé, R. J. Donato et al., "Anode biofilm transcriptomics reveals outer surface components essential for high density current production in *Geobacter sulfurreducens* fuel cells," *PLoS One*, 4, e5628 (2009).

64. R. M. Donlan and J. W. Costerton, "Biofilms: Survival mechanisms of clinically relevant microorganisms," *Clin. Microbiol. Rev.*, 15, 167–193 (2002).

65. M. Simões, M. O. Pereira and M. J. Vieira, "Effect of mechanical stress on biofilms challenged by different chemicals," *Water Res.*, 39, 5142–5152 (2005).

66. J. W. Costerton, Z. Lewandowski, D. E. Caldwell, D. R. Korber and H. M. Lappin-Scott, "Microbial biofilms," *Annu. Rev. Microbiol.*, 49, 711–745 (1995).

67. T. F. Mah and G. A. O'Toole, "Mechanisms of biofilm resistance to antimicrobial agents," *Trends Microbiol.*, 9, 34–39 (2001).

68. T. F. Mah, "Biofilm-specific antibiotic resistance," *Future Microbiol.*, 7, 1061–1072 (2012).

69. A. K. John, M. Schmaler, N. Khanna and R. Landmann, "Reversible daptomycin tolerance of adherent *staphylococci* in an implant infection model. Antimicrob," *Agents Chemother.*, 55, 3510–3516 (2011).

70. Y. Qu, A. J. Daley, T.S. Istivan, D. A. Rouch and M. A. Deighton, "Densely adherent growth mode, rather than extracellular polymer substance matrix build-up ability, contributes to high resistance of *Staphylococcus epidermidis* biofilms to antibiotics," *J. Antimicrob. Chemother.*, 65, 1405–1411 (2010).

71. S. Y. Queck, M. Weitere, A. M. Moreno, S. A. Rice and S. Kjelleberg, "The role of quorum sensing mediated developmental traits in the resistance of *Serratia marcescens* biofilms against protozoan grazing," *Environ, Environ. Microbiol.*, 8, 1017–1025 (2006).

72. S. S Branda, J. E. Gonzalez-Pastor, S. Ben-Yehuda, R. Losick and R. Kolter, "Fruiting body formation by *Bacillus subtilis*," *Proc. Natl. Acad. Sci. USA*, 98, 11621–11626 (2001).

73. H. Vlamakis, C. Aguilar, R. Losick and R. Kolter, "Control of cell fate by the formation of an architecturally complex bacterial community," *Genes Dev.*, 22, 945–953 (2008).

74. J. A. J. Haagensen, M. Klausen, R. K. Ernst, S. I. Miller, A. Folkesson, T. Tolker-Nielsen and S. Molin, "Differentiation and distribution of colistin and sodium dodecyl sulfate tolerant cells in *Pseudomonas aeruginosa* biofilms," *J. Bacteriol.*, 189, 28–37 (2007).

75. B. Purevdorj-Gage, W. J. Costerton and P. Stoodley, "Phenotypic differentiation and seeding dispersal in non-mucoid and mucoid *Pseudomonas aeruginosa* biofilms, *J. Gen. Microbiol.*, 151, 1569–1576 (2005).

76. M. F. Copeland and D. B. Weibel, "Bacterial swarming: A model system for studying dynamic self-assembly," *Soft Matter*, 5, 1174–1187 (2009).

77. J. E. Patrick and D. B. Kearns, "Swarming motility and the control of master regulators of flagellar biosynthesis," *Mol. Microbiol.*, 83, 14–23 (2012).

78. M. T. Butler, Q. Wang and R. M. Harshey, "Cell density and mobility protect swarming bacteria against antibiotics," *Proc. Nat. Acad. Sci. USA*, 107, 3776–3781 (2010).

79. L. Wang, C. Zhang, F. Gong, H. Li, X. Xie, C. Xia, J. Chen, Y. Song, A. Shen and J. Song, "Influence of *Pseudomonas aeruginosa* pvdQ gene on altering antibiotic susceptibility under swarming conditions," *Curr. Microbiol.*, 63, 377–386 (2011).

80. C. J. Ingham, O. Kalisman, A. Finkelshtein and E. Ben-Jacob, "Mutually facilitated dispersal between the nonmotile fungus *Aspergillus fumigatus* and the swarming bacterium *Paenibacillus vortex*," *Proc. Natl. Acad. Sci. USA*, 108, 19731–19736 (2011).
81. L. Chai, H. Vlamakis and R. Kolter, "Extracellular signal regulation of cell differentiation in biofilms," *MRS Bull*, 36, 374–379 (2011).
82. K. Hamze, S. Autret, K. Hinc, S. Laalami, D. Julkowska, R. Briandet, M. Renault et al., "Single-cell analysis in situ in a *Bacillus subtilis* swarming community identifies distinct spatially separated subpopulations differentially expressing hag (flagellin), including specialized swarmer," *Microbiology*, 157, 2456–2469 (2011).
83. T. Matsuyama, Y. Takagi, Y. Nakagawa, H. Itoh, J. Wakita and M. Matsushita, "Dynamic aspects of the structured cell population in a swarming colony of *Proteus mirabilis*," *J. Bacteriol.*, 182, 385–393 (2000).
84. H. H. Tuson, M. F. Copeland, S. Carey, R. Sacotte and D. B. Weibel, "Flagellum density regulates *Proteus mirabilis* swarmer cell motility in viscous environments," *J. Bacteriol.*, 195, 368–377 (2013).
85. W. L. Ng and B. L. Bassler, "Bacterial quorum-sensing network architectures," *Annu. Rev. Genet.*, 43, 197–222 (2009).
86. M. Frederix and J. A. Downie, "Quorum sensing: Regulating the regulators," *Adv. Microb. Physiol.*, 58, 23–80 (2011).
87. S. T. Flickinger, M. F. Copeland, E. M. Downes, A. T. Braasch, H. H. Tuson, Y. J. Eun and D. B. Weibel, "Quorum sensing between *Pseudomonas aeruginosa* biofilms accelerates cell growth," *J. Am. Chem. Soc.*, 133, 5966–5975 (2011).
88. G. E. Dilanji, J. B. Langebrake, P. De Leenheer and S. J. Hagen, "Quorum activation at a distance: Spatiotemporal patterns of gene regulation from diffusion of an autoinducer signal," *J. Am. Chem. Soc.*, 134, 5618–5626 (2012).
89. J. D. Shrout, T. Tolker-Nielsen, M. Givskov and M. R. Parsek, "The contribution of cell-cell signaling and motility to bacterial biofilm formation," *MRS Bull.*, 36, 367–373 (2011).
90. J. S. Madsen, M. Burmolle, L. H. Hansen and S. J. Sorensen, "The interconnection between biofilm formation and horizontal gene transfer," *FEMS Immunol. Med. Microbiol.*, 65, 183–195 (2012).
91. K. L. Meibom, M. Blokesch, N. A. Dolganov, C. Y. Wu and G. K. Schoolnik, "Chitin induces natural competence in *Vibrio cholerae*," *Science*, 310, 1824–1827 (2005).
92. D. T. Pathak, X. Wei, A. Bucuvalas, D. H. Haft, D. L. Gerloff and D. Wall, "Cell contact-dependent outer membrane exchange in *myxobacteria*: Genetic determinants and mechanism," *PLoS Genet.*, 8, e1002626 (2012).
93. P. V. Krasteva, J. C. Fong, N. J. Shikuma, S. Beyhan, M. V. Navarro, F. H. Yildiz and H. Sondermann, "Vibrio cholerae VpsT regulates matrix production and motility by directly sensing cyclic di-GMP," *Science*, 327, 866–868 (2012).
94. N. C. Caiazza, J. H. Merritt, K. M. Brothers and G. A. O'Toole, "Inverse regulation of biofilm formation and swarming motility by *Pseudomonas aeruginosa* PA14," *J. Bacteriol.*, 189, 3603–3612 (2007).
95. K. M. Blair, L. Turner, J. T. Winkelman, H. C. Berg and D. B. Kearns, "A molecular clutch disables flagella in the *Bacillus subtilis* biofilm," *Science*, 320, 1636–1638 (2008).
96. C. Allison, N. Coleman, P. L. Jones and C. Hughes, "Ability of *Proteus mirabilis* to invade human urothelial cells is coupled to motility and swarming differentiation," *Infect. Immun.*, 60, 4740–4746 (1992).
97. S. Bakalova, A. Doycheva, I. Ivanova, V. Groudeva and R. Dimkov, "Bacterial microflora of contaminated metalworking fluids," *Biotechnol. Biotechnol. Eq.* 21, 437–441 (2007).
98. K. Bruellhoff, J. Fiedler, M. Muller, J. Groll and R. E. Brenner, "Surface coating strategies to prevent biofilm formation on implant surfaces," *Int. J. Artif. Organs*, 33, 646–653 (2012).

99. G. Kang and Y. Cao, "Development of antifouling reverse osmosis membranes for water treatment: A review," *Water Res.*, 46, 584–600 (2012).

100. C. E. Marcato-Romain, Y. Pechaud, E. Paul, E. Girbal-Neuhauser and V. Dossat-Létisse, "Removal of microbial multi-species biofilms from the paper industry by enzymatic treatments," *Biofouling*, 28, 305–314 (2012).

101. M. P. Schultz, J. A. Bendick, E. R. Holm and W. M. Hertel, "Economic impact of biofouling on a naval surface ship," *Biofouling*, 27, 87–98 (2011).

102. R. M. Goulter-Thorsen, E. Taran, I. R. Gentle, K. S. Gobius and G. A. Dykes, "CsgA production by *Escherichia coli* O157:H7 alters attachment to abiotic surfaces in some growth environments," *Appl. Environ. Microbiol.*, 77, 7339–7344 (2011).

103. P. Lejeune, "Contamination of abiotic surfaces: What a colonizing bacterium sees and how to blur it," *Trends Microbiol.*, 11, 179–184 (2003).

104. S. Y. Wong, L. Han, K. Timachova, J. Veselinovic, M. N. Hyder, C. Ortiz, A. M. Klibanov and P. T. Hammond, "Drastically lowered protein adsorption on microbicidal hydrophobic/hydrophilic polyelectrolyte multilayers," *Biomacromol.*, 13, 719–726 (2012).

105. L. D. Melo, R. R. Palombo, D. F. S. Petri, M. Bruns, E. M. A. Pereira and A. M. Carmona-Ribeiro, "Structure-activity relationship for quaternary ammonium compounds hybridized with poly (methyl methacrylate)," *ACS Appl. Mater. Interfaces*, 3, 1933–1939 (2011).

106. J. C. Tiller, C. J. Liao, K. Lewis and A. M. Klibanov, "Designing surfaces that kill bacteria on contact," *Proc. Natl. Acad. Sci. USA.*, 98, 5981–5985 (2001).

107. T. de Wouters, C. Jans, T. Niederberger, P. Fischer and P. A. Rühs, "Adhesion potential of intestinal microbes predicted by physicochemical characterization methods," *PLoS One*, 10, 1–17 (2015).

108. K. Bazaka, R. J. Crawford and E. P. Ivanova, "Do bacteria differentiate between degrees of nanoscale surface roughness?" *J. Biotechnol.*, 6, 1103–1114 (2011).

109. G. Cheng, Z. Zhang, S. Chen, J. D. Bryers and S. Jiang, "Inhibition of bacterial adhesion and biofilm formation on zwitterionic surfaces," *Biomaterials*, 28, 4192–4199 (2007).

110. Z. Zhang, T. Chao, S. Chen and S. Jiang, "Super low fouling sulfobetaine and carboxybetaine polymers on glass slides," *Langmuir*, 22, 10072–10077 (2006).

111. Z. Zhang, S. Chen, Y. Chang and S. Jiang, "Surface grafted sulfobetaine polymers via atom transfer radical polymerization as super low fouling coatings," *J. Phys. Chem. B.*, 110, 10799–10804 (2006).

112. L. K. Ista, S. Mendez, V. H. Pérez-Luna and G. P. López, "Synthesis of poly(N-isopropylacrylamide) on initiator-modified self-assembled monolayers," *Langmuir*, 17, 2552–2555 (2001).

113. T. Okano, A. Kikuchi, Y. Sakurai, Y. Takei and N. Ogata, "Temperature-responsive poly(N-isopropyl acrylamide) as a modulator for alteration of hydrophilic/hydrophobic surface properties to control activation/inactivation of platelets," *J. Control. Release*, 36, 125–133 (1995).

114. W. Senaratne, L. Andruzzi and C. K. Ober, "Self-assembled monolayers and polymer brushes in biotechnology: Current applications and future perspectives," *Biomacromoles*, 6, 2427–2448 (2005).

115. M. R. Nejadnik, H.C. van der Mei, W. Norde and H. J. Busscher, "Bacterial adhesion and growth on a polymer brush-coating," *Biomaterials*, 29, 4117–4121 (2008).

116. G. Gao, K. Yu, J. Kindrachuk, D. E. Brooks, R. E. Hancock and J. N. Kizhakkedathu, "Antibacterial surfaces based on polymer brushes: Investigation on the influence of brush properties on antimicrobial peptide immobilization and antimicrobial activity," *Biomacromoles*, 12, 3715–3727 (2011).

117. R. S. Kane, P. Deschatelets and G. M. Whitesides, "Kosmotropes form the basis of protein-resistant surfaces," *Langmuir*, 19, 2388–2393 (2003).

118. P. H. Yancey, M. E. Clark, S. C. Hand, R. D. Bowlus and G. N. Somero, "Living with water stress: Evolution of osmolyte systems," *Science*, 217, 1214–1219 (1982).

119. K. K. Chung, J. F. Schumacher, E. M. Sampson, R. A. Burne, P. J. Antonelli and A. B. Brennan, "Impact of engineered surface microtopography on biofilm formation of *Staphylococcus aureus*," *Biointerphases*, 2, 89–94 (2007).

120. A. K. Epstein, A. I. Hochbaum, P. Kim and J. Aizenberg, "Control of bacterial biofilm growth on surfaces by nanostructural mechanics and geometry," *Nanotechnology*, 22, 494007 (2011).

121. A. K. Epstein, T. S. Wong, R. A. Belisle, E. M. Boggs and J. Aizenberg, "Liquid-infused structured surfaces with exceptional anti-biofouling performance," *Proc. Nat. Acad. Sci. USA*, 109, 13182–13187 (2012).

122. S. K. Lower, R. Yongsunthon, N. N. Casillas-Ituarte, E. S. Taylor, A. C. DiBartola, B. H. Lower, T. J. Beveridge, A. W. Buck and V. G. Fowler, "A tactile response in *Staphylococcus aureus*," *Biophys. J.*, 99, 2803–2811 (2010).

123. K. Otto and M. Hermansson, "Inactivation of ompX causes increased interactions of type 1 fimbriated *Escherichia coli* with abiotic surfaces," *J. Bacteriol.*, 186, 226–234 (2004).

124. M. R. Couturier and M. Stein, "*Helicobacter pylori* produces unique filaments upon host contact in vitro," *Can. J. Microbiol.*, 54, 537–548 (2008).

125. H. Chisholm, Ed. '*Chemotaxis*', *Encyclopedia Britannica* (11th Ed.), Cambridge University Press (UK), p. 77 (1911).

126. J. R. O'Connor, N. J. Kuwada, V. Huangyutitham, P. A. Wiggins and C. S. Harwood, "Surface sensing and lateral subcellular localization of WspA, the receptor in a chemosensory-like system leading to c-di-GMP production," *Mol. Microbiol.*, 86, 720–729 (2012).

127. Z. T. Güvener and C. S. Harwood, "Subcellular location characteristics of the *Pseudomonas aeruginosa* GGDEF protein, WspR, indicate that it produces cyclic-di-GMP in response to growth on surfaces," *Mol. Microbiol.*, 66, 1459–1473 (2007).

128. J. W. Hickman, D. F. Tifrea and C. S. Harwood, "A chemosensory system that regulates biofilm formation through modulation of cyclic diguanylate levels," *Proc. Natl. Acad. Sci. USA*, 102, 14422–14427 (2005).

129. S. L. Kuchma, K. M. Brothers, J. H. Merritt, N. T. Liberati, F. M. Ausubel and G. A. O'Toole, "BifA, a cyclic-Di-GMP phosphodiesterase, inversely regulates biofilm formation and swarming motility by *Pseudomonas aeruginosa* PA14," *J. Bacteriol.*, 189, 8165–8178 (2007).

130. R. Hengge, "Principles of c-di-GMP signaling in bacteria," *Nat. Rev. Microbiol.*, 7, 263–273 (2009).

131. R. Simm, M. Morr, A. Kader, M. Nimtz and U. Romling, "GGDEF and EAL domains inversely regulate cyclic di-GMP levels and transition from sessility to motility," *Mol. Microbiol.*, 53, 1123–1134 (2004).

132. S. S. Justice, D. A. Hunstad, L. Cegelski and S. J. Hultgren, "Morphological plasticity as a bacterial survival strategy," *Nat. Rev. Microbiol.*, 6, 162–168 (2008).

133. D. J. Horvath, B. Li, T. Casper, S. Partida-Sanchez, D. A. Hunstad, S. J. Hultgren, and S. S. Justice, "Morphological plasticity promotes resistance to phagocyte killing of uropathogenic *Escherichia coli*," *Microb. Infect.*, 13, 426–437 (2011).

134. K. R. Finer, K. M. Larkin, B. J. Martin and J. J. Finer, "Proximity of Agrobacterium to living plant tissues induces conversion to a filamentous bacterial form," *Plant Cell Rep.*, 20, 250–255 (2001).

135. H. H. Tuson and D. B. Weibel, "Bacteria-surface interactions," *Soft Matter*, 14, 4368–4380 (2013).

136. S. Salvetti, K. Faegri, E. Ghelardi, A. B. Kolst and S. Senesi, "Global gene expression profile for swarming *Bacillus cereus* bacteria," *Appl. Environ. Microbiol.*, 77, 5149–5156 (2011).

137. Q. Wang, J. G. Frye, M. McClelland and R. M. Harshey, "Gene expression patterns during swarming in *Salmonella typhimurium*: Genes specific to surface growth and putative new motility and pathogenicity genes," *Mol. Microbiol.*, 52, 169–187 (2004).

138. J. W. Costerton, P. S. Stewart and E. P. Greenberg, "Bacterial biofilms: A common cause of persistent infections," *Science*, 21, 284, 1318–1322 (1999).

139. A. Roosjen, H. C. Van Der Mei, H. J. Busscher and W. Norde, "Microbial adhesion to poly (ethylene oxide) brushes: Influence of polymer chain length and temperature," *Langmuir*, 20, 10949–10955 (2004).

140. D. Cunliffe, C. A. Smart, C. Alexander and E. N. Vulfson, "Bacterial adhesion at synthetic surfaces," *Appl. Environ. Microbiol.*, 65, 4995–5002 (1999).

141. K. Page, M. Wilson and I. P. Parkin, "Antimicrobial surfaces and their potential in reducing the role of the inanimate environment in the incidence of hospital-acquired infections," *J. Mater. Chem.*, 19, 3819–3831 (2009).

142. A. S. Lynch and G. T. Robertson, "Bacterial and fungal biofilm infections," *Ann. Rev. Med.*, 59, 415–428 (2008).

143. J. P. Bearinger, S. Terrettaz, R. Michel, N. Tirelli, H. Vogel, M. Textor and J. A. Hubbell, "Chemisorbed poly (propylene sulphide)-based copolymers resist biomolecular interactions," *Nat. Mater.*, 2, 259–264 (2003).

144. V. A. Liu, W. E. Jastromb and S. N. Bhatia, "Engineering protein and cell adhesivity using PEO-terminated triblock polymers," *J. Biomed. Mater. Res.*, 60, 126–134 (2002).

145. N. Xia, C. J. May, S. L. McArthur and D. G. Castner, "Time-of-flight secondary ion mass spectrometry analysis of conformational changes in adsorbed protein films," *Langmuir*, 18, 4090–4097 (2002).

146. G. R. Llanos and M. V. Sefton, "Does polyethylene oxide possesses a low thrombogenicity?" *J. Biomater. Sci., Polym.*, 4, 381–400 (1993).

147. S. I. Jeon, J. H. Lee, J. D. Andrade and P. G. De Gennes, "Protein-surface interactions in the presence of polyethylene oxide: I. Simplified theory," *J. Colloid Interface Sci.*, 142, 149–158 (1991).

148. A. Hucknall, S. Rangarajan and A. Chilkoti, "In pursuit of zero: Polymer brushes that resist the adsorption of proteins," *Adv. Mater*, 21, 2441–2446 (2009).

149. R. G. Chapman, E. Ostuni, S. Takayama, R. E. Holmlin, L. Yan and G. M. Whitesides, "Surveying for surfaces that resist the adsorption of proteins," *J. Am. Chem. Soc.*, 122, 8303–8304 (2000).

150. L. Deng, M. Mrksich and G. M. Whitesides, "Self-assembled monolayers of alkane thiolates presenting tri(propylene sulfoxide) groups resist the adsorption of protein," *J. Am. Chem. Soc.*, 118, 5136–5137 (1996).

151. P. H. Yancey, M. E. Clark, S. C. Hand, R. D. Bowlus and G. N. Somero, "Living with water stress: Evolution of osmolyte systems," *Science*, 24, 1214–1222 (1982).

152. M. V. Athawale, J. S. Dordick and S. Garde, "Osmolyte trimethylamine-N-oxide does not affect the strength of hydrophobic interactions: Origin of osmolyte compatibility," *Biophys. J.*, 89, 858–866 (2005).

153. S. N. Timasheff, "Control of protein stability and reactions by weakly interacting cosolvents: The simplicity of the complicated." In: *Advances in Protein Chemistry* E. Di Cera, F. M. Richards, D. S. Eisenberg and P. S. Kim, Eds., pp. 51, Academic Press, San Diego, CA (1998).

154. I. Baltzer-Mattsby, M. Sandin, B. Aldstrom, S. Allenmark, M. Edebo, E. Falsen, K. Pedersen, N. Rodin, R. Thompsan and L. Edebo, "Microbial growth and accumulation in industrial metal-working fluids," *Appl. Environ. Microbiol.*, 55, 2681–2689 (1989).

155. M. Veillette, P. S. Thorne, T. Gordon and C. Duchaine, "Six month tracking of microbial growth in a metalworking fluid after system cleaning and recharging," *Ann. Occup. Hyg.*, 48, 541–546 (2004).

156. F. J. Passman, R. E. Kauffman, G. L. Munson, "The use of non-chemical technologies to control microbial contamination in aviation fuels," In: 8th International Fuels Colloquium, W. J. Bartz, Ed., Technische Akademie Esslingen, Ostfildern, Germany (2011).

157. C. Cheng, D. Philips and R. Akhaddar, "Bacterial microflora of contaminated metalworking fluids," *Water Res.*, 39, 4051–4063 (2005).

158. V. Gast, A. Whiteley and I. Thnopsam, "Temporal dynamics and degradation activity of a bacterial inoculum for treating waste metal-working fluid," *Environ. Microbiol.*, 6, 254–263 (2004).

Section III

Green, Nano-, and Biotribology

8 Tribological Behavior of Nanoadditives in Lubricants

Guangbin Yang, Shengmao Zhang,
Pingyu Zhang, and Zhijun Zhang

CONTENTS

LIST OF ABBREVIATION

C_xS	alkanethiols with different chain lengths
DDP	dialkyldithiophosphate
DDP-Cu	DDP-modified Cu
DIOS	diisooctyl sebacate
DTC	dialkyl dithiocarbamate
EDS	energy dispersive spectrometry
LP	liquid paraffin
MADE	maleic anhydride dodecyl ester
$MPEGOCS_2K$	methoxylpolyethyleneglycol xanthate potassium
nano-ZnO	Nanoscale ZnO
OA	oleic acid

OM	oleylamine
PAO	poly α-olefin
P_B value	maximum non-seizure load
P_D value	sintering load
PyDDP	pyridinium di-n-octadecyldithiophosphate
SAED	selected area electron diffraction
SEM	scanning electron microscopy
TEM	transmission electron microscopy
THA	tetradecyl hydroxamic acid
WSD	wear scar diameter
ZDDP	zinc dialkyldithiophosphate

8.1   INTRODUCTION

About one third to one half of the world's primary energy is dissipated in mechanical friction, and 80% of machinery component failures is caused by wear [1–3]. To deal with these issues, researchers have made great efforts to develop various lubricant additives thereby saving energy and increasing the reliability of machinery components by effectively reducing friction and wear. Among the many types of lubricant additives, nanomaterials are of particular significance, since they exhibit advantages such as low dosage, better anti-friction property, and wear resistance performance [3–8]. In recent years, many investigations have been carried out on the use of nanoparticles as lubricating oil additives, including the following nanomaterials: nanolayered inorganic nanoparticles [9], metal nanoparticles [10,11], borate and carbonate nanomaterials [12,13], nanostructured oxides and sulfides [14–16], hard nanoparticles [17,18], rare earth compound nanoparticles [19,20], and so on.

Since the early 1990s, we studied nanoadditives, and considerable research has been carried out in this field. In this chapter, we shall focus on the work of our group over the last two decades on the use of nanoparticles in liquid lubricant systems. We introduce the topic of in situ surface modification technology and tribology of nanoadditives in the chapter. In situ surface modification technology was used to solve problems of dispersion and agglomeration of nanoparticles in lubricant base fluids. Several types of nanoadditives from metals, as well as their alloys, oxides, and sulfides, were synthesized by in situ surface modification method with a variety of organic molecules, and we discuss their application in the field of tribology.

8.2   IN SITU SURFACE MODIFICATION TECHNOLOGY

Over the past decades, nanoscale materials have been extensively investigated and widely employed in many scientific and technological fields due to their fascinating physical and chemical characteristics [21–23]. Many inorganic nanoparticles have been shown to have promising lubricant additive properties since they provide outstanding ability in reducing friction and wear when formulated into lubricants [24]. However, inorganic nanoparticles (in particular, naked nanoclusters) have high surface activity and poor compatibility with lubricating oils and are easy to aggregate, which is the main obstacle to their applications [25]. Surface chemical modification has been shown to be an effective way to solve such problems. Surface chemical modification of nanoparticles usually refers to the surface treatment of the already prepared nanoparticles obtained by the separation, and makes the nanoparticles acquire a good dispersion property in the medium; but the effects are often limited. On the other hand, in situ surface chemical modification of nanoparticles is chemical modification by forming a surface modification layer on the surface of nanoparticles through a variety of chemical bonds between a surface modification agent and the surface atoms (ions) of nanoparticles in the early stages of the formation of nanoparticles. The surface modification layer can effectively prevent the particles from further growth, or aggregating. Therefore, in situ surface chemical modification can effectively control the nanoparticle size, improve its dispersion, and provide chemical stability in the medium.

In this chapter, the preparation of in situ surface-modified oil-soluble nanoparticles is summarized. Model and formation mechanism of the organic molecules on the surface of nanoparticles based on type of chemical bonds are proposed. The approach lays the theoretical foundation for the design and development of nanomaterials with good dispersibility in liquid media.

8.2.1 IN SITU SURFACE MODIFICATION IN TERMS OF COVALENT BONDS

This method mainly applies to the surface modification of oxide nanoparticles [26–30] by coupling agents and metal nanoparticles by organometallic compounds. An example [26] is the in situ surface modification of nanosilica as shown in Figure 8.1: First, sodium metasilicate is hydrolyzed to form silicic acid in the presence of HCl. Second, condensation polymerization occurs via anhydration of hydroxyls of silicic acids in three dimensions and the Si and O are bonded to form nanosilica clusters. Large numbers of hydroxyls are left on the surface of the nanoclusters. Third, hexamethyldisilazane is hydrolyzed to produce trimethylsilyl and ammonia. As soon as the nanoclusters are formed, the trimethylsilyl reacts rapidly with the hydroxyl of nanosilica to form the modification layer on the surface of the clusters. Here, the trimethylsilyls reacts with a majority of active sites on SiO_2 and result in a steric hindrance, which prevents SiO_2 from further growth to cause it to agglomerate. These reactions are competitive to each other. By controlling the reaction conditions, nanosilica particles "capped" with organic compounds are obtained.

8.2.2 IN SITU SURFACE MODIFICATION IN TERMS OF IONIC BONDS

This method is mainly observed during surface modification of nanoparticles from ionic compounds [31,32]. Often, organic compounds with the same ions (anions or cations) as the nanoparticles are used as modifiers. As an example, the formation mechanism of $Mg(OH)_2$ nanocrystallines [31] modified with stearic acid is shown in Figure 8.2. Generally, an excess of water molecules are adsorbed on the surface of $Mg(OH)_2$ colloid by hydrogen bonds. When the colloid is mixed with butanol and

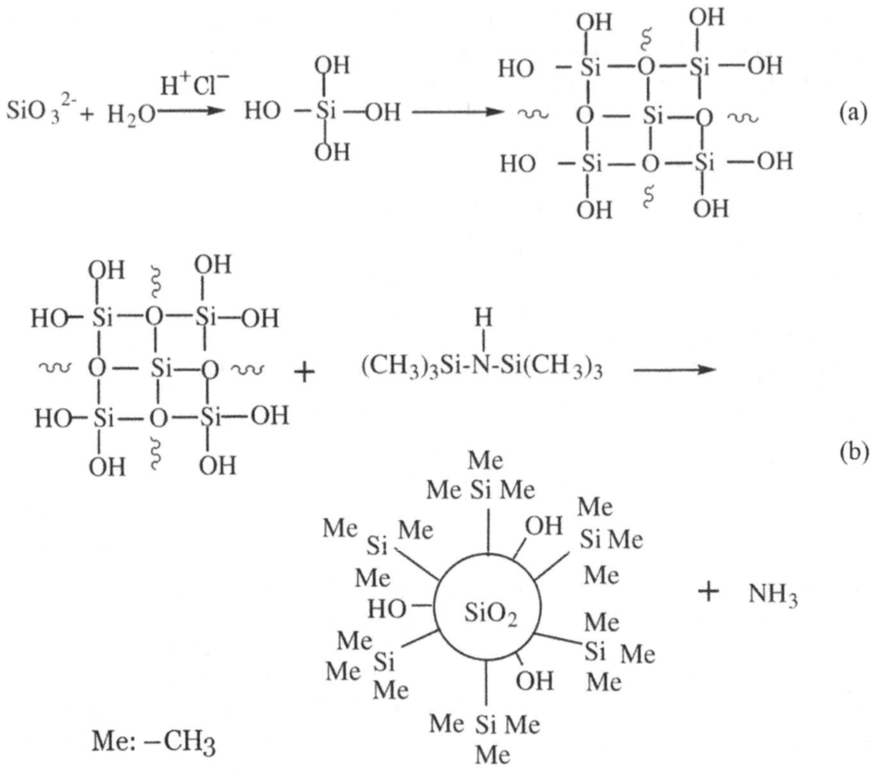

FIGURE 8.1 Schematic diagram for in situ surface modification process using covalent bond formation, (a) Formation of nanosilica clusters, and (b) The modification layer of nanosilica clusters was formed by reacting with the modifier.

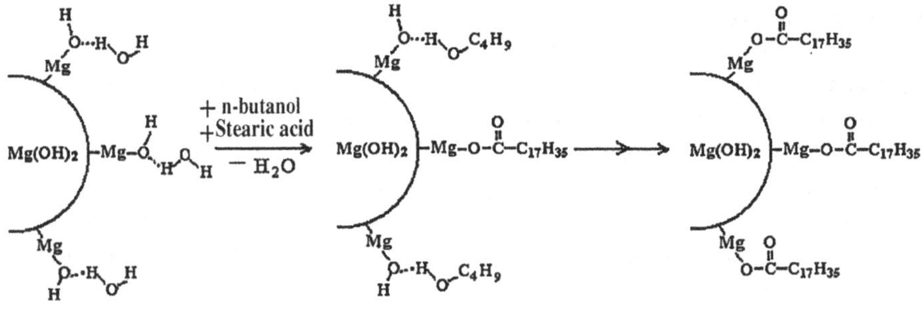

FIGURE 8.2 The schematic of in situ surface-modified Mg(OH)$_2$ using ion bond formation.

stearic acid, and the temperature of the slurry is raised gradually, the water molecules attached by hydrogen bond with the free hydroxyls on the colloid surfaces are displaced by butanol and HOOCC$_{17}$H$_{35}$. At the azeotropic temperature, all the water and some of the butanol molecules are distilled off as azeotropic mixture, and adsorbed HOOCC$_{17}$H$_{35}$ molecules react with the free hydroxyls on the colloid surfaces to form ionic bonds. When the temperature is raised further to the boiling point of butanol, the residual butanol is completely removed and replaced by $-OOCC_{17}H_{35}$ to achieve surface modified $Mg(OH)_2$ nanocrystallines.

8.2.3 IN SITU SURFACE MODIFICATION IN TERMS OF COORDINATION BONDS

This method applies mainly during the surface modification of nanoparticles composed of transition metal ions [33–36]. The first step is selection of organic compound modifiers that can react with transition metal elements to form a stable complex. As an example, Figure 8.3 shows a schematic diagram of surface-modified MoS_2 with dialkyldithiophosphate (DDP) by coordination bonds [33]. The mechanism of formation of DDP-modified MoS_2 nanoclusters is complex because of the multivalent state of Mo and the acidity of Mo^{4+}. The following mechanism is proposed: At first Na_2MoO_4 reacted at a certain temperature with $NH_2OH \cdot HCl$ to produce Mo^{4+}. The addition of Na_2S and pyridinium di-n-octadecyldithiophosphate (PyDDP) produced MoDDP, MoS_3^{2-}, and other molybdenum species, respectively. When HCl was injected into the reaction mixture slowly, it caused MoS_3^{2-} to decompose slowly into MoS_2. It is well known that the initial MoS_2 crystallite is extremely active; it has a large number of defect sites and dangling bonds on its surface. So the initial MoS_2 will react with MoDDP or DDP in solution to form a surface modification layer that stops the growth and further condensation of MoS_2. These bare MoS_2 and other molybdenum compounds are then removed by filtration and recrystallization processes.

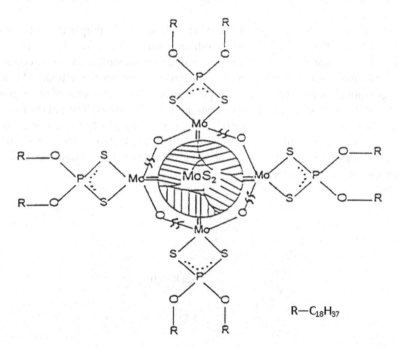

FIGURE 8.3 Schematic of surface modification of MoS_2 by dialkyldithiophosphate using coordination bond formation.

8.2.4 IN SITU SURFACE MODIFICATION IN TERMS OF WEAK CHEMICAL BONDS

Weak chemical bonds, such as hydrogen bonds, van der Waals interaction and dangling bonds, and so on, are mainly used for in situ surface modification of oxide and metal nanoparticles [37,38]. An example of hydrogen bond formation between nanoparticles and modifier is the bonding of amino or ammonium groups to the SiO_2 surfaces (Figure 8.4). First, there are abundant hydroxyl groups on the surface of SiO_2 through ion exchange or acid–base reaction. Second, SiO_2 nanoparticles reacted with the surface modifier of alkyl diamine to form functionalized surface-modified silica nanoparticles through hydrogen bond formation. Third, bonding of another molecular of carboxylic acid to the particle surface was through the interaction between the amino and carboxyl groups. Thus, SiO_2 nanoparticles with bimolecular layer modification were obtained involving hydrogen bond formation.

From the above discussions, the oxidation stability of inorganic nanoparticles and their dispersion stability in organic solvents improved by in situ surface modification with organic compound, which is due to the strong interaction between the functional groups in the modifying agent and the nanoparticles, especially, the coordination bond and the covalent bond. This strong interaction is helpful to effectively inhibit agglomeration between the nanoparticles, and render the nanoparticles to have an affinity for the solvents. Furthermore, the surface modification of nanoparticles with organic compound molecules containing reactive groups can not only improve the dispersion stability of nanoparticles but also achieve the functional characteristics by secondary bonding of active organic functional groups on the nanoparticles.

Therefore, choosing the appropriate modifier is key to the preparation of surface-modified nanoparticles. Only with the modifier, which can chemically adsorb on the surface of the nanometer cores and form a dense monomolecular layer, could the modified nanoparticles be stable and be prepared by appropriate methods. The modifier is composed of both polar and non-polar groups. The structures of polar groups and non-polar groups affect the modification of nanoparticles. The polar groups of the modifier, such as carboxyl groups, mercapto groups, hydroxyl groups, amino groups, and so on, generally determine the ability to adsorb onto the nanoparticles. Appropriate surface modifiers can provide surface-modified nanoparticles with good dispersibility and stability.

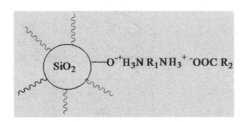

FIGURE 8.4 Bimolecular layer modification of SiO_2 nanoparticles involving hydrogen formation.

8.3 TRIBOLOGY OF NANOADDITIVES

8.3.1 TRIBOLOGY OF METALS NANOPARTICLES

Metal nanoparticles are the most common and widely used nanomaterials. The performance of nanoparticles can be adjusted by changing their size, the surface morphology, uniformity, and dispersion property. The main problems and difficulties in the application of nanoparticles include controllability of size, dispersion in media, and chemical and physical stability. Oil-soluble Cu nanoparticles were synthesized by in situ surface modification technique. During nanoparticles formation, different organic modifiers were coated on the surface of the Cu nanoparticles by strong chemical bonds, so that a series of multifunctional lubricating additives with stable structure, excellent oxidation stability, and excellent dispersion in oil media were obtained. Zhou and coworkers [39,40] reported a series of investigations about the preparation and tribological properties of Cu nanoparticles with O,O'-dialkyldithiophosphate (DDP) containing S and P as the capping agent. The DDP-modified Cu nanoparticles were prepared by means of a redox surface modification technique. The DDP-modified Cu nanoparticles dispersed well in some organic solvents and had good oxidation stability in air. It was assumed that the hydrophobic groups on the surface of DDP-Cu nanoparticles contributed to the improvement of the dispersion property. As shown in Table 8.1, the DDP-modified Cu nanoparticles were evaluated as additives in liquid paraffin (LP) base oil and showed different antiwear property depending on their particle size. The difference in the antiwear behaviors between unmodified and DDP-modified Cu nanoparticles is explained by their difference in interactions with the surface of the friction pairs during the friction process. It was observed that DDP-modified Cu nanoparticles with sizes of 2 nm and 5 nm could effectively improve the antiwear property of LP. In contrast, DDP-modified Cu nanoparticles of 12 nm size decreased the antiwear property of LP, while unmodified Cu nanoparticles (with an average size of 40 nm) as additives in LP caused instant seizure under the same test conditions. It is reasonable to assume that the smaller the size of Cu nanoparticles, the lower its melting point and the better the ductility of nanometer Cu core. Melting point depression of nanoparticles as a function of size has been proposed by Zhang and Kim [41,42]. It was confirmed that the melting point depresses with decreasing size when the size is smaller than the critical size. However, the melting points of nanoparticles with larger than critical size are almost the same as that of bulk materials. Consequently, Cu nanoparticles of smaller size were more likely to interact with

TABLE 8.1

Wear Scar Diameter of Dialkyldithiophosphate-Modified Cu Nanoparticles

		Particle Size of Cu Nanoparticles (nm)		
	Liquid Paraffin	**2**	**5**	**12**
WSD (mm)	0.448	0.261	0.393	0.602

Note: WSD: wear scar diameter.

the surfaces of the friction pairs to form a surface protective film. Such a surface protective film contributed primarily to improved antiwear property. In the case of DDP-modified copper nanoparticles with larger particle sizes and of those not modified at all, the weaker interaction among the copper nanoparticles and the surfaces of the friction pairs was not beneficial to the formation of the surface protective film. Ultimately, unmodified copper nanoparticles of too large particle size acted as abrasives during the friction process, resulting in a decreased antiwear property [39].

Zhou et al. [40] also found that DDP-modified Cu nanoparticles as an oil additive had better friction reduction and antiwear properties than zinc dialkyldithiophosphate (ZDDP), especially at high applied load. They could also improve the load-carrying capacity of the base oil. Table 8.2 gives the relationship between tribological properties and the applied load with the lubrication by LP containing 4 wt% modified Cu nanoparticles. The results showed that relatively lower wear scar diameter (WSD) and lower friction coefficient were observed for LP containing Cu nanoparticles as compared to those of LP containing ZDDP. Moreover, the system can be lubricated effectively for loads of up to 1000–1500 N by adding Cu nanoparticles to the base oil, while for ZDDP, the system will scuff at a load of 800 N. Additionally, Cu nanoparticles in the base oil can remarkably reduce the friction coefficient, especially in the load range 1000–1500 N. However, at loads in the range 500–900 N, the system will scuff and severe wear will occur. The reason why this occurs may be that at the low load (<500 N), the tribochemical reaction film formed by S and P elements is responsible for the good antiwear property, and at the high load (>900 N), Cu nanoparticles can deposit and melt on the rubbed surface to form a layer of compact protection film, but at loads of 500–900 N, the tribological properties of the reaction film are poor and the deposited Cu nanocores cannot melt to form a good boundary film comparable to that formed at the higher load, so the tribological performance is poor.

Unfortunately, phosphorus and sulfur elements are harmful to the environment and humans as well. This means it is imperative to seek novel P- and/or S-free surface modifying agents for Cu nanoparticles so as to make an environmentally acceptable

TABLE 8.2
The Relationship between Tribological Properties and the Applied Load for Liquid Paraffin with 4% DDP-Modified Cu Nanoparticles versus ZDDP

	Load (N)							
	300	400	500	600	700	1000	1200	1500
WSD (mm)								
Cu-DDP	0.40	0.45	/	/	/	0.95	1.025	0.94
ZDDP	0.48	0.51	0.66	0.72	1.02	/	/	/
Friction coefficient								
Cu-DDP	0.093	0.080	/	/	/	0.080	0.065	0.048
ZDDP	0.100	0.101	0.110	0.112	0.120	/	/	/

Note: WSD: wear scar diameter; DDP: dialkyldithiophosphate; ZDDP: zinc dialkyldithiophosphate; and Cu-DDP: DDP-modified Cu nanoparticles.

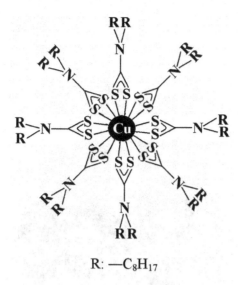

R: —C_8H_{17}

FIGURE 8.5   Proposed structure of dialkyl dithiocarbamate–modified Cu nanoparticles.

nano-Cu lubricant additive. With that in mind, Yang et al. [10,35] used P-free N,N'-dialkyl dithiocarbamate (DTC) as modifier to synthesize surface-modified Cu nanoparticles. The structure of DTC-modified Cu nanoparticles is shown in Figure 8.5. A proper amount of as-prepared DTC-modified Cu nanoparticles was dispersed in poly α-olefin (PAO) and centrifuged at 12,000 rpm for 10 minutes, and no precipitate was observed at the bottom of the centrifuge tube. Furthermore, no sign of sedimentation was observed even after the dispersion of DTC-modified Cu nanoparticles in PAO was stored under ambient conditions for 6 months. Figure 8.6 shows pictures of the dispersions with 0.5 wt% and 1.0 wt% DTC-capped Cu nanoparticles in PAO after 6 months of storage. They are free of color change or precipitation during storage in ambient conditions. This indicates that as-synthesized DTC-modified Cu nanoparticles are highly stable in ambient conditions, due to the excellent capping effect of the bidentate DTC.

The influence of DTC-modified Cu nanoparticles concentration on WSD is shown in Figure 8.7. The WSD is reduced from 0.665 mm with lubrication of LP without additive to 0.460 mm with the lubrication of LP containing 1.5 wt% DTC-modified Cu nanoparticles. This indicates that Cu nanoparticles as a lubricant additive have excellent antiwear property, which might be closely related to the small size, low melting point, and high reactivity of copper nanoparticles. Particularly in the form of a stable dispersion in oils, surface-modified copper nanoparticles are easy to enter between the rubbing surfaces and deposit there to form surface protective film under local high temperature and high contact pressure, resulting in significantly increased antiwear property. Moreover, surface-modified Cu nanoparticles in LP can fill up micropits and grooves of the friction surface to play the role of self-repair, resulting in increased antiwear property. The mechanism is proposed in Figure 8.8.

Furthermore, Cu nanoparticles surface-modified by alkanethiols with different chain lengths (C_xS) were synthesized using the ligand exchange method in a

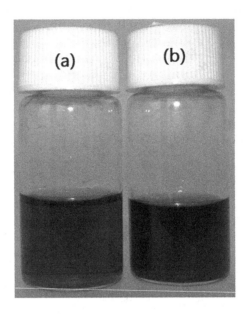

FIGURE 8.6 Photographs of dispersions of dialkyl dithiocarbamate–modified Cu nanoparticles after 6 months of storage at ambient conditions in PAO: (a) 0.5 wt% and (b) 1.0 wt%.

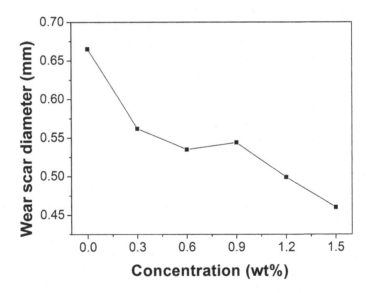

FIGURE 8.7 Wear scar diameter as a function of concentration of Cu nanoparticles modified by dialkyl dithiocarbamate in liquid paraffin (four-ball, 300 N, 1450 rpm, 30 min).

two-phase system. As-prepared C_xS-modified Cu nanoparticles as additive in LP possess excellent antiwear and friction-reduction performance because of the deposition of nano-Cu with low melting point on worn steel surface, leading to the formation of a self-repairing protective layer [43].

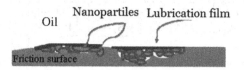

FIGURE 8.8 Mechanism of friction reduction between friction surfaces in oil with nanoparticles.

Environmentally friendly oil-soluble Cu nanoparticles without phosphorus and sulfur elements [44,45] were prepared using in situ surface modification technique with tetradecyl hydroxamic acid (THA), oleic acid, and oleylamine as modifiers. As-synthesized THA-capped Cu nanoparticles as additive in LP showed excellent antiwear property for steel–steel contact, indicating promising application as environmentally acceptable lubricating oil additive. Also the load-carrying capacity of THA-capped Cu nanoparticles has been investigated. Table 8.3 gives the data of WSD obtained during the maximum non-seizure load (P_B value) and sintering load (P_D value) test process. It shows the relationship between tribological properties and the applied load for lubrication with LP containing 0.2 wt% THA-modified Cu nanoparticles. It can be concluded from the results that THA-modified Cu nanoparticles in LP have excellent tribological properties, especially in the load range 800–2100 N. At the same time, the friction system will scuff at intermediate loads of 430–700 N. This is attributed to the tribochemical reaction film formed by the organic modifying agent with active O and N elements, which mainly contributes to improving the tribological properties at the low load (<430 N) and then at the high load (>700 N). Cu nanoparticles can deposit by the friction shearing force and high pressure at the interface; after that, the bare Cu nanocores melt with the high temperature of friction surface to form a layer of compact protection film. Nevertheless, the deposited Cu nanoparticles cannot melt to form a good boundary film at loads of 500–900 N, which is not high enough, thus resulting in the poor tribological properties of THA-modified Cu nanoparticles.

Compared to the oil-based lubricating fluid, water-based lubricating fluid has certain advantages, such as incombustibility, excellent cooling ability, environmental compatibility, and low cost. This has led to rapid progress in the development and use of water-based lubricating fluids in recent years. However, the synthesis of high-performance water-based additives is a prerequisite for the development of new water-based lubricating fluids. Zhang and coworkers [36,46,47] have synthesized

TABLE 8.3

Wear Scar Diameter Obtained in the P_B and P_D Value Test Process for Liquid Paraffin with 0.2 wt% Tetradecyl Hydroxamic Acid–Modified Cu Nanoparticles

Load (N)	300	392	411	430	598	696	804	1049	1490	2078	2205
WSD (mm)	0.25	0.31	0.31	/	/	/	0.81	0.91	0.87	1.03	Sintering

Note: WSD: wear scar diameter; P_B value is maximum non-seizure load; and P_D value is sintering load.

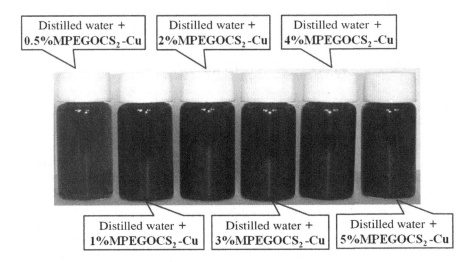

FIGURE 8.9 Digital camera images of Cu nanoparticles modified by methoxylpolyethyl-eneglycol xanthate potassium (MPEGOCS$_2$-Cu) dispersed in distilled water.

water-soluble Cu nanoparticles with excellent friction-reducing and antiwear proper-ties, as well as outstanding extreme pressure performance. As-prepared Cu nanopar-ticles modified by methoxylpolyethyleneglycol xanthate potassium (MPEGOCS$_2$K) have a diameter of 3 ± 1 nm. They are readily soluble in water, and were dissolved in distilled water at different concentrations. After 1 month of storage, the digital camera images shown in Figure 8.9 were obtained. There is no settling at the bot-tom of the bottles. This indicates that as-prepared Cu nanoparticles modified by MPEGOCS$_2$K can be homogeneously dispersed in distilled water in the test concen-tration range, indicating good dispersion capability in water [36]. This is critical to the tribological application of Cu nanoparticles as a water-based additive.

Cu nanoparticles surface-modified by MPEGOCS$_2$K as water-based lubricant additives can significantly improve the tribological properties and load-carrying capacity of distilled water. Figure 8.10 compares the P_B values of distilled water, distilled water containing MPEGOCS$_2$K modifier, and distilled water containing Cu nanoparticles modified by MPEGOCS$_2$K. It can be seen that the addition of both the modifier and Cu nanoparticles modified by MPEGOCS$_2$K can increase the P_B value of distilled water. Surface-modified Cu nanoparticles by MPEGOCS$_2$K addi-tive were better than the modifier alone in terms of the ability to improve the load-carrying capacity of distilled water. Besides, the as-prepared Cu nanoparticles as an additive in distilled water (mass fraction 5%) can increase the P_B value from 88 N of distilled water to 696 N. The reason is that surface-capped Cu nanoparticles as an additive to distilled water experience tribochemical reaction in the sliding of steel against steel to form boundary lubrication film consisting of Cu, FeS, and FeSO$_4$ on the rubbed steel surface. This greatly improves the tribological properties of distilled water, thereby reducing friction and wear of the steel–steel pair.

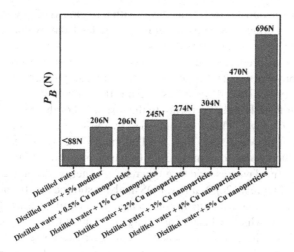

FIGURE 8.10 P_B values of distilled water, distilled water containing surface modifier of methoxylpolyethyleneglycol xanthate potassium, and distilled water containing Cu nanoparticles modified by methoxylpolyethyleneglycol xanthate potassium.

In order to further improve the oxidation resistance of Cu nanoparticles, a water-soluble nanosized Cu/SiO$_2$ composite was prepared via the sol–gel method. Copper nanocrystals were prepared in aqueous solution via the reduction of copper (II) ions by hydrazine hydrate and coating of as-obtained Cu nanoparticles with silica (SiO$_2$) [46]. Figure 8.11 shows the transmission electron microscopy (TEM) images of Cu/SiO$_2$ nanocomposite. It is observed that Cu metal nanoparticles are well embedded in network-like silica (Figure 8.11a), which allows as-prepared Cu/SiO$_2$ nanocomposite to form suspension in distilled water. Cu nanoparticles appear as spheres and have an average particle diameter of 20 nm (Figure 8.11b). In the meantime, a careful observation of the TEM image in Figure 8.11b indicates that the

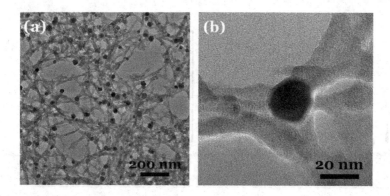

FIGURE 8.11 TEM images of as-prepared Cu/SiO2 nanocomposite with different magnifications, (a) 20,000× and (b) 200,000×.

Cu nanocore is surrounded by a silica layer with a thickness of about 2 nm. Such a silica coating can effectively prevent Cu nanoparticles from oxidation, resulting in increased oxidation resistance of Cu nanocores.

Furthermore, as-prepared Cu/SiO_2 nanocomposite as a lubricant additive can effectively improve the friction-reducing and antiwear property as well as load-carrying capacity of distilled water. As a result, friction and wear of the steel–steel contact are significantly reduced owing to the formation of the protective and lubricious deposited Cu layer. This was confirmed by the scanning electron microscopy (SEM) images of the wear scar (Figure 8.12a and b). The worn steel surface lubricated with distilled water containing Cu/SiO_2 nanocomposite additive is covered by a metallic protective layer, as evidenced by relevant energy dispersive spectrometry (EDS) analysis (Figure 8.12c). Namely, EDS analysis showed that the atomic concentration of Cu in the wear scar surface lubricated with distilled water-Cu/SiO_2 additive is up to 36.70%, which indicates that a large amount of Cu has been deposited on worn steel surface. This is rational, since Cu nanoparticles possessing good wetting behavior and a much lower melting point than bulk Cu are able to form a protective

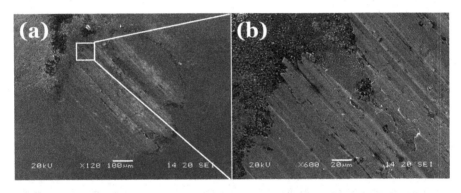

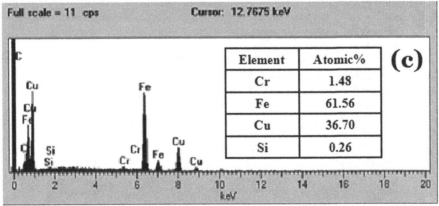

FIGURE 8.12 (a and b) SEM micrographs and (c) EDS pattern (corresponding to image (b) of worn steel ball surfaces lubricated with distilled water containing 1.0 wt% Cu/SiO_2 nanocomposite (load: 150 N; speed: 1450 rpm; time: 30 min; room temperature).

and lubricious layer on steel sliding surface under friction-induced high flash temperature, thereby increasing contact area, providing fast relaxation and reduction of contact stress, and avoiding direct steel–steel contact.

In addition to metal nano-copper, many soft metals materials with low melting points were used as lubricant additives. Examples include silver [48,49], indium [50,51], tin [52,53], lead [54], nickel [11,55], and bismuth [56,57]. In particular, bismuth has attracted much interest as a potential thermoelectric and "green" lubricant material. Zhao and coworkers [56] reported a simple preparation procedure for bismuth nanoparticles and evaluated their tribological properties on a four-ball tribometer. In addition, an improved method based on the in situ formation of a surfactant layer of stearic acid chemisorbed on the surface of metal bismuth nanoparticles was proposed by Zhao and coworkers [57]. The method allows for control the oxidation of bismuth nanoparticles during the synthesis process. Liu et al. [58] reported a green and simple solvent thermolysis route for the synthesis of surface-capped bismuth nanoparticles. As-synthesized Bi nanoparticles additives in synthetic ester base oils exhibit excellent tribological properties, especially good antiwear property, due to the deposition of bismuth nanoparticles with a low melting point on worn steel surface leading to the formation of a self-repairing protective layer.

Nanosized particles of nickel have promising applications in the fields of catalysis, magnetic recording, medical diagnosis, and conduction. Dandelion-like 3D nickel nanostructures were prepared by reducing nickel chloride with hydrazine hydrate in glycol solution at a relatively mild temperature of 100°C. The prepared Ni samples had obvious shape anisotropy and were composed of fine nanocrystallites, while they had significantly enhanced ferromagnetic properties compared to bulk Ni and Ni nanoparticles [59]. Chen et al. [11] reported that nickel-based nanolubricants containing size-tunable monodispersed nickel nanoparticles were in situ synthesized in PAO via a simple one-step thermal decomposition method with $Ni(HCOO)_2 \cdot 2H_2O$ as the Ni source, PAO as base oil, as well as oleylamine (OM) and oleic acid (OA) as the surface-capping agents. TEM images as well as the size distribution of Ni nanoparticles with different sizes are shown in Figure 8.13. It can be seen that as-prepared Ni nanoparticles have sphere-like morphology, and their average diameter is tunable from about 7.5 to 13.5 and 28.5 nm, depending on the varying dosage of PAO from 80 to 40 and 20 mL. More importantly, as-synthesized Ni nanoparticles show no sign of obvious aggregation, which indicates that OM and OA are able to well adjust the nucleation and growth processes of Ni nanocrystalline by influencing chemical reaction driving force and affording monodispersed Ni nanoparticles with different sizes.

The effect of particle size on tribological properties of as-synthesized Ni-based nanolubricants was evaluated with a four-ball friction and wear tribometer. The nickel-based nanolubricants exhibit good antiwear behavior even at a low Ni concentration of 0.05 wt%. Moreover, the antiwear property of Ni nanoparticles in PAO was closely related to their size and concentration. Thus, at an additive concentration of 0.05 wt%, sample Ni1 with a smaller diameter of 7.5 nm (Figure 8.13a) is more effective than samples Ni2 with diameter of 13.5 nm (Figure 8.13b) and Ni3 with diameter of 28.5 nm (Figure 8.13c) in reducing WSD in the steel–steel contact, which is consistent with the well-known Hall–Petch effect and Archard's law mostly used in adhesion and abrasive wear conditions [60].

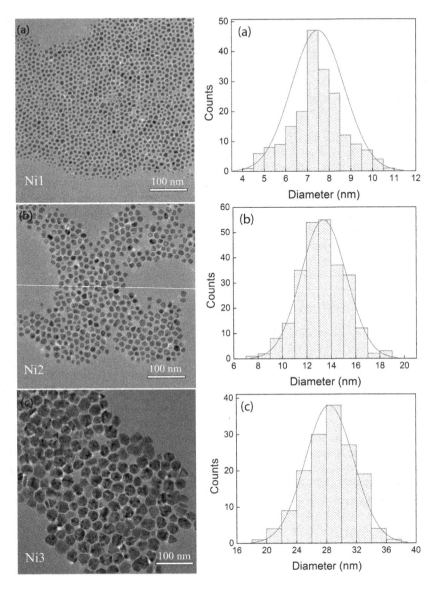

FIGURE 8.13 TEM images (left) and size distributions (right) of surface-capped Ni nanoparticles with different sizes.

Figure 8.14 shows typical SEM images of worn surfaces of steel balls lubricated with various lubricant systems. It can be seen that the WSD of the steel balls lubricated with PAO containing different sizes Ni nanoparticles was smaller than that lubricated with pure PAO. Particularly, the smallest WSD is obtained under lubrication with PAO containing Ni1 with particles size of 7.5 nm (Figure 8.14c). In addition, the worn surface of the steel ball lubricated with PAO + Ni1 is smooth and shows only shallow scratch grooves as compared to those lubricated with PAO + Ni2

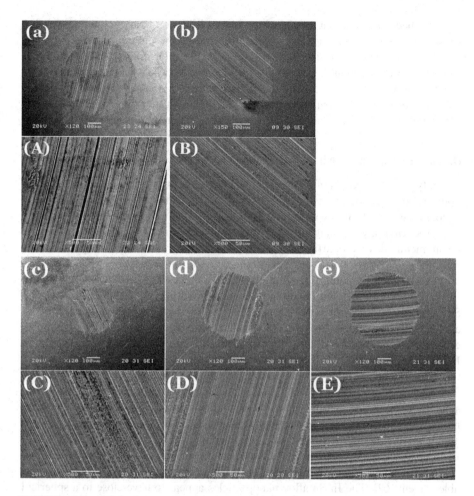

FIGURE 8.14 SEM micrographs of wear scars lubricated with (a, A) pure PAO, (b, B) PAO containing 0.05 wt% surface modifiers, (c, C) PAO containing 0.05 wt% Ni1 nanoparticles with particle size of 7.5 nm, (d, D) PAO containing 0.05 wt% Ni2 nanoparticles with particle size of 13.5 nm, and (e, E) PAO containing 0.05 wt% Ni3 nanoparticles with particle size of 28.5 nm (four-ball friction and wear tester, 1,450 rev/min, 300 N, 30 min).

(Figure 8.14d) or PAO + Ni3 (Figure 8.14d). In the latter two cases, the worn steel surfaces are dominated by obvious scuffing and adhesion wear. Ni nanocores with a smaller size are easier to fill in micropits and grooves on worn steel surfaces to form a compact protective layer thereon. Therefore, the chemically adsorbed and deposited layer of fine nickel particles together with the boundary lubricating film jointly result in significantly improved tribological properties of PAO base stock.

Among all the metal nanoparticles that our group has synthesized, the overall performance of Cu nanoparticles is the best. The preparation conditions and parameters of large-scale production were explored, and copper nanoparticles have been produced on an industrial scale.

It has been realized that metals with a low melting point can melt on the friction surface into liquid under the flash temperature caused by rubbing, and exhibit a low friction coefficient. Accordingly, it is expected that their corresponding nanoparticles should be excellent lubricants. It is reasonable to expect that the nanoparticles of low melting point metals, particularly in the form of a stable dispersion in oils, are able to enter between the rubbing surfaces and melt on it to form a liquid lubricating film, thus exhibiting an enhanced lubricating performance.

8.3.2 Tribology of Metal Alloy Nanoparticles

When two or more kinds of metals are melted together to form an alloy, the melting point, rigidity, conductivity, and extensibility will change considerably. Therefore, nanoalloy materials are expected to have more fascinating properties in electronics, catalysis, tribology and magnetism compared to nanoparticles of single metals. As a result, metal alloy nanoparticles, especially low melting point metal alloy nanomaterials, have been the focus of intensive research in recent years [61,62].

The low melting point alloys mainly refer to those with two or more metals of Bi, Pb, Sn, In, and Cd, with melting points below 400 C (Bi: 271.4 C; Pb: 327.4 C: Sn: 231.8 C; In: 156.4 C; Cd: 320.9 C), that form bimetal, trimetal, or multi-metal alloys. Nanoalloy metals are anticipated to have excellent tunable tribological properties as the composition of the alloys is varied. To date, nanoalloys of Bi-In [61,63], In-Sn [62], Pb-Bi [64], Sn-Bi [65], and Sn-Cd [66] dispersed in oils have been synthesized and their tribological properties investigated.

In 2004, Zhao and coworker [61] used the direct solution-dispersing method to synthesize Bi-In low melting point metallic alloy nanocrystals. For this approach, molten metal was dispersed in a solvent without using complex apparatus, then cooled and centrifuged, and the product was obtained. Figure 8.15 shows TEM images of Bi-In dendritic nanocrystals obtained at different temperatures. Using this method, In-Sn alloy nanoparticles were prepared by dispersing molten In-Sn alloy in a suitable solvent [62]. The In-Sn alloy nanoparticles appear to have close to a spherical shape with an average diameter of 60 nm (Figure 8.16a). In–Sn alloy nanoparticles were evaluated as additives in lubricating oil on a four-ball tribometer. Figure 8.16b shows that the In–Sn alloy nanoparticles have excellent antiwear performance over a wide concentration range. Furthermore, the In-Sn alloy nanoparticles performed better than the monometallic In or Sn particles, which might be attributed to the structure of In-Sn alloy.

It is well known that lead is an important lubricating material and bismuth, with very similar properties to lead, is considered as an ecologically green "lubricating metal." Pb-Bi bimetal nanoparticles were prepared from alloy ingot by solution dispersion [64]. The Pb-Bi alloy nanoparticles had a spherical shape with an average particle diameter of 50 nm. The formation mechanism of Pb-Bi nanoparticles involves diffusion and oxidation of Pb atoms, solvent adsorption at the droplet surface, and dispersion of alloy droplets by the shear force. A possible pathway for the formation of the nanoparticles is shown in Figure 8.17. First, large alloy droplets are dispersed as small droplets in the solvent by stirring, but these small droplets are thermodynamically unstable and tend to coagulate to reduce total energy of the system.

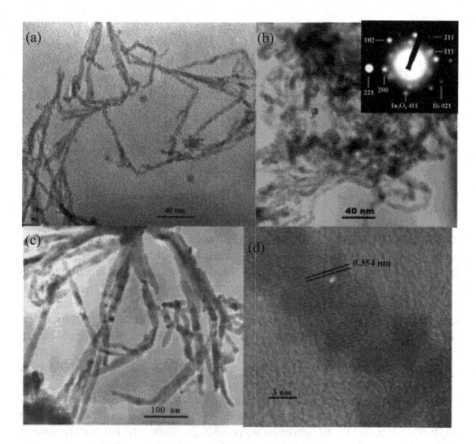

FIGURE 8.15 TEM images of various Bi-In dendritic nanocrystals prepared at (a) 220°C, (b) 200°C, and (c) 185°C; (d) HRTEM image of an individual Bi-In dendritic nanocrystal.

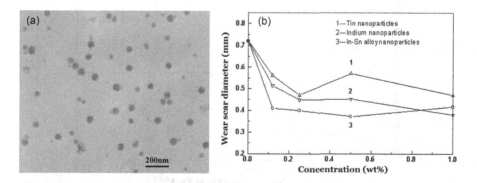

FIGURE 8.16 (a) A typical TEM image of solution dispersion synthesized In-Sn alloy nanoparticles, and (b) variations of wear scar diameter with concentration of tin, indium, and In-Sn alloy nanoparticles in paraffin oil under 300 N, 1450 rpm, 30 min.

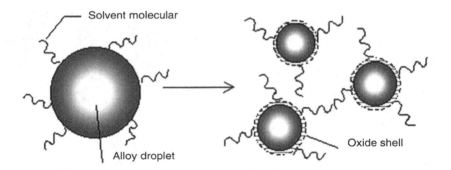

FIGURE 8.17 Schematic representation of the mechanism of formation for Pb–Bi nanoparticles.

At the reaction temperature of 200 C, Pb in the droplet surface has high activity and easily forms PbO sheath through reaction with oxygen, which necessarily prevents the aggregation of alloy droplet. As the oxidation of surface Pb proceeds, more Pb atoms within the droplet diffuse to the surface, which results in alloy segregation and forms Bi-rich phase (Bi) and Pb_7Bi_3 phase. In addition, the PbO sheath seems to adsorb solvent molecules, which improves the stability of alloy droplets in the solvent. In the presence of PbO layer, alloy droplets can be gradually separated into nanosized alloy droplets, resulting in nanoparticles when the temperature is reduced.

The tribological properties of Pb-Bi nanoparticles as lubricating additives in oil were evaluated on a four-ball tribometer. Figure 8.18 shows the variation of the WSD with the concentration of Pb-Bi nanoparticles in paraffin oil under a load of 300 N. When the concentration is zero, the coordinate represents WSD of paraffin oil. It is evident that Pb-Bi nanoparticles provide excellent antiwear performance in a broad concentration range. Even at a very low concentration, Pb-Bi nanoparticles

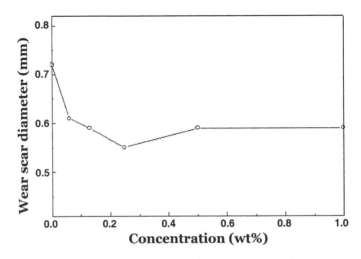

FIGURE 8.18 Variation of wear scar diameter as a function of Pb-Bi bimetal nanoparticles concentration in paraffin oil at 300 N load.

can efficiently improve the antiwear properties of paraffin oil. The best antiwear properties were obtained at 0.25 wt% concentration of Pb-Bi nanoparticles in the base oil. The reason that 0.25 wt% is the best concentration may be that the Pb-Bi nanoparticles have appropriate content in base oil to achieve the best performance. When the concentration increased, the WSD was slightly higher, perhaps because abrasive wear occurred.

Figure 8.19 illustrates the relationship between the WSD and the applied load during lubrication test with base oil containing 0.25 wt% Pb-Bi nanoparticles. For comparison, the results with just the base oil and the base oil with 0.5 wt% Pb nanoparticles are also given. For the base oil, the WSD is 0.72 mm at a load of 300 N, and the system scuffing load is below 400 N. It is clear that the Pb-Bi nanoparticles give excellent antiwear properties under different loads. Lubrication with base oil containing Pb-Bi nanoparticles gave smaller WSD than base oil containing 0.5 wt% Pb nanoparticles. At 300 N load, the WSD of 0.25 wt% Pb-Bi nanoparticles was only 0.55 mm, which is much lower than with 0.5 wt% Pb nanoparticles (0.6 mm). At the load of 500 N, the WSD of 0.25 wt% Pb-Bi nanoparticles was 0.64 mm, which is still below that with the 0.5 wt% Pb nanoparticles (0.72 mm). At 600 N, the WSD of 0.25 wt% Pb-Bi nanoparticles was 0.70 mm, which is close to that of the pure base oil at 300 N. These results show that Pb-Bi nanoparticles can be used as a lubricating additive at mild loads.

Chen and coworkers [65,66] reported a new ultrasonic-assisted solution dispersing method to prepare low melting point metal alloy nanoparticles for use as oil additives. In this method, bulk alloys of low melting point metals were placed in oils and then directly dispersed in the oils by ultrasonic irradiation at room temperature. In addition to Sn-Bi [65] and Sn-Cd [66] bimetal alloy, Pb-Sn-Cd trimetal alloy [67] and Bi-Pb-Sn-Cd tetrametal alloy [68] have been successfully prepared using this method. The tribological properties of these alloy nanoparticles as additives in LP

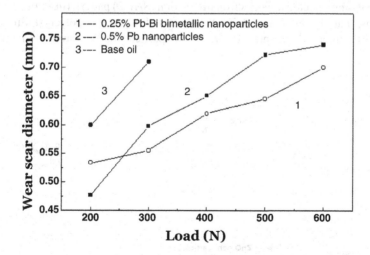

FIGURE 8.19 Relationship between wear scar diameter and applied load for lubrication with (1) 0.25 wt% Pb–Bi nanoparticles, (2) 0.5 wt% Pb nanoparticles, and (3) base oil without nanoparticles.

were evaluated using a four-ball tribometer. The alloy nanoparticles as additives in LP base oil showed good friction-reducing and antiwear properties. Moreover, the alloy nanoparticles showed better lubricating properties than the monometal particles, hence the alloy nanoparticles can be used as good lubricating oil additives.

The alloy nanoparticles as lubricating oil additives have small size and low melting point. They can form a high-strength composite lubrication film on the friction surface under the high load, thereby reducing the friction coefficient and wear. At the same time, alloy nanoparticles with strong reactivity can form "micro-solid solution" on the rubbing surface so as to repair the wear damage. The alloy nanoparticles with small size can also effectively fill the microcracks caused by fatigue wear and prevent its propagation, thereby reducing fatigue wear.

8.3.3 TRIBOLOGY OF OXIDE NANOPARTICLES

Among the various nanolubricants, oxides with a compact and stable structure as well as good thermal stability may be promising green lubricant additives [69–72]. Nanoscale ZnO (nano-ZnO) is an example of the so-called nanolubricants, as it has a large surface area, high surface energy, strong adsorption, high diffusion, easy sintering, low melting point, and good tribological properties. Particularly, nano-ZnO, free of sulfur and phosphorus, is environment-friendly. Oleic acid–modified ZnO nanoparticles were prepared via a facile in situ one-step route for directly preparing oxide nanoparticles from an organometallic precursor [69]. The formation of OA-modified ZnO is illustrated in Figure 8.20.

Figure 8.21 shows TEM images with different magnifications of OA-modified ZnO nanoparticles. It can be seen that the OA-modified ZnO nanoparticles show no sign of obvious aggregation (Figure 8.21a–c), and they display distinct lattice stripes with a lattice spacing of 0.32 nm (Figure 8.21d). This, in association with corresponding selected area electron diffraction (SAED) pattern (inset Figure 8.21c), confirms that the as-prepared ZnO nanoparticles possess polycrystalline structure. Furthermore, the as-prepared ZnO nanoparticles have an average diameter of

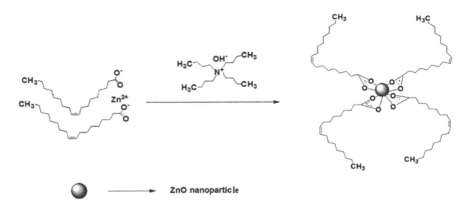

FIGURE 8.20 Schematic diagram showing the procedure for the preparation of ZnO nanoparticles modified by oleic acid.

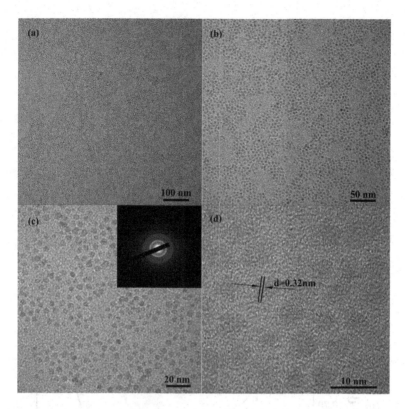

FIGURE 8.21 TEM images with different magnifications and SAED pattern (inset) of aspreparedoleic acid modified ZnO nanoparticles, (a) 40,000×, (b) 80,000×, (c) 200,000× and(d) 800,000×.

4.04 nm. This indicates that the surface modification of ZnO nanoparticles by oleic acid helps to inhibit their aggregation and retain a relatively narrow size distribution. The reason might be that OA molecules anchored on the surface of ZnO nanoparticles are able to reduce the surface energy of ZnO nanoparticles thereby preventing them from agglomeration. In the meantime, oleic acid molecules anchored on the surface of ZnO nanoparticles also help to decrease the collision rate of ZnO crystal nucleus, thereby retarding the growth of ZnO nanoparticles.

The tribological properties of OA-modified ZnO nanoparticles in non-polar PAO and polar diisooctyl sebacate (DIOS) lubricant base oils were investigated. Figure 8.22 illustrates the friction coefficient and WSD of the steel balls lubricated with PAO containing different concentrations of OA-modified ZnO nanoparticles (load: 392 N; speed: 1200 rev min^{-1}; time: 60 min; temperature: 75 C). It can be seen that the OA-modified ZnO nanoparticles as an additive in PAO can effectively reduce the friction coefficient and WSD of the steel–steel pair. Namely, when the OA-modified ZnO nanoparticles are added into PAO at a concentration of 1.2 wt%, the best antiwear performance and friction-reducing ability are obtained: the friction coefficient is reduced from 0.106 to 0.096 and the WSD is reduced from 0.850 mm to

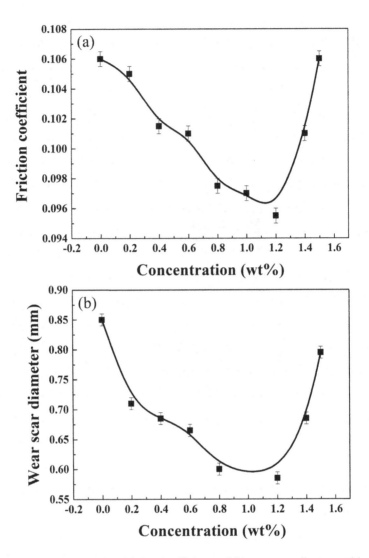

FIGURE 8.22 Variations of (a) friction coefficient and (b) wear scar diameter with concentration of oleic acid–modified ZnO nanoparticles in PAO (load: 392 N; speed: 1200 rev/min; time: 60 min; temperature: 75°C).

0.585 mm (corresponding to reductions of 9.9% and 31.2%, respectively). However, higher concentrations of OA-modified ZnO nanoparticles in PAO were less effective in reducing the friction coefficient and WSD. It can be inferred that at a too low additive concentration, the lubricant can hardly form a tribofilm with good coverage to prevent direct contact of the sliding steel pairs, thereby leading to relatively high friction coefficient. When the additive concentration is too high, the as-prepared OA-modified ZnO nanoparticles would tend to form large aggregates and can hardly fill the valleys between the asperities of the sliding steel pair, thereby also yielding a relatively high friction coefficient along with enhanced scratching and shearing.

Therefore, it is suggested that the concentration of the OA-modified ZnO nanoparticles in PAO be kept at 1.2 wt% in order to effectively reduce the friction and wear of the steel–steel sliding pair.

The friction coefficient and WSD of the steel balls lubricated with DIOS containing different concentrations of as-prepared ZnO nanoparticles are depicted in Figure 8.23. Here the friction and wear tests were carried out under the same test conditions as those for PAO. It can be seen that both the friction coefficient and WSD of the steel–steel sliding pair gradually decreased with increasing concentration of

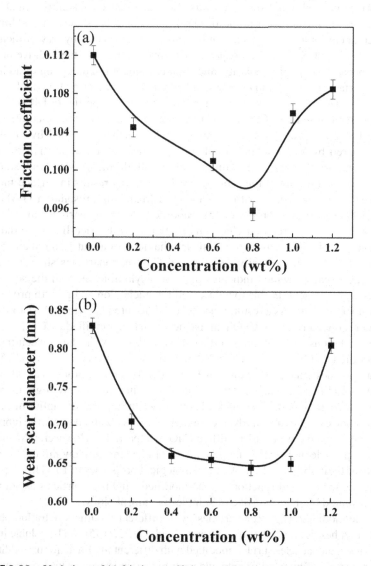

FIGURE 8.23 Variations of (a) friction coefficient and (b) wear scar diameter for steel–steel sliding as a function of oleic acid–modified ZnO concentration in diisooctyl sebacate base oil (load: 392 N; speed: 1200 rev/min; time: 60 min; temperature: 75°C).

as-prepared ZnO nanoparticles in DIOS to a minimum value. Further increase of the concentrations of as-prepared ZnO nanoparticles in DIOS resulted in larger friction coefficient and WSD. The lowest friction coefficient and WSD were obtained at 0.80 wt% concentration of OA-modified ZnO nanoparticles in DIOS, which resulted in the friction coefficient decreasing from 0.112 to 0.096 and the WSD decreasing from 0.830 mm to 0.645 mm (corresponding to reduction of 14.5% and 22.3%, respectively). From these results, it can be said that ZnO nanoparticles in PAO and DIOS gave similar friction coefficient and WSD of the steel–steel sliding pair under boundary lubrication condition. However, it was also found that some additives in DIOS led to a decrease in the antiwear properties [73]. This could be because the additive and DIOS compete for adsorption sites on the rubbed steel surface by way of their functional groups, DIOS adsorption capacity was much more than the additive on metal surface because of its high polarity, and smaller amount of adsorption of additive in friction surface, resulting in DIOS base oil continuous lubricating film not be formed on the metal surface, thereby hindering the tribological properties of DIOS base oil.

It has been found that CuO nanoparticles are potential lubricant additives with good tribological properties. CuO nanoparticles surface modified by oleylamine were prepared by a unique and simple one-pot thermal decomposition route [70]. In the synthesis process, oleylamine acts as both the solvent and stabilizer. This allows the probing of the capping mechanism that results in surface-modified rice-like CuO nanoparticles without S and P. TEM and high-resolution TEM images as well as the size distribution of OM-modified CuO nanoparticles are shown in Figure 8.24. The as-prepared CuO nanoparticles exhibit rice-like or spindle-like morphology and have an average length and diameter of about 12.8 nm and 5.5 nm, respectively. More importantly, as-prepared CuO nanoparticles show no sign of obvious aggregation, which indicates that the oleylamine, as both the solvent and surface-capping agent, is able to adjust well the nucleation and growth processes of CuO nanocrystalline. As a result, it has induced oriented growth of the nanocrystalline and produced rice-like CuO nanoparticles surface modified by OM.

The variations of friction coefficient and WSD with the concentration of OM-modified CuO nanoparticles in LP base oil are displayed in Figure 8.25. For a comparison, the data for LP containing OM under the same conditions are also given. Both OM-modified CuO nanoparticles and oleylamine lead to only slight increase of the friction coefficient of base stock LP, but they can improve the antiwear property of LP to some extent. Particularly, the smallest WSD is obtained under the lubrication of LP containing 0.5 wt% OM-modified CuO nanoparticles. However, OM-modified CuO nanoparticle and surface modifying agent oleylamine show minor differences in terms of their ability to improve the tribological properties of LP. This is possibly because the tribological function of OM-modified CuO nanoparticles as additive in LP is prevented by their rice-like or spindle-like morphology.

Surface-modified TiO$_2$ nanoparticles, with different modifiers (tetrafluorobenzoic acid, 2-ethyl hexoic acid, cis-9-octadecenoic acid) [71,72,74,75] can be soluble in LP or water. TiO$_2$ nanoparticles surface-modified with different modifiers for use additives in LP or in water, exhibit good friction and wear reduction as well as load-carrying properties. This makes surface-modified TiO$_2$ nanoparticles an environmentally acceptable lubricant additive with good dispersion stability in lubricating base fluids.

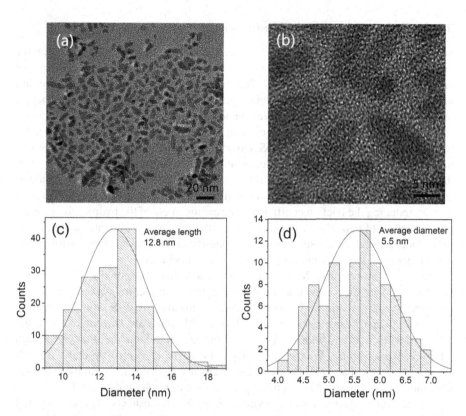

FIGURE 8.24 (a) TEM and (b) HRTEM images as well as size distributions (c and d) of oleylamine-modified CuO nanoparticles.

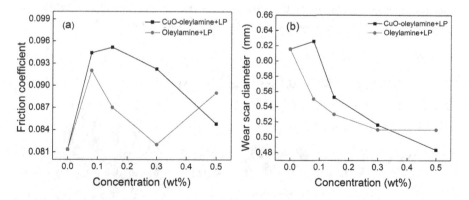

FIGURE 8.25 Relationship between (a) friction coefficient (b) wear scar diameter and concentration of oleylamine-modified CuO in liquid paraffin (four-ball tribometer, 1450 rev/min, 300 N, 30 min; the data for liquid paraffin containing oleylamine under the same conditions are also given for a comparison).

8.3.4 Tribology of Sulfide Nanoparticles

Sulfide nanoparticles have been widely used in lubricants as friction modifiers for many years because they can efficiently reduce friction and wear in the boundary lubrication regime [33,34,76]. MoS_2, due to its special lamellar-type structure and lows hear strength, is a very useful lubricant additive that provides excellent friction reduction, especially in high-pressure contacts. However, it is only used in colloidal suspension due to its insolubility in organic media. MoS_2 nanoparticles modified by DDP are much more stable than MoS_2 particles in air, and they have good dispersity in oils [33,34], which not only reduces the friction coefficient but also increases the antiwear property. This may be the result of the spheroid structure of nanocluster MoS_2 particles, which changes the sliding friction into partial rolling friction, thereby reducing the friction coefficient. The arrangement of the hydrocarbon chains on the surface of nanocluster can also reduce the friction coefficient, owing to "the brush mechanism." At the same time, the cohesive effect between long hydrocarbon chains increases the antiwear property of MoS_2 nanoclusters.

Tungsten disulfide, like MoS_2 and graphite, also has traditional solid lubricants lamellar-type structure, since it shears easily. For example, tungsten disulfide has a characteristic anisotropic layered structure consisting of W-S atoms covalently bonded in planar hexagonal arrays. Each of its W atoms are surrounded by a trigonal prism of S atoms within a lamella, while adjacent lamellae interact through relatively weak van der Waals forces. This allows the layers to slip under a small shear force. Jiang et al. [77] reported that ultrathin WS_2 nanosheets modified by oleylamine were synthesized by high-temperature solution-phase method. Figure 8.26 shows schematically the route to synthesizing WS_2 nanosheets. The as-prepared OM-modified WS_2 nanosheets exhibit an average particle size of 6–8 nm and belong to 2 H polytype of platelets. TEM images of OM-modified WS_2 nanosheets are shown in Figure 8.27.

OM-modified WS_2 nanosheets as an additive in PAO base oil are effective in reducing the friction coefficient and WSD of steel–steel pairs over a wide temperature range. Figure 8.28 shows the variations of friction coefficient and WSD with temperature under the lubrication of PAO oil containing 2.0 wt% of WS_2 nanosheets

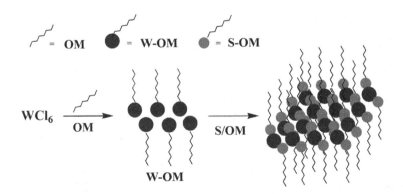

FIGURE 8.26 Synthesis of ultrathin WS_2 nanosheets modified by oleylamine.

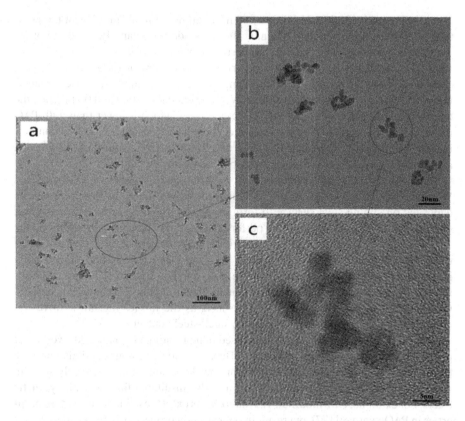

FIGURE 8.27 TEM images of oleylamine-modified WS$_2$ nanosheets under different magnifications (a) 25,000×, (b) 100,000×, and (c) 500,000×.

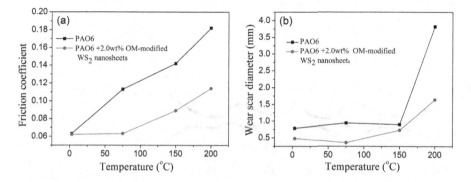

FIGURE 8.28 Variation of (a) friction coefficient and (b) wear scar diameter with temperature of PAO6 alone and PAO6 containing 2.0 wt% oleylamine-modified WS$_2$ nanosheets.

(load: 392; speed: 1200 rev/min; time: 60 min; temperature: from ambient temperature to 200°C). It is seen that WS_2 nanosheets as additives can effectively improve the friction-reducing and antiwear properties as well as load-carrying capacity of PAO base oil. However, the WSD and friction coefficient of the steel–steel pairs considerably increase with increasing temperature. For example, the friction coefficient of PAO oil containing 2.0 wt% of WS_2 nanosheets is 0.089 and 0.114 when the temperature reaches 150°C and 200°C, respectively, which is lower than with PAO alone. Similarly, the WSDs of the steel–steel contact lubricated by PAO oil containing 2.0 wt% of WS_2 nanosheets at 150°C and 200°C are 0.731 mm and 1.643 mm, respectively, which is smaller than that lubricated with PAO alone.

Figure 8.29 shows the schematic of the tribological model of OM-modified WS_2 nanosheets in PAO6 base oil. WS_2 nanosheets with lamellar structure can be deposited on sliding steel surface to form a surface protective and lubricious layer, thereby improving the lubrication performance of the base oil. In addition, the OM-modified WS_2 nanosheets can participate in tribochemical reaction to form a tribochemical reaction film composed of WO_3 and iron oxides, which significantly improve the tribological properties of the base oil. The friction-reducing and antiwear property of the tribochemical reaction product might be inferior to the surface-modified WS_2 nanosheets, but WS_2 nanosheets as the lubricating oil additive can reduce the overall friction and wear of the steel–steel contact.

Among the multifunctional additives used in lubricating oils, nanoadditives are of particular significance because they are effective at lower dosages and also provide better tribological properties than conventional lubricant additives widely used in mineral oil. However, nanoadditives are usually unsuitable for use with synthetic esters. For example, OM-modified WS_2 nanosheets exhibit excellent lubricating properties in PAO6 base oil [77], but result in poor friction properties in the synthetic ester oil DIOS. Therefore, it is imperative to manipulate the surface chemical properties of nanoadditives like OM-modified WS_2 nanosheets in order to improve their compatibility with synthetic ester oil and to effectively enhance the tribological properties. For example, amphiphilic maleic anhydride dodecyl ester (MADE) surfactant was used to tune the surface properties of OM-capped WS_2 nanoparticles [78] to improve dispersion stability of WS_2 nanoparticles in synthetic ester base oil and increase its

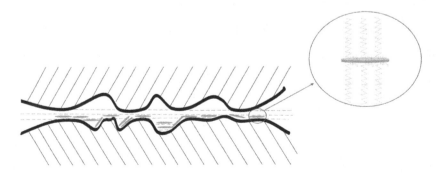

FIGURE 8.29 Tribological model for lubrication with oleylamine-modified WS_2 nanosheets in PAO6.

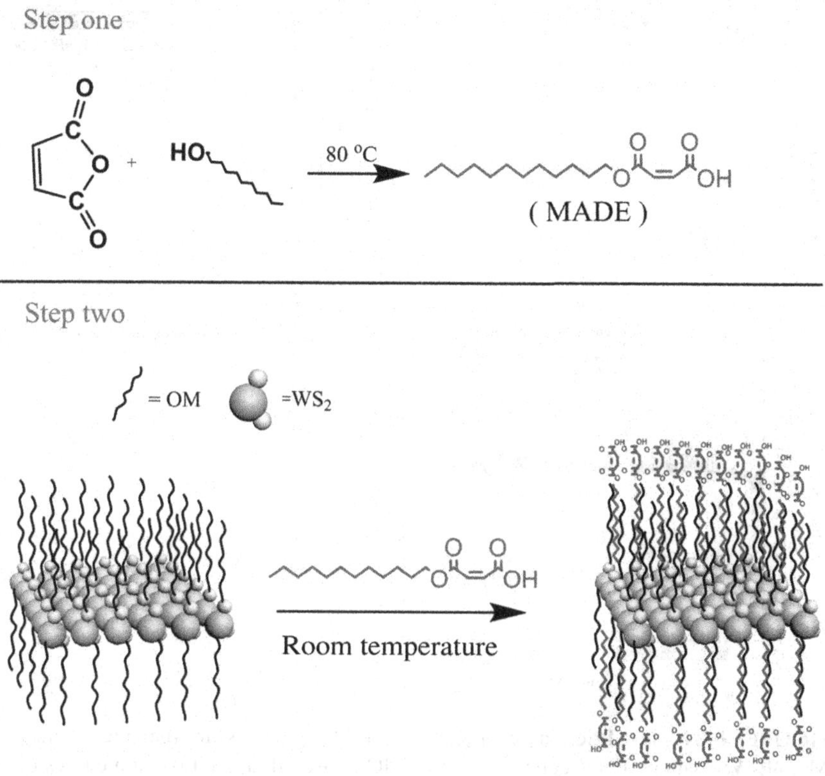

Step one

(MADE)

Step two

Room temperature

FIGURE 8.30 Schematic diagram showing the synthesis of OM/MADE-capped WS$_2$ nanoparticles (OM: oleylamine; MADE: maleic anhydride dodecyl ester).

compatibility with the base oil as well. Figure 8.30 schematically shows the method for synthesizing the target ultrathin OM/MADE-capped WS$_2$ nanoparticles.

The variations in the friction coefficient and WSD with the concentration of OM/MADE-capped WS$_2$ in DIOS are shown in Figure 8.31a and b. OM/MADE-capped WS$_2$ as an additive in DIOS can effectively reduce the friction coefficient and WSD of the steel–steel pairs. Thus, when 2.0 wt% of OM/MADE-capped WS$_2$ nanoparticles were added into DIOS, the best antiwear performance and friction-reducing ability were obtained: the WSD and the friction coefficient were reduced by 39.6% and 56.9%, respectively. Nevertheless, when OM/MADE-capped WS$_2$ nanoparticles were added in DIOS base oil with a concentration above 2.0 wt%, both the friction coefficient and WSD increased slightly with increasing additive concentration. Thus, it can be inferred that when the lubricant additive was added into the base oil at too low a concentration, it was unable to form a tribofilm to cover well the sliding steel surface. As a result, a relatively high friction coefficient was observed, due to direct contact of the sliding surfaces. On the other hand, when too many surface-capped WS$_2$ nanoparticles were added into the base stock, the concentration was too high, forming large aggregates. As a result, the interasperity valleys of the sliding surfaces

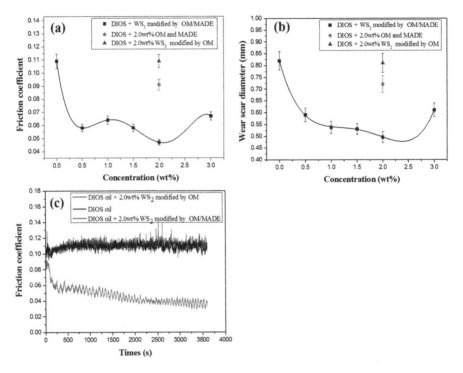

FIGURE 8.31 (a) Friction coefficient and (b) wear scar diameter versus OM/MADE-WS$_2$ nanoparticles concentration in DIOS base oil, and (c) friction curves of DIOS, DIOS with 2.0 wt% of OM-modified WS$_2$ and DIOS with 2.0 wt% of OM/MADE-modified WS$_2$ (OM: oleylamine; MADE: maleic anhydride dodecyl ester; DIOS: diisooctyl sebacate).

could not be filled well by the lubricant, and the friction coefficient increased owing to enhanced scratching and shearing [77,79].

OM/MADE-capped WS$_2$ nanoparticles added in DIOS base oil are advantageous over OM-capped WS$_2$ nanoparticles. Thus, when 2.0 wt% of OM-capped WS$_2$ was added in DIOS, the friction coefficient remained nearly unchanged, but the WSD rose relative to that with the lubrication by DIOS alone. Also, a friction coefficient of 0.091 and a WSD of 0.720 mm were obtained during lubrication with DIOS containing 2.0 wt% OM and MADE. Both the friction coefficient and WSD were larger than with DIOS containing 2 wt% of OM/MADE-capped WS$_2$ nanoparticles. Figure 8.31c shows the friction curves of DIOS and DIOS with OM-capped WS$_2$ or OM/MADE-capped WS$_2$. Under the lubrication by DIOS with 2.0 wt% of OM/MADE-capped WS$_2$ nanoparticles, the friction coefficient reduced gradually and eventually stabilized at about 0.036, while OM-capped WS$_2$ nanoparticles displayed no friction-reducing property.

This might be related to the polarity of DIOS. Namely, the OM-capped WS$_2$ nanoparticles exhibit weak polarity attributed to OM, which makes it possible for the additives rather than the nonpolar PAO base oil to be readily adsorbed on the sliding steel ball surfaces to reduce friction and wear. Unlike the nonpolar PAO, DIOS contains the polar group –COO$^-$, and can easily adsorb on the sliding steel

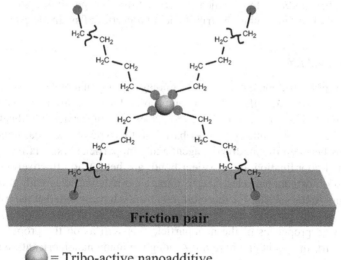

= Tribo-active nanoadditive

= Coordination group

= Functional group for adsorbing on friction pair

FIGURE 8.32 Strategy designing a nanoadditive compatible with synthetic ester oils.

balls surfaces than the OM-capped WS$_2$ nanoparticles. As a result, the tribological function of the lubricant additive is retarded and an increase friction coefficient, even larger than that under the lubrication with DIOS base oil alone, was obtained when OM-capped WS$_2$ nanoparticles were introduced in DIOS base oil [80,81].

The fact that OM-capped WS$_2$ nanoparticle additives have no influence on the lubricating performance of DIOS, while OM/MADE-capped WS$_2$ nanoparticles result in greatly improved friction-reducing and antiwear properties, implies that the surface properties of nanoadditive have a significant influence on their tribological properties. This observation can be applied to develop a strategy for the design of nanoadditives suitable for synthetic ester base oil (Figure 8.32). Thus, three aspects, namely, nanoadditive with triboactivity, surface modifier with coordination group, and surface modifier with polar functional groups, should be considered for the development of high-performance nanoadditives suitable for synthetic ester base oil. First, the nanoadditive should be tribologically active and participate in tribochemical reactions during the friction process, thereby contributing to reduced friction and wear. Second, the surface modifier should contain a coordination group that helps it bind to the surface of the nanoadditive, as well as an alkyl group to improve the dispersion stability of the nanoadditive in base oil. Third, the surface modifier should contain a functional group at the alkyl terminal in order to promote its adsorption on sliding metallic surface, thereby contributing to reduced friction and wear. With the assistance of the triboactivity as well as the coordination group and functional group of the surface modifiers, nanoadditives will be able to form physicochemically adsorbed films and/or tribochemical reaction films on the sliding steel surfaces, thereby effectively improving the tribological properties of synthetic ester base oil.

8.4 SUMMARY

With the rapid development of nanotechnology, many nanolubricants from metals, alloys, oxides, and sulfides have been proposed and prepared by in situ surface modification with a variety of organic molecules. In situ surface modification technique can be classified into five types based on the type of chemical bond. Strong interactions between the modifying agent and nanoparticles, especially in the form of combined coordination and covalent bond, are helpful to effectively inhibit the agglomeration of nanoparticles, and to impart the nanoparticles with good dispersion stability in the solvent. Furthermore, test results have shown that the tribological properties of nanoparticles are very strongly dependent on the chemical, physical, type, and size properties of the nanoparticles, as well as on the properties of the surface-modifying agent and base oil. Among the many nanolubricants synthesized by our group, Cu nanoparticles offer the highest advantage of performance to price ratio. Also the preparation conditions and parameters for large-scale production are explored, and copper nanoparticles have been produced on an industrial scale.

ACKNOWLEDGMENTS

The authors gratefully acknowledge the financial supports of the Ministry of Science and Technology of China (project of "973" Plan, grant No. 2013CB632303) and the National Natural Science Foundation of China (grant Nos. 21671053 and 51405132).

REFERENCES

1. M. Amiri and M. M. Khonsari, "On the Thermodynamics of Friction and Wear-A Review," *Entropy*, 12, 1021–1049 (2010).
2. Y. Mo, D. Tao, X. Wei and Q. Li, "Research on Friction-Coatings with Activated Ultra-Thick Tin-Base," In: *Advanced Tribology*, J. Luo, Y. Meng, T. Shao and Q. Zhao (Eds.), pp. 915–919. Springer, Heidelberg, Germany (2010).
3. H. Huang, H. Hu, S. Qiao, L. Bai, M. Han, Y. Liu and Z. Kang, "Carbon Quantum Dot/ CuS_x Nanocomposites Towards Highly Efficient Lubrication and Metal Wear Repair," *Nanoscale*, 7, 11321–11327 (2015).
4. S. Ingole, A. Charanpahari, A. Kakade, S. S. Umare, D. V. Bhatt and J. Menghani, "Tribological Behavior of Nano TiO_2 as an Additive in Base Oil," *Wear*, 301, 776–785 (2013).
5. P. Ye, X. Jiang, S. Li and S. Li, "Preparation of $NiMoO_2S_2$ Nanoparticle and Investigation of its Tribological Behavior as Additive in Lubricating Oils," *Wear*, 253, 572–575 (2002).
6. Z. Zhang, Q. Xue and W. Liu, "Study on Lubricating Mechanisms of $La(OH)_3$ Nanocluster Modified by Compound Containing Nitrogen in Liquid Paraffin," *Wear*, 218, 139–144 (1998).
7. S. Chen, W. Liu and L. Yu, "Preparation of DDP-coated PbS Nanoparticles and Investigation of the Antiwear Ability of the Prepared Nanoparticles as Additive in Liquid Paraffin," *Wear*, 218, 153–158 (1998).

8. X. Hou, J. He, L.Yu, Z. Li, Z. Zhang and P. Zhang, "Preparation and Tribological Properties of Fluorosilane Surface-Modified Lanthanum Trifluoride Nanoparticles as Additive of Fluoro Silicone Oil," *Appl. Surf. Sci.*, 316, 515–523 (2014).

9. C. P. Koshy, P. K. Rajendrakumar and M. V. Thottackkad, "Evaluation of the Tribological and Thermo-Physical Properties of Coconut oil Added with MoS_2 Nanoparticles at Elevated Temperatures," *Wear*, 330–331, 288–308 (2015).

10. G. Yang, S. Chai, X. Xiong, S. Zhang, L. Yu and P. Zhang, "Preparation and Tribological Properties of Surface Modified Cu Nanoparticles," *Trans. Nonferrous Metal. Soc. China*, 22, 366–372 (2012).

11. Y. Chen, Y. Zhang, S. Zhang, L. Yu, P. Zhang and Z. Zhang, "Preparation of Nickel-Based Nanolubricants via a Facile In Situ One-Step Route and Investigation of Their Tribological Properties," *Tribol. Lett.*, 51, 73–83 (2013).

12. Z. F. Jia and Y. Q. Xia, "Hydrothermal Synthesis, Characterization, and Tribological Behavior of Oleic Acid-Capped Lanthanum Borate with Different Morphologies," *Tribol. Lett.*, 41, 425–434 (2011).

13. Y. Pu, F. Kang, J. F. Chen, X. F. Zeng and J. X. Wang, "Synthesis of Transparent Oil Dispersion of Monodispersed Calcium Carbonate Nanoparticles with High Concentration," *AIChE J.*, 63: 3663–3669 (2017).

14. L. Peña-Parás, J. Taha-Tijerina, L. Garza, D. Maldonado-Cortés, R. Michalczewski and C. Lapray, "Effect of CuO and Al_2O_3 Nanoparticle Additives on the Tribological Behavior of Fully Formulated Oils," *Wear*, 332–333: 1256–1261 (2015).

15. Y. Guo, L. Zhang, G. Zhang, D. Wang, T. Wang and Q. Wang, "High Lubricity and Electrical Responsiveness of Solvent-Free Ionic SiO_2 Nanofluids," *J. Mater. Chem. A*, 6, 2817–2827 (2018).

16. R. C. Zhang, D. Qiao, X. Q. Liu, Z. G. Guo, M. R. Cai and L. Shi, "A Facile and Effective Method to Improve the Dispersibility of WS_2 Nanosheets in PAO8 for the Tribological Performances," *Tribol. Int.*, 118, 60–70 (2018).

17. M. Abdullah, M. Abdollah, N. Tamaldin, H. Amiruddin, N. Mat Nuri, C. Gachot and H. Kaleli, "Effect of Hexagonal Boron Nitride Nanoparticles as an Additive on the Extreme Pressure Properties of Engine Oil," *Ind. Lubr. Tribol.*, 68, 441–445 (2016).

18. S. Ma, S. Zheng, D. Cao and H. Guo, "Anti-Wear and Friction Performance of ZrO_2 Nanoparticles as Lubricant Additive," *Particuology*, 8, 468–472 (2010).

19. M. Zhang, X. B. Wang and W. M. Liu, "Tribological Behavior of LaF_3 Nanoparticles as Additives in Poly-Alpha-Olefin," *Ind. Lubr. Tribol.*, 65, 226–235 (2013).

20. R. D. Liu, X. C. Wei, D. H. Tao and Y. Zhao, "Study of Preparation and Tribological Properties of Rare Earth Nanoparticles in Lubricating Oil," *Tribol. Int.*, 43, 1082–1086 (2010).

21. B. Chen, K. Gu, J. Fang, J. Wu, J. Wang and N. Zhang, "Tribological Characteristics of Monodispersed Cerium Borate Nanospheres in Biodegradable Rapeseed Oil Lubricant," *Appl. Surf. Sci.*, 353, 326–332 (2015).

22. D. L. Jiang, X. L. Hu, R. Wang and D. Q. Yin, "Oxidation of Nanoscale Zero-Valent Iron under Sufficient and Limited Dissolved Oxygen: Influences on Aggregation Behaviors," *Chemosphere*, 122, 8–13 (2015).

23. V. Y. Rudyak and S. L. Krasnolutskii, "Dependence of the Viscosity of Nanofluids on Nanoparticle Size and Material," *Phys. Lett. A*, 378, 1845–1849 (2014).

24. Z. Tang and S. Li, "A Review of Recent Developments of Friction Modifiers for Liquid Lubricants (2007–present)," *Curr. Opin. Solid State Mater. Sci.*, 18, 119–139 (2014).

25. G. Yang, J. Zhao, L. Cui, S. Song, S. Zhang, L. Yu and P. Zhang, "Tribological Characteristic and Mechanism Analysis of Borate Ester as a Lubricant Additive in Different Base Oils," *RSC Adv.*, 7, 7944–7953 (2017).

26. X. Li, Z. Cao, F. Liu, Z. Zhang and H. Dang, "A Novel Method of Preparation of Superhydrophobic Nanosilica in Aqueous Solution," *Chem. Lett.*, 35, 94–95 (2006).

27. X. Li, Z. Cao, Z. Zhang and H. Dang, "Surface-Modification in Situ of Nano-SiO$_2$ and Its Structure and Tribological Properties," *Appl. Surf. Sci.*, 252, 7856–7861 (2006).

28. F. Yan, X. Zhang, F. Liu, X. Li and Z. Zhang, "Adjusting the Properties of Silicone Rubber Filled with Nanosilica by Changing the Surface Organic Groups of Nanosilica," *Compos. B Eng.*, 75, 47–52 (2015).

29. P. Liu, S. Guo, M. Lian, X. Li and Z. Zhang, "Improving Water-Injection Performance of Quartz Sand Proppant by Surface Modification with Surface-Modified Nanosilica," *Colloid Surf. A*, 470, 114–119 (2015).

30. H. Yao, H. Pan, R. Li, X. Li and Z. Zhang, "Unsaturated Polyester Resin/Epoxy-Functionalised Nanosilica Composites Constructed by in Situ Polymerization," *Micro Nano Lett.*, 10, 427–431 (2015).

31. B. Li, Y. Zhang, Y. Zhao, Z. Wu and Z. Zhang, "A Novel Method for Preparing Surface-Modified Mg(OH)$_2$ Nanocrystallines," *Mater. Sci. Eng. A*, 452–453, 302–305(2007).

32. L. Wang, M. Zhang, X. Wang and W. Liu, "The Preparation of CeF$_3$ Nanocluster Capped with Oleic Acid by Extraction Method and Application to Lithium Grease," *Mater. Res. Bull.*, 43, 2220–2227 (2008).

33. Z. Zhang, J. Zhang and Q. Xue, "Synthesis and Characterization of a Molybdenum Disulfide Nanocluster," *J. Phys. Chem.*, 98, 12973–12977 (1994).

34. Z. Zhang, Q. Xue and J. Zhang, "Synthesis, Structure and Lubricating Properties of Dialkyldithiophosphate Modified Mo-S Compound Nanoclusters," *Wear*, 209, 8–12 (1997).

35. G. Yang, Z. Zhang, S. Zhang, L. Yu and P. Zhang, "Synthesis and Characterization of Highly Stable Dispersions of Copper Nanoparticles by a Novel One-Pot Method," *Mater. Res. Bull.*, 48, 1716–1719 (2013).

36. C. Zhang, S. Zhang, S. Song, G. Yang, L. Yu, Z. Wu, X. Li and P. Zhang, "Preparation and Tribological Properties of Surface-Capped Copper Nanoparticle as a Water-Based Lubricant Additive," *Tribol. Lett.*, 54, 25–33 (2014).

37. M. Li, K. Wong and S. Mann, "Organization of Inorganic Nanoparticles Using Biotin-Streptavidin Connectors," *Chem. Mater.*, 11, 23–26 (1999).

38. R. Subbiah, M. Veerapandian and K.S. Yun, "Nanoparticles: Functionalization and Multifunctional Applications in Biomedical Sciences," *Curr. Med. Chem.*, 17, 4559–4577 (2010).

39. J. Zhou, J. Yang, Z. Zhang, W. Liu and Q. Xue, "Study on the Structure and Tribological Properties of Surface-Modified Cu Nanoparticles," *Mater. Res. Bull.*, 34, 1361–1367 (1999).

40. J. Zhou, Z. Wu, Z. Zhang, W. Liu and Q. Xue, "Tribological Behavior and Lubricating Mechanism of Cu Nanoparticles in Oil," *Tribol. Lett.*, 8, 213–218 (2000).

41. J. Zhang, Y. Zheng, D. Zhao, S. Yang, L. Yang, Z. Liu, R. Zhang, S. Wang, D. Zhang and L. Chen, "Ellipsometric Study on Size-Dependent Melting Point of Nanometer-Sized Indium Particles," *J. Phys. Chem. C*, 120, 10686–10690 (2016).

42. E. Kim and B. Lee, "Size Dependency of Melting Point of Crystalline Nano Particles and Nano Wires: A Thermodynamic Modeling," *Met. Mater. Int.*, 15, 531–537 (2009).

43. G. Yang, Z. Zhang, S. Zhang, L. Yu, P. Zhang and Y. Hou, "Preparation and Characterization of Copper Nanoparticles Surface-Capped by Alkanethiols," *Surf. Interface Anal.*, 45, 1695–1701 (2013).

44. X. Xiong, Y. Kang, G. Yang, S. Zhang, L. Yu and P. Zhang, "Preparation and Evaluation of Tribological Properties of Cu Nanoparticles Surface Modified by Tetradecyl Hydroxamic Acid," *Tribol. Lett.*, 46, 211–220 (2012).

45. Y. Zhang, Y. Xu, G. Yang, S. Zhang, P. Zhang and Z. Zhang, "Synthesis and Tribological Properties of Oil-Soluble Copper Nanoparticles as Environmentally Friendly Lubricating Oil Additives," *Ind. Lubr. Tribol.*, 67, 227–232 (2015).

46. C. Zhang, S. Zhang, L. Yu, Z. Zhang, Z. Wu and P. Zhang, "Preparation and Tribological Properties of Water-Soluble Copper/Silica Nanocomposite as a Water-Based Lubricant Additive," *Appl. Surf. Sci.*, 259, 824–830 (2012).

47. Y. H. Xu, G. B. Yang, S. M. Zhang, P. Y. Zhang and Z. J. Zhang, "Preparation of Water-Soluble Copper Nanoparticles and Evaluation of Their Tribological Properties," *Tribology*, 32, 165–170 (2012).

48. L. Sun, X. Tao, Y. Zhao and Z. Zhang, "Synthesis and Tribology Properties of Stearate-Coated Ag Nanoparticles," *Tribol. Trans.*, 53, 174–178 (2010).

49. L. Sun, Z. J. Zhang, Z. S. Wu and H. X. Dang, "Synthesis and Characterization of DDP Coated Ag Nanoparticles," *Mater. Sci. Eng. A*, 379, 378–383 (2004).

50. Y. Zhao, Z. Zhang and H. Dang, "A Novel Solution Route for Preparing Indium Nanoparticles," *J. Phys. Chem. B*, 107, 7574–7576 (2003).

51. Z. Li, X. Tao, Y. Cheng, Z. Wu, Z. Zhang and H. Dang, "A Simple and Rapid Method for Preparing Indium Nanoparticles from Bulk Indium via Ultrasound Irradiation," *Mater. Sci. Eng. A*, 407, 7–10 (2005).

52. Y. Zhao, Z. Zhang and H. Dang, "Preparation of Tin Nanoparticles by Solution Dispersion," *Mater. Sci. Eng. A*, 359, 405–407 (2003).

53. Z. Li, X. Tao, Y. Cheng, Z. Wu, Z. Zhang and H. Dang, "A Facile Way for Preparing Tin Nanoparticles from Bulk Tin via Ultrasound Dispersion," *Ultrason. Sonochem.*, 14, 89–92 (2007).

54. Y. Zhao, Z. Zhang and H. Dang, "Fabrication and Tribological Properties of Pb Nanoparticles," *J. Nanopart. Res.*, 6, 47–51 (2004).

55. Y. Liu, L. Xin, Y. Zhang, Y. Chen, S. Zhang and P. Zhang, "The Effect of Ni Nanoparticles on the Lubrication of a DLC-Based Solid–Liquid Synergetic System in All Lubrication Regimes," *Tribol. Lett.*, 65, 31 (2017).

56. Y. Zhao, Z. Zhang and H. Dang, "A Simple Way to Prepare Bismuth Nanoparticles," *Mater. Lett.*, 58, 790–793 (2004).

57. Y. B. Zhao, Z. J. Zhang, Z. S. Wu and H. X. Dang, "Preparation of Stearic Acid Modified Bismuth Nanoparticles by Solution Dispersion," *Chinese J. Inorg. Chem.*, 19, 997–1000 (2003).

58. H. Liu, S. Zhang, P. Zhang and Y. Zhang, "Preparation of Bismuth Nanoparticles via a Thermal Decomposition Process and Evaluation of their Tribological Properties in Synthetic Ester Oils," In *14th World Congress on Mechanism and Machine Science*, Department of Mechanical Engineering, National Taiwan University, Taipei, Taiwan (2015).

59. J. T. Tian, C. H. Gong, L. G. Yu, Z. S. Wu and Z. J. Zhang, "Synthesis of Dandelion-Like Three-Dimensional Nickel Nanostructures via Solvothermal Route," *Chinese Chem. Lett.*, 19, 1123–1126 (2008).

60. L. Wang, Y. Gao, T. Xu and Q. Xue, "A Comparative Study on the Tribological Behavior of Nanocrystalline Nickel and Cobalt Coatings Correlated with Grain Size and Phase Structure," *Mater. Chem. Phys.*, 99, 96–103 (2006).

61. Y. Zhao, Z. Zhang, W. Liu, H. Dang and Q. Xue, "Controlling Synthesis of BiIn Dendritic Nanocrystals by Solution Dispersion," *J. Am. Chem. Soc.*, 126, 6854–6855(2004).

62. Y. Zhao, Z. Zhang and H. Dang, "Synthesis of In–Sn Alloy Nanoparticles by a Solution Dispersion Method," *J. Mater. Chem.*, 14, 299–302 (2004).

63. G. Xiao, Y. Zhao, X. Meng, Z. Wu and Z. Zhang, "Shape-Controlled Synthesis of BiIn Alloy Nanostructures," *J. Alloy Compd.*, 437, 329–331 (2007).

64. Y. Zhao, J. Liu, L. Cao, Z. Wu, Z. Zhang and H. Dang, "Synthesis and Characterization of Pb–Bi Bimetal Nanoparticles by Solution Dispersion," *Mater. Chem. Phys.*, 99, 71–74 (2006).

65. H. Chen, Z. Li, Z. Wu and Z. Zhang, "A Novel Route to Prepare and Characterize Sn–Bi Nanoparticles," *J. Alloy Compd.*, 394, 282–285 (2005).

66. H. J. Chen, Z. W. Li, X. J. Tao, Z. S. Wu and Z. J. Zhang, "Sonochemical Preparation and Characterization of Sn-Cd Nanoparticles," *Chem. Res.*, 16 (3), 34–37 (2005) (in Chinese).

67. H. J. Chen, Z. W. Li, Z. S. Wu and Z. J. Zhang, "Preparation and Characterization of Ternary Pb-Sn-Cd Alloy Nanoparticles," *J. Chem. Indus. Eng.*, 56, 1590–1593 (2005).

68. H. J. Chen, Z. W. Li, B. B. Hu, Z. S. Wu, P. Y. Zhang, Z. J. Zhang and H. X. Dong, "Preparation of Bi-Pb-Sn-Cd Nanoalloy Particles and its Tribological Properties as an Additive in Liquid Paraffin," *Tribology*, 25, 169–172 (2005).

69. L. Wu, Y. Zhang, G. Yang, S. Zhang, L. Yu and P. Zhang, "Tribological Properties of Oleic Acid–Modified Zinc Oxide Nanoparticles as the Lubricant Additive in Poly-Alpha Olefin and Diisooctyl Sebacate Base Oils," *RSC Adv.*, 6, 69836–69844 (2016).

70. H. Liu, Y. Zhang, S. Zhang, Y. Chen, P. Zhang and Z. Zhang, "Preparation and Evaluation of Tribological Properties of Oil-Soluble Rice-Like CuO Nanoparticles," *Ind. Lubr. Tribol.*, 67, 276–283 (2015).

71. Q. Xue, W. Liu and Z. Zhang, "Friction and Wear Properties of a Surface-Modified TiO_2 Nanoparticle as an Additive in Liquid Paraffin," *Wear*, 213, 29–32 (1997).

72. W. Ye, T. Cheng, Q. Ye, X. Guo, Z. Zhang and H. Dang, "Preparation and Tribological Properties of Tetrafluorobenzoic Acid Modified TiO_2 Nanoparticles as Lubricant Additives," *Mater. Sci. Eng. A*, 359, 82–85 (2003).

73. F. Jin, G. Yang, Y. Peng, S. Zhang, L. Yu and P. Zhang, "Preparation of Borate Ester and Evaluation of its Tribological Properties as an Additive in Different Base Oils," *Lubr. Sci.*, 28, 505–519 (2016).

74. Y. Gao, R. Sun, Z. Zhang and Q. Xue, "Tribological Properties of Oleic Acid—Modified TiO_2 Nanoparticle in Water," *Mater. Sci. Eng. A*, 286 149–151 (2000).

75. Y. Gao, G. Chen, Y. Oli, Z. Zhang and Q. Xue, "Study on Tribological Properties of Oleic Acid–Modified TiO_2 Nanoparticle in Water," *Wear*, 252, 454–458 (2002).

76. Z. Li, X. Tao, Z. Wu, P. Zhang and Z. Zhang, "Preparation of In_2S_3 Nanoparticle by Ultrasonic Dispersion and its Tribology Property," *Ultrason. Sonochem.*, 16, 221–224 (2009).

77. Z. Jiang, Y. Zhang, G. Yang, K. Yang, S. Zhang, L. Yu and P. Zhang, "Tribological Properties of Oleylamine-Modified Ultrathin WS_2 Nanosheets as the Additive in Polyalpha Olefin over a Wide Temperature Range," *Tribol. Lett.*, 61, 24 (2016).

78. Z. Jiang, Y. Zhang, G. Yang, J. Ma, S. Zhang, L. Yu and P. Zhang, "Tribological Properties of Tungsten Disulfide Nanoparticles Surface-capped by Oleylamine and Maleic Anhydride Dodecyl Ester as Additive in Diisooctylsebacate," *Ind. Eng. Chem. Res.*, 56, 1365–1375 (2017).

79. A. Moshkovith, V. Perfiliev, I. Lapsker, N. Fleischer, R. Tenne and L. Rapoport, "Friction of Fullerene-Like WS_2 Nanoparticles: Effect of Agglomeration," *Tribol. Lett.*, 24, 225–228 (2006).

80. A. Naveira Suarez, M. Grahn, R. Pasaribu and R. Larsson, "The Influence of Base Oil Polarity on the Tribological Performance of Zinc Dialkyl Dithiophospate Additives," *Tribol. Int.*, 43, 2268–2278 (2010).

81. P. Waara, J. Hannu, T. Norrby and Å. Byheden, "Additive Influence on Wear and Friction Performance of Environmentally Adapted Lubricants," *Tribol. Int.*, 34, 547–556 (2001).

9 Application of Tribology Test for Quality Assessment of Fabric Softeners Based on Cationic Surfactants

Marta Ogorzalek and Tomasz Wasilewski

CONTENTS

9.1 INTRODUCTION

Fabric softeners are household chemicals used in the final rinse of the laundry wash cycle. Their purpose is to reduce the adverse effects of laundry washing by modifying the surface of fibers and providing the washed fabric with a long-lasting fresh scent. Fabric softeners are milky white suspensions of cationic surface active agents in water. In addition, fabric softener formulations are enriched with a fragrance composition, stabilizers, emulsifiers, and dispersants (most commonly nonionic surfactants). Other ingredients include small amounts of viscosity modifiers (e.g., inorganic salts), dyes, and preservatives [1–6].

Fabric softening is the primary functional property of this type of household chemical, and a determinant of their quality. Furthermore, fabrics treated with fabric softeners should have a reduced water retention capacity (water runs off faster, which shortens drying time) and should not attract static electricity [4,7–9]. The beneficial macroscopic properties of fabrics outlined above are achieved via the adsorption of cationic surfactant molecules (added to the final rinse bath) on fabrics. The adsorption layer forming on the surface of fibers is a specific type of deposit which smooths out the fiber surface, neutralizes negatively charged active centers, and facilitates ironing of fabrics [10–15]. It is an effect of a decrease in the friction coefficient between individual fibers, and between the fibers and the soleplate (surface in contact with the fabric during ironing), of the dry iron [16].

The simplest commonly used method of determining the performance of fabric softeners is "fabric hand"—that is, based on fabric feel (degree of softness) by hand [17]. It is a subjective evaluation of the mechanical properties of fabrics using the sense of touch (friction coefficient of fabric against fabric and fabric against skin contacts) [7–9,17]. Methods of sensory evaluation of fabric softness after the rinse process on the basis of fabric hand were proposed by H. Heine and O. Viertel in the 1980s [18,19]. The method proposed by H. Hein involves rinsing a standard fabric in the solution of the test liquid, followed by repeated comparisons (using the hand method) with reference fabrics. A similar test was developed by Viertel assessing the roughness of a cotton towel after the rinse process on a scale of 1–5. Both methods (Hein's hand evaluation and Viertel's tactile of hand evaluation) are comparative in nature and reflect a set of subjective impressions of the testers [18,19]. A sensory evaluation of textile products is the only method available to consumers to describe the level of performance of fabric softeners. However, from a scientific point of view such methods fail to deliver repeatable and measurable values defining the softening properties. In addition, tests of this type are relatively time-consuming and labor-intensive.

Consequently, it is necessary to develop methods that would make it possible to reduce testing time and eliminate subjective evaluations performed by testing personnel. It is important to note that the evaluation of fabric hand comprises a number of fabric properties including elasticity, compressibility, stretchability, plasticity, and friction coefficient [7–9,16,17,20–23]. Scientific research on fabric softeners includes studies investigating the relationship between the friction coefficient in the fabric against fabric system and the sensation of softness [16,20–23]. Danmei and Stylios [20] analyzed the mechanical properties (including the friction coefficient) of a fabric modified by a surfactant—a softening agent. The tests were performed using the Kawabata Evaluation System (KES). It is the first system for the evaluation of fabric hand value, introduced by Professor Sueo Kawabata of Kyoto University in the 1970s. The results showed that the friction coefficient of fabric rinsed with fabric softeners was lower than the reference fabric (control). Similar results were reported by Nair et al. [21] who examined the friction coefficient of fabric against fabric system using an apparatus developed by Central Institute for Research on Cotton Technology (CIRCOT) and attached to an Instron tensile tester. In another study, Parvinzadeh et al. [22] analyzed the friction coefficient of fabric treated with a cationic softener in a conventional bath and treated with ultrasound. The friction coefficient decreased in both cases. The test was conducted performed using Advanced Friction Tester (AFT) from Hanatek Instruments, UK.

The present study was an attempt to assess a ball-and-disc tribometer (T-11) for a rapid and comprehensive evaluation of fabric softener performance. Furthermore, the study explored the relationships between sensory properties and a mechanical parameter (friction coefficient). It is anticipated that the proposed method of determining the friction coefficient in the fabric against fabric and fabric against steel systems by means of T-11 tribometer will represent a significant advance in fabric softener quality evaluation.

9.2 EXPERIMENTAL

9.2.1 MATERIALS

9.2.1.1 Preparation of Fabric Softener

Based on the literature data [1–6] and previous experience [16,23,24], a model fabric softener formulation was developed, containing commonly used cationic surfactant Bis (acyloxyethyl) hydroxyethyl methylammonium methosulphate (Praepagen TQ from Clariant International Ltd, Muttenz, Switzerland) at a concentration of 5 wt%. This surfactant is classified as a quaternary ammonium salt. The commercial product contains approximately 90 wt% of a cationic surface active agent and 10 wt% of isopropanol. The compound is commonly used mainly because of its relatively low price and high efficiency as a fabric softener. In addition, the formulation contains a preservative (Acticide MBS from Thor GmbH, Speyer, Germany) consisting of benzisothiazolinone and methylisothiazolinone. The fabric softener was prepared by adding the appropriate amount of the cationic surface active agent Praepagen TQ to water at a temperature of about 50°C. The mixture was stirred for 15 minutes, allowed to cool with the stirring to below 30°C, and finally the preservative (Acticide MBS) was added. Product was diluted with water used to make aqueous fabric softener solutions at several concentrations (20, 2, 0.2, 0.02, and 0.002 wt%) and tested.

9.2.1.2 Preparation of Cotton Fabric for Tests

A cotton fabric with properties listed in Table 9.1 was used for testing.

The cotton fabric was subjected to a washing process at a temperature of close to 95°C within 15 minutes in order to remove impurities resulting from production. The wash bath contained 3 g of sodium alkyl benzene sulphate (Pasta ABS Na, Brenntag, Kędzierzyn-Koźle, Poland), 2 g of sodium hydrogen carbonate, and 1 L of distilled water. The ratio between the weight of the fabric and the weight of the rinsing bath was 1:20. The fabric was manually rinsed five times in room-temperature water. Next, the fabric was cut into appropriate pieces, marked, and rinsed in previously prepared fabric softener solutions for 15 minutes.

TABLE 9.1
Properties of Cotton Fabric Used in Tests

Fabric type	Weight, g/m²	Yarn Linear Density, Tex	Number of Threads/10 cm	Tensile Strength, %	Shrinkability, %
Cotton	133	Warp–25	Warp–256	Warp–34.4	Warp–5.0
		Weft–30	Weft–192	Weft–34.5	Weft–3.2

9.2.2 METHODS

9.2.2.1 Determination of Friction Coefficient on T-11 Tribometer

The evaluation of friction coefficient in contact conditions of the ball-and-disc type was performed using T-11 tribometer produced at the Institute for Sustainable Technologies—National Research Institute in Radom, Poland. The upper element of the friction couple consisted of steel ball, 6.35 mm in diameter. The lower element of the friction couple comprised steel discs (25.4 mm in diameter, 8 mm thick) made from 100Cr6 steel (surface roughness $R_a = 0.043$ μm). Tests were conducted by rotating the disc against a ball tested at a constant radius from the center of the disc. Tests were performed under two configurations: fabric against fabric (with fabric placed on the ball and disc and appropriately fixed) and fabric against steel (with fabric placed on the disc only). The tests were carried out under the following conditions: load—10 N, time—180 s, track radius—0.01 m, linear speed—0.05 m/s, and rotational speed—48 rpm. A schematic of the T-11 ball-on-disk tribometer is shown in Figure 9.1. Using the measured friction force, the formula below was used to calculate the friction coefficient:

$$\mu = \frac{F_T}{P} \tag{9.1}$$

where μ is the friction coefficient; F_T is the friction force, N; and P is the pressure, N. The friction force was recorded every second during the test. Average values of data from 30-second time of test are reported.

9.2.2.2 Evaluation of Softening Effect

Samples of cotton fabrics measuring 30 × 30 cm were rinsed for 15 minutes at room temperature in the following solutions: aqueous solution of the fabric softener tested, 8 wt% solution of disodium salt of ethylenediaminetetraacetic acid (POCH S.A., Gliwice, Poland)—pattern hard; 8% solution of polyoxyethylenealkylamine (Rokamin SR8, PCC Group, Brzeg Dolny, Poland)—pattern soft; 8 wt% solution of polyoxyethyleneglycol (Polikol 1500, PCC Group, Brzeg Dolny, Poland)—first pattern intermediate; and 8 wt% solution of nonylphenylpolyoxyethyleneglycol ether (Rokafenol N8, PCC Group, Brzeg Dolny, Poland)—second pattern intermediate. Following drying and appropriate marking, the hand of the fabric samples was evaluated. Determination of softening properties was described by Zieba et al. [23].

9.2.2.3 Determination of Friction Coefficient in "Iron Test"

Fabric samples measuring 30 × 30 cm were placed on a 1 m long movable countertop. Next, the dry iron with a steel soleplate (surface in contact with the fabric during ironing), weighing 978.60 g, was placed on the test fabric. Schematic representation of the procedure employed for measuring the friction coefficient in the fabric against steel system (iron test) is shown in Figure 9.2. Based on measurement data (maximum height h from which the iron slides down the surface of the test fabric), the friction coefficient (μ) was calculated from the following formula:

$$\mu = \frac{T}{N} = \frac{Q\sin\alpha}{Q\cos\alpha} = tg\alpha = \frac{h}{L} \qquad (9.2)$$

where μ is the friction coefficient; Q is the gravity force, N; T, N is the components of force Q, N; sin, cos, and tg are the trigonometric functions; α is the angle; h is the height, mm; and L is the length of the base of inclined surface, mm.

9.3   RESULTS AND DISCUSSION

9.3.1   TRIBOLOGICAL TESTS IN FABRIC AGAINST FABRIC CONTACT

Tests were performed to evaluate the fabric softening effect using a sensory method, that is, the fabric hand evaluation system. The method involves a comparative evaluation of a fabric sample (Table 9.1) rinsed in the test product (Sample 1), fabrics rinsed in distilled water (Sample 2), and fabrics rinsed in references aqueous solution (Samples 3—pattern hard, 4—pattern soft, 5—first pattern intermediate, 6—second pattern intermediate). Applying the sensory method, the tester analyzes the level of softness by touching and rubbing the fabric. It can thus be stated that the analysis is based on the friction coefficient (μ) between two fabrics or a fabric and human skin. The evaluation was performed by a team of 10 testers (designated A to J), each of whom was tasked with comparing the samples arranged in pairs. Based on their own perception, the testers determined which of the two samples in a pair was softer. The softer fabric in each pair was assigned one point. If a tester perceived both samples to be equally soft, each of them was given half a point. In addition, the remaining fabrics (2–6) were of necessity evaluated in relation to one another. In this manner every fabric sample received a maximum of 10 points. The maximum score that could be awarded to a test product was 50, which corresponds to the maximum softening effect. By way of illustration, Table 9.2 lists the results of measurements for evaluating the softening effect in one of the test fabrics after being rinsed in 20% aqueous solution of fabric softener, where the numbers represent the following:

1. fabric sample rinsed in the test product;
2. fabric sample rinsed in distilled water;
3. fabric sample rinsed in pattern hard (8 wt% aqueous solution of disodium salt of ethylenediaminetetraacetic acid);
4. fabric sample rinsed in pattern soft (8 wt% aqueous solution of polyoxyethylenealkylamine);
5. fabric sample rinsed in first pattern intermediate (8 wt% aqueous solution of polyoxyethyleneglycol); and
6. fabric sample rinsed in second pattern intermediate (8 wt% aqueous solution of nonylphenylpolyoxyethyleneglycol ether).
   The testers are represented by the letters A–J.

TABLE 9.2

An Example of Results Obtained in Manual Measurements Evaluating the Softening Effect Noted for One Test Fabric after Rinsing in 20 wt% Aqueous Fabric Softener Solution

Sample	A	B	C	D	E	F	G	H	I	J	Total [points]
	Probants — Softening effect [points]										
(1)	1	0	0.5	1	1	1	0	1	0.5	1	7
(2)	0	1	0.5	0	0	0	1	0	0.5	0	3
(1)	1	1	1	1	1	1	1	1	1	1	10
(3)	0	0	0	0	0	0	0	0	0	0	0
(1)	1	0	1	0.5	0.5	0.5	1	0	0.5	1	6
(4)	0	1	0	0.5	0.5	0.5	0	1	0.5	0	4
(1)	1	0.5	0.5	1	1	1	0.5	1	0.5	1	8
(5)	0	0.5	0.5	0	0	0	0.5	0	0.5	0	2
(1)	0.5	1	0.5	1	1	1	0.5	1	0.5	1	8
(6)	0.5	0	0.5	0	0	0	0.5	0	0.5	0	2
(2)	1	1	1	1	1	1	1	1	1	1	10
(3)	0	0	0	0	0	0	0	0	0	0	0
(2)	0	0	0.5	1	0	0	0.5	0	0	1	3
(4)	1	1	0.5	0	1	1	0.5	1	1	0	7
(2)	0.5	0.5	0.5	0.5	0	0.5	0.5	0.5	0	0.5	4
(5)	0.5	0.5	0.5	0.5	1	0.5	0.5	0.5	1	0.5	6
(2)	0.5	0.5	1	0	0	0.5	0.5	0.5	0	0	4
(6)	0.5	0.5	0	1	1	0.5	0.5	0.5	1	1	6
(3)	0	0	1	0	0	0	1	0	0	0	2
(4)	1	1	0	1	1	1	0	1	1	1	8
(3)	0	0	0	0	0	0	0	0	0	0	0
(5)	1	1	1	1	1	1	1	1	1	1	10
(3)	0	0	0	0	0	0	0	0	0	0	0
(6)	1	1	1	1	1	1	1	1	1	1	10
(4)	1	0	1	0.5	0	1	1	0	0	0.5	5
(5)	0	1	0	0.5	1	0	0	1	1	0.5	5
(4)	1	0	1	0	0	1	1	0	0	0	4
(6)	0	1	0	1	1	0	0	1	1	1	6
(5)	0.5	1	0.5	0	0	0.5	0.5	1	0	0	4
(6)	0.5	0	0.5	1	1	0.5	0.5	0	1	1	6

The results obtained in the evaluation of softness of fabrics rinsed in all tested solutions of fabric softener are listed in Table 9.3.

Tests determining the effectiveness of fabric softening by aqueous fabric softener solutions showed that an increase in the concentration of the product in water, in the opinion of testers, was accompanied by a substantial increase in the softness of the test fabrics. The smallest differences, 17–18 points, were observed for the fabrics

TABLE 9.3

Results of Measurements for Evaluating the Softening Effect Produced by Aqueous Fabric Softener Solutions

Concentration of Aqueous Solutions of Fabric Softener [wt%]	Softening Effect [Points]					
	Sample (1)	Sample (2) Water	Sample (3) Pattern Hard	Sample (4) Pattern Soft	Sample (5) First Pattern Intermediate	Sample (6) Second Pattern Intermediate
0.002	17	29	3	34	32	35
0.02	18	28	4	33	33	34
0.2	22	29	2	32	31	34
2	33	24	2	28	30	33
20	39	24	2	28	27	30

that were rinsed in solutions at the lowest concentrations (0.002 wt% and 0.02 wt%). The fabric rinsed in 20 wt% solution of the test formulation was assigned 39 points, which is 50% more than the fabric rinsed in the solution with the lowest concentration.

In the next stage, tribological tests were conducted on fabrics prepared in the same manner as in the sensory evaluation of softness. Friction coefficient measurements were carried out in the fabric against fabric contact. The friction system consisted of a steel disc and a ball to which the test fabric was fixed (Figure 9.1).

The disk was then subjected to rotational motion (in constant measurement conditions) under the load of 10 N. The measured friction coefficients as a function of time for fabrics rinsed in solutions of the fabric softener under study are shown in Figure 9.2. The data in grey circles are the results for fabrics rinsed with distilled water.

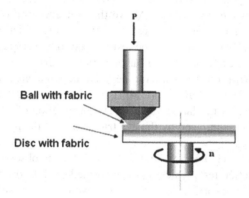

FIGURE 9.1 Schematic of a ball (fabric) against disc (fabric) friction configuration of the T-11 tribometer.

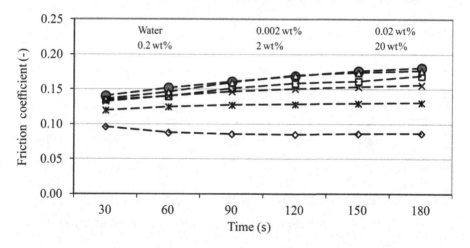

FIGURE 9.2 Friction coefficient versus time and concentration of aqueous fabric softener solution. T-11 tribometer. Fabric against fabric at various contact. Test time 180 s, load 10 N, sliding speed 0.05 m/s.

Change in the friction coefficient in the presence of the fabric rinsed in distilled water increases monotonically as a function of time. Initially, the value of μ is approximately 0.14, and increases to 0.18 at the end. The values for the fabrics rinsed in the fabric softener solutions of concentrations 0.002, 0.02, and 0.2 wt% exhibited similar results as those rinsed in distilled water. The friction coefficients for the fabrics rinsed in fabric softener solution of 2 and 20 wt% were constant over the entire test time and were approximately 0.13 and 0.08, respectively.

To facilitate the analysis of results, an arithmetic average of friction coefficients determined from three test runs was calculated. The relationship between the average friction coefficient and the concentration of the fabric softener solution is shown in Figure 9.3.

The analysis shows that the average friction coefficient for the fabrics rinsed in the 0.002 wt% and 0.02 wt% solutions were comparable to the values for the fabrics rinsed in distilled water ($\mu = 0.16$). Above the concentration of 0.02 wt% fabric softener, there was an observed decrease in the average friction coefficient: for the fabric rinsed in 20% solution of the test product the average μ value was 0.09. This tendency is a result of adsorption of cationic surfactant molecules (added to the final rinse bath) on fibers having negative active centers on their surfaces. According to Wahle and Falkowski [4], cationic surface active agents occupy active centers found on the fabric surface. The orientation of adsorbed molecules on the fabric surface is induced by the electric charge or the hydrophobic effect of these molecules.

The tendencies of change in the friction coefficient observed for the fabrics analyzed in the friction tests which were performed on T-11 tribometer are similar to those noted in the sensory tests carried out to evaluate fabric softness on the basis of fabric hand. The above findings justify the conclusion that tribological tests are suitable for a rapid and quantitative evaluation of a fabric softener performance, that can be correlates with sensory tests results of fabric softness.

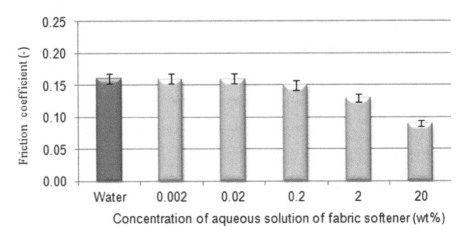

FIGURE 9.3 Average friction coefficient versus concentration of aqueous fabric softener solution. T-11 tribometer. Fabric against fabric contact. Test time 180 s, load 10 N, sliding speed 0.05 m/s.

9.3.2 TRIBOLOGICAL TESTS IN FABRIC AGAINST STEEL CONTACT

One of the most desirable properties of fabrics after the rinsing process is their ability to reduce the friction coefficient between the fibers and the dry iron soleplate (surface in contact with the fabric during ironing). Therefore, tests were performed to determine the correlations between the test methods employed (iron test—manual method, T-11 tribometer) and the concentration of the fabric softener under study.

The "iron test" is a manual method developed for the purpose of quality assessment of fabric softeners. After rinsing, the test fabric is placed on a movable countertop and dry iron with a known weight is placed on it. The end of the countertop is raised at a uniform speed. After overcoming the friction force, at a certain height h, the dry iron begins to slide down the surface of the test fabric (Figure 9.4). The data obtained from the test are used to calculate the friction coefficient using Equation 9.2.

The measured friction coefficient (μ) between the steel surface (iron soleplate) and the fabrics rinsed in the aqueous solutions (0.002, 0.02, 0.2, 2, and 20 wt%) of the fabric softener under study are given in Figure 9.5.

The fabrics rinsed in 0.002 and 0.02 wt% solutions of fabric softener gave similar friction coefficient ($\mu = 0.21$) as the fabric rinsed with distilled water. Further increase in the concentration of fabric softener solutions decreased the friction coefficient value by approximately 0.02. Increasing the concentration of the fabric softeners in the rinse bath to 0.1 and 1 wt% decreased the friction coefficient value between the fabric and steel to 0.18 and 0.17, respectively.

Measurements of the friction coefficient in fabric against steel configuration were also conducted using the T-11 tribometer. The friction couple consisted of a steel ball and a steel disc to which the test fabric was fixed (Figure 9.6).

Tests were conducted between rotating the disk and fixed ball under a load of 10 N. The measured friction coefficient as a function of time for fabrics rinsed with fabric softener solutions is shown in Figure 9.7. The data in grey circles are the results for fabrics rinsed with distilled water.

The results for the fabrics rinsed in the fabric softener solutions showed a similar tendency of friction coefficient as a function of time as those rinsed in distilled

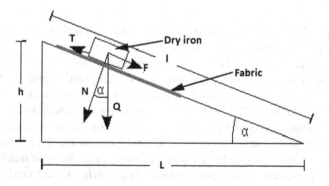

FIGURE 9.4 Schematic representation of the procedure employed for measuring the friction coefficient in the fabric against steel system—the "iron test."

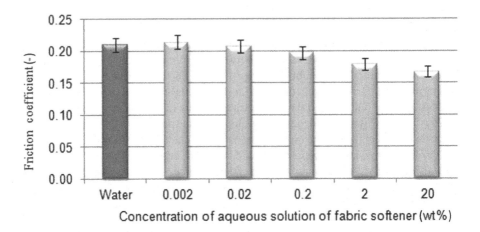

FIGURE 9.5 Average friction coefficient versus concentration of aqueous fabric softener solution. Fabric against steel configuration on the "iron test."

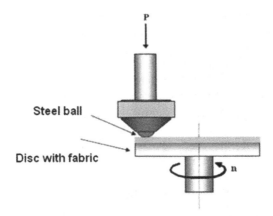

FIGURE 9.6 Schematic of a ball against disc (fabric) friction configuration on the T-11 tribometer.

water. The μ values were in the range 0.08–0.12. The arithmetic means of the friction coefficients for consecutive tests runs are shown in Figure 9.8.

The value of the friction coefficient determined for the fabric rinsed in distilled water was 0.12. The friction coefficient values determined for the fabrics rinsed in 0.002, 0.02, and 0.2 wt% aqueous solutions of the fabric softener were similar to the value obtained for the fabric rinsed in distilled water. The lowest friction coefficient 0.08 was observed for the friction couple comprising the fabric rinsed in 20% solution of the fabric softener.

The friction tests results from fabric against steel on the T-11 tribometer and fabric against steel on the iron test (original method) reveal the same trend of friction coefficient dependence on the concentration of the fabric softener used. The above findings provide evidence that the friction method based on T-11 tribometer may be regarded as a comprehensive quality assessment method for fabric softener products.

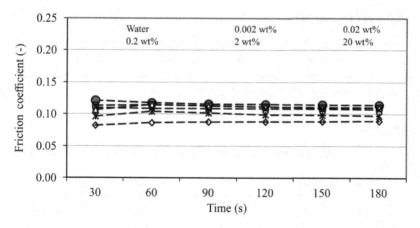

FIGURE 9.7 Friction coefficient versus time and fabric softener concentration. T-11 tribometer. Fabric against steel contact. Test time 180 s, load 10 N, sliding speed 0.05 m/s.

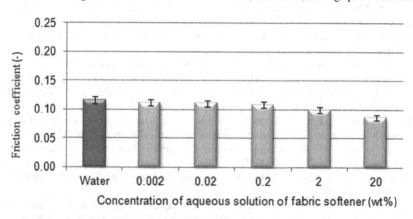

FIGURE 9.8 Average friction coefficient versus concentration of fabric softeners. T-11 tribometer. Fabric against fabric contact. Test time 180 s, load 10 N, sliding speed 0.05 m/s.

9.4 CONCLUSIONS

The following conclusions are drawn:

- There is a correlation between the results obtained with manual tests (evaluation of softening effect, iron test) and those of tribometer tests (T-11 tribometer).
- Tribometer tests offer the possibility to differentiate the properties of fabrics rinsed in fabric softener solutions above 0.02 wt%, concentration.
- Tribological tests markedly reduce the duration of experiments and make it possible to carry out these tests in different measurement conditions.

The results of the study detailed above show that friction tests with tribometer are suitable for evaluating fabric softener performance (softness, ease of ironing).

ACKNOWLEDGMENTS

This research was funded by the Ministry of Science and Higher Education from subsidies for statutory activity. Project no. 3086/35/P entitled "Development of formulations and technologies for the manufacture of innovative cosmetics, pharmacy supplies, household and industrial chemicals."

REFERENCES

1. U. Zoller, *Handbook of Detergents, Part E: Applications*. CRC Press, Boca Raton, FL (2008).
2. E. Smulders, *Laundry Detergenst*. Wiley, New York (2002).
3. C. Vidrago, M. J. Abreu, G. Soares and H. Carvalho, "Cost and Efficiency Analysis of Commercial Softeners in the Sewability Behavior of Cotton Fabrics," *J Eng Fiber Fabr*, 10, 21–28 (2015).
4. B. Wahle and J. Falkovski, "Softeners in Textile Processing. Part 1: An Overview," *Color Technol*, 32, 118–124 (2002).
5. D. S. Murphy, "Fabric Softener Technology: A Review," *J Surfactants Deterg*, 18, 199–204 (2015).
6. M. Levinson, "Rinse–Added Fabric Softener Technology at the Close of the Twentieth Century," *J Surfactants Deterg*, 2, 223–235 (1999).
7. G. Agarwal, L. Koehl, A. Perwuelz and K. S. Lee, "Interaction of Textile Parameters, Wash-Ageing and Fabric Softener with Mechanical Properties of Knitted Fabrics and Correlation with Textile-Hand. I. Interaction of Textile Parameters with Laundry Process," *Fiber Polym*, 12, 670–678 (2011).
8. G. Agarwal, L. Koehl, A. Perwuelz and K. S. Lee, "Interaction of Textile Parameters, Wash-Ageing and Fabric Conditioner with Mechanical Properties and Correlation with Textile-Hand. II. Relationship between Mechanical Properties and Textile-Hand," *Fiber Polym*, 12, 795–800 (2011).
9. S. I. Ali and S. Begum, "Fabric Softeners and Softness Perception," *Ergonomics*, 37, 801–806 (1994).
10. G. Agarwal, A. Perwuelz, L. Koehl and K. S. Lee, "Interaction between the Surface Properties of the Textiles and the Deposition of Cationic Softener," *J Surfactants Deterg*, 15, 97–105 (2012).
11. T. Igarashi, N. Morita, Y. Okamoto and K. Nakamura, "Elucidation of Softening Mechanism in Rinse Cycle Fabric Softeners. Part 1: Effect of Hydrogen Bonding," *J Surfactants Deterg*, 19, 183–192 (2016).
12. T. Igarashi, K. Nakamura, M. Hoshi, T. Hara, H. Kojima, M. Itou, R. Ikeda and Y. Okamoto, "Elucidation of Softening Mechanism in Rinse-Cycle Fabric Softeners. Part 2: Uneven Adsorption—The Key Phenomenon to the Effect of Fabric Softeners," *J Surfactants Deterg*, 19, 759–773 (2016).
13. M. S. Mohammadi, "Colloidal Stability of Di-Chain Cationic and Ethoxylated Nonionic Surfactant Mixtures Used in Commercial Fabric Softeners," *Colloids Surf A*, 288, 96–102 (2006).
14. D. Gupta, "Softening Treatments for Technical Textiles." In: *Advances in the Dyeing and Finishing of Technical Textiles*, M. L. Gulrajani (Ed.), pp. 154–173. Elsevier, Sawston, UK (2013).
15. P. S. Purohit and P. Somasundaran, "Modification of Surface Properties of Cellulosic Substrates by Quaternized Silicone Emulsions," *J Colloid Interface Sci*, 426, 235–240 (2014).

16. T. Wasilewski and M. Ogorzałek, "Influence of Isopropanol on Quality of Fabric Softeners," *Pol J Commod Sci*, 2, 89–98 (2009).
17. H. Behery, *Effect of Mechanical and Physical Properties on Fabric Hand*. CRC Press, Washington, DC (2005).
18. H. Hein, "Problems Regarding the Production of Concentrated Household Rinse Softeners," *Tenside Det*, 18, 243–256 (1981).
19. O. Viertel, "Ein Beitrag zur Prüfung von Weichspülmitteln," *Fette, Seifen, Anstrichmittel*, 75.1, 56–60 (1973).
20. S. Danmei and G. K. Stylios, "Cotton fabric mechanical properties affected by post-finishing processes," *Fiber Polym*, 13, 1050–1057 (2012).
21. A. U. Nair, B. A. Patwardhan and R. P. Nachane, "Studies on Friction in Cotton Textiles: Part II—A Study on the Relationship between Physical Properties and Frictional Characteristics of Chemically Treated Cotton Fabrics," *Indian J Fiber Text*, 38, 366–374 (2013).
22. M. Parvinzadeh, N. Memari, M. Shaver, B. Katozian, S. Ahmadi and I. Ziadi, "Influence of Ultrasonic Waves on the Processing of Cotton with Cationic Softener," *J Surfactants Deterg*, 13, 135–141 (2010).
23. M. Zięba, A. Małysa, T. Wasilewski and M. Ogorzałek, "Effects of Chemical Structure of Silicone Polyethers Used as Fabric Softener Additives on Selected Utility Properties of Cotton Fabric," *Autex Res J*. doi:10.1515/aut-2018-0009.
24. T. Wasilewski, M. Ogorzałek, E. Klimaszewska, E. Rój and M. Zalewska, "Influence of Plant Extract Obtained under Supercritical Carbon Dioxide Conditions on Applicable Properties of Fabric Softeners," *Pol J Commod Sci*, 2, 104–112 (2016).

10 A Green Approach to Tribology

Nadia G. Kandile and David R.K. Harding

CONTENTS

10.1 INTRODUCTION

Green adds another dimension to the control of tribology. Use of green lubricants to deal with friction, adhesion, and wear is of course the ideal goal. Green in itself poses challenges, as does tribology itself. Issues such as raw material availability, production, life-cycle expectations, and cost play important roles in defining ideal green lubricants. As with all lubricants, disposal of used materials is a major issue.

The world's population continues to grow. At the same time, petroleum reserves are dwindling. The desire for improved lifestyles for all people is an objective of the United Nations [1] (Figure 10.1).

Arable land for foodstuffs is decreasing as cities expand. Marginal land, however, can be used for growing nonedible green products. An important objective for a green world is the current search for green lubricants that can be as good as or better than petroleum lubricants. Improved, readily replaceable green lubricants will not only reduce pollution but also allow the world's limited supplies of raw materials, metals in particular, to survive much longer.

This chapter addresses the issue of green lubrication in relation to extending the lifetime of all solid materials. Human use of lubricants began with water, natural oils (e.g., olive oil), and animal fats. As we now look to green lubricants, have we come full circle?

10.1.1 TRIBOLOGY HISTORY

Tribology carries with it serious economic issues that are directly related to the particular application of lubrication in question. One might say that the lubrication issues in a large passenger jet plane are of considerably more importance (and worth greater expense) than those of a home lawn mower. Tribology is a multidisciplinary

FIGURE 10.1 The Sustainable Development Goals (SDGs). (From The Sustainable Development Goals (SDGs) Are 17 Goals to Transform Our World by 2030, September 2016, https://ohorizons.org/blog/sustainable-development-goals?gclid=EAIaIQobChMI8ZKe1ajJ2 AIVWR0rCh24tAGQEAAYAiAAEgLYI_D_BwE.)

topic with the overall objective of increasing the lifetime of moving surfaces while keeping costs as low as possible. Now given the limited supply of nonrenewable lubricants, green lubrication is growing in importance as a critical issue.

The use of wheels dates back to at least 3500 BC, where of course the surface interactions are between wheel hub and the axle and lubricants are used to reduce wear and remove heat. Pictures exist showing two-wheeled harvest carts with studded wheels, circa 1338 AD [2]. The moving of large objects in ancient times relied on water or oils such as olive oil. A very large number of slaves were needed to move an Egyptian colossus, El-Bersheh, on a sled, circa 1880 BC [3]. Such a task is estimated to have a coefficient of friction estimated at close to 0.23 [3]. A tomb in Egypt that was dated several thousand years BC provides evidence of the use of lubricants. A chariot in this tomb still contained some of the original animal-fat lubricant in its wheel bearings.

In China many years ago, large stones were moved on an ice path. Figure 10.2 [4]. Li et al. [4] have also detailed other large object movement methods and these are listed in Table 10.1.

The European Industrial Revolution occurred from about the mid-1700s to the mid-1800s. Overall this period saw hand processing replaced by machinery production coupled with chemical production and large-scale iron (later steel) processing. This started a major upswing in the need for expanded and focused tribology and led to an increased need for the involvement of scientists from many disciplines.

FIGURE 10.2 (a) The Forbidden City, Beijing, China; (b) large stone carving; (c) ●⸳⸳⸳⸳● the 70 km ice road trip for the 300 tons of large stone moved from the Dashiwo quarry to the Forbidden City. (From Li, J. et al., *Proc. Natl. Acad. Sci. USA*, 110, 20023–20027, 2013.)

Bartz describes the history of tribological lubricants in conjunction with the development of friction and heat-producing moving equipment involving, for example, wheels and axles [5]. In each case, of course, the earlier lubricant is carried forward with/without modification. Hence, the following list summarizes developments only for each era:

- Prehistoric (up to about 3500 BC)—oil, fat, bitumen.
- Early civilization (after 3500 BC)—water, oil, mud.
- Greek-Roman era (900 BC–400 AD)—light oil from crude oil, bitumen advances—vegetable and animal oils.
- Middle Ages (400–1450 AD)—nothing new—but continued use of vegetable and animal lubricants.
- Renaissance (1450–1600)—nothing new, although lubricant acknowledged as friction-reducing agent.
- Early industrial revolution (1600–1750)—lubricant application became a major issue—porcine tallow, lard oil, vegetable oils came more into focus as did friction and wear theories.
- Industrial Revolution (1750–1850)—mineral oil is now on the lubricant list—lubricants patented: graphite/pig tallow used for greasing bearings, olive oil/lime in water; palm oil, tallow, or graphite improved the lubricants' performance.

TABLE 10.1

Heavy Objects Transported with Sledges Using Different Lubrication Methods

Case	Year	Object	Location	m, tons	s, km	Method	Lubricant	Power	CoKF
1	ca. 1880 BC	Statue of Tehuti-Hetep	El-Bersheh, Egypt	60	Unknown	Sliding on wooden planks	Water	172 men	0.23
2	ca. AD 400	A stone	Japan	14	Unknown	Long rollers on a dirt road	Oil	36 men	0.21
3	AD 1557	A stone	Beijing, China	123	70	Sliding on ice	?	A team of men	?
4	AD 1934	Mining equipment	Far north, Canada	28	212.4	Sliding on ice	None	A tractor	0.36
5	AD 1999	Cape Hatter as Lighthouse	North Carolina	4,400	0.884	Steel roller on steel tracks	Soap bar	Five 30-ton jacks	<0.03

Source: Li, J. et al., *Proc. Natl. Acad. Sci. USA*, 110, 20023–20027, 2013.

Note: CoKF: coefficient of kinetic friction, m: mass of the object in tons, s: distance of transport in kilometers.

- "Technical progress" (1850–1925)—mineral oil takes over—patented materials include light and heavy oils, compressor oils, and the first additive composites involving graphite, emulsifiers, thickeners.
- 1925 to present day (2018)—development of additives, mineral oil improvements, synthetic base oils and, more recently, a return to natural-sourced (plant/animal) base lubricant materials.

Tribology—the study of friction, wear, and adhesion—affects a significant part of our lives. Lubrication, a critical component of reducing friction, wear, and adhesion allows our car engines to function, cutting tools to perform efficiently, and our joints to operate smoothly. Tribology is a multifaceted science.

10.2 TRIBOLOGICAL LUBRICANTS

Lubrication is essential for the proper and reliable operation of mechanical systems. It is a crucial component of reducing friction, wear, and undesired adhesion between moving surfaces in order to ensure high-quality performance. Lubricants can be liquid, solid, and gaseous. Liquid, solid, and semisolid lubricants are the most widely used forms. Lubricants are also used to remove particulate waste by-products, as well as heat, as quickly as possible from the surface-surface interactions friction zone.

10.2.1 CLASSIFICATION OF LUBRICANTS

Lubricants can be classified into four categories: liquid lubricants, greases, solid lubricants, and gas lubricants (such as air, nitrogen, hydrogen, and helium) [6]. Compared to solid lubrication, fluid lubrication has long-term endurance, low mechanical noise, promotion of thermal conductance, and very low friction in the elastohydrodynamic regime. Solid lubrication is used where liquid lubricants cannot function, such as under very high temperatures, and where volatility is forbidden. Gas lubricants are usually used in some special cases [7]. Also lubricants can be classified based on the following criteria [8,9]:

1. **Physical appearance**
   a. **Solid.** The film of a solid material is composed of organic or inorganic compounds, such as graphite, molybdenum disulphide, and cadmium disulphide.
   b. **Semisolid liquid.** It is suspended in a solid matrix of thickener and additives, such as grease. Examples are oils such as petroleum, vegetable, animal, and synthetic oils. Base oil resources are natural oils derived from animal fats and vegetable oils. Refined oils are oils derived from crude or petroleum reserves, such as paraffinic, naphthenic, and aromatic oils.
   c. **Synthetic oils.** Oils synthesized as end products of reactions that are tailored per requirement; examples are synthetic esters, silicones, and polyalphaolefins.

2. **Applications**
   a. **Automotive oils** are used in the automobile and transportation industry; examples are engine oils, transmission fluids, gearbox oils, as well as brake and hydraulic fluids.
   b. **Industrial oils** are used for industrial purposes; examples are machine oils, compressor oils, metalworking fluids, and hydraulic oils. Special oils used for special purposes according to specific operations; examples are process oils, white oils, and instrumental oils.
   c. **Special oils** are used for special purposes according to specific operations; examples are process oils, white oils, and instrumental oils.

Basic applications of lubricants.

Transport vehicles: all moving parts need to be lubricated from the engine to the brakes. Even with the evolution of the electrical automotive engine, all moving parts will still need lubricating.

Industrial equipment: often large and needing high-density, lifelong lubrication

Specialty application lubricants: can even involve human applications, for example—bone joints.

10.2.2 Green Philosophy

In his book *Green Philosophy*, Roger Scruton (2011) maintains that the environment is also an urgent political problem. He contends that everyone should shoulder responsibility for achieving a green world. While a green future is by no means guaranteed, Scruton discusses possibilities for a safer future for our planet and us [10].

10.2.3 Green Chemistry

Green chemistry is defined as the design of chemical products and processes that reduce or eliminate the use or generation of hazardous substances. Green chemistry technologies provide a number of benefits, including reduced waste, safer products, reduced use of energy and resources, and improved competitiveness of chemical manufacturers and their customers.

Green chemistry, known as *sustainable chemistry*, is based on a set of principles used in the design, development, and implementation of chemical products and processes that reduce or eliminate risks and enable scientists to protect and benefit the economy, people, and the planet. Green chemistry uses renewable, biodegradable materials that do not persist in the environment. Green chemistry uses catalysis and biocatalysis to improve efficiency and conducts reactions at low or ambient temperatures. Green chemistry is a proven systems approach. Green chemistry offers a strategic pathway to build a sustainable future.

Green chemistry consists of chemicals and chemical processes designed to reduce or eliminate negative environmental impacts. The use and production of these chemicals may involve reduced waste products, nontoxic components, and improved efficiency.

Anastas and Warner proposed the following of 12 principles to provide a road map for chemists in implementing green chemistry [11]:

1. Prevent waste
2. Atom economy
3. Less hazardous synthesis
4. Design benign chemicals
5. Use benign solvents and auxiliaries
6. Design for energy efficiency
7. Use renewable feedstocks
8. Reduce derivatives
9. Catalysis (vs. stoichiometric)
10. Design for degradation
11. Real-time analysis for pollution prevention
12. Inherently benign chemistry for accident prevention

Recently, the green chemistry concept has been introduced and developed in the lubrication field, such as for biodegradable, environment-friendly, and natural lubrication.

10.2.4 Green Technology

Green technology is defined as technology with less impact on the environment than traditional technology and contributes to green applications such as renewable energy (e.g., friction and wear issues in wind turbines). Keywords regarding green technology include life cycle assessment, green manufacturing, ecological design, green chemistry, industrial symbiosis (employs principles of ecological systems to industrial systems), and ecological modernization (true green technology, disruptive technologies, e.g., new bio-inspired approaches without the use of plastics or metals), and integrating concepts and frameworks at the interface of technology, society, and the environment [12].

10.2.5 Sustainable Technologies

The concept of "green" is not the same as the concept of "sustainable." Green refers to "better for the environment than conventional" or "related to green applications." The concept of sustainability emerged in the 1980s, when the United Nations published the Brundtland report (World Commission on Environment and Development 1987) [13] and defined sustainable development as "development that meets the needs of the present without compromising the ability of future generations to meet their own needs." The Brundtland report states:

> In essence, sustainable development is a process of change in which the exploitation of resources, the direction of investments, the orientation of technological development, and institutional change are all in harmony and enhance both current and future potential to meet human needs and aspirations.
>
> Sustainability is now a widely used word, and depending on which group uses it, it can denote very different concepts.

Examples of sustainable technologies with relevance to tribology are alternative fuels (such as biodiesel, bioalcohol, nonedible vegetable oil, and non-fossil methane and natural gas), electric cars, energy recycling, environmental technologies (such as renewable energy, water purification, air purification, sewage treatment, environmental remediation, solid waste management, and energy conservation), hydropower, synthetic crude oil, soft energy technologies, water power engines, wave power, and windmills. For more in-depth information on the connections and interdependencies of technology, globalization, and sustainable development, the reader is referred to Ashford and Hall [14].

10.2.6 GREEN TRIBOLOGY

Tribological advances are aimed at improving lubrication in order to control friction, reduce wear, and eliminate undesired adhesion and equipment failure. This in turn leads to increased equipment lifetime and energy conservation. All of which should lead to a better society with reduced pollution.

The term "green tribology" is defined by H. P. Jost as the science and technology of the tribological aspects of ecological balance and of environmental and biological impacts [15]. Green tribology is also known as ecotribology. Since its inception, green tribology has been a key subject in major international and world tribology conferences. The basic goals in green tribological efforts include reduction of material loss, reduction in energy loss, and life-cycle extension. Si-wei Zhang introduced green tribology as an international concept in 2009 [16]. Green tribology is the science and technology of tribological aspects of the ecological balance and their influence on the environment and living nature [17,18]. Tribology must proceed in consensus with the most important worldwide rules and regulations concerning the environment and energy.

Green tribology can be linked to two other green areas: green engineering and green chemistry. The U.S. Environmental Protection Agency (EPA) defined green engineering as "the design, commercialization and use of processes and products that are technically and economically feasible while reducing the generation of pollution at the source and minimizing the risk to human health and the environment" [19]. While many scientists use "green chemistry," there are also critics who argue that the term is no more than a public relations label, since some chemists use it without relating it to the green chemistry principles proposed by Anastas and Warner [11], as pointed out by Linthorst [20]. Green tribology may have to deal with the same problem.

The three tiers of green engineering assessment in design involve (1) process research and development; (2) conceptual/preliminary design; and (3) detailed pollution prevention, process heat/energy integration, and process mass integration [21].

Green tribology is now with us as we recognize the need for continued tribological input into our lifestyles as mineral oil deposits decrease. While green in one sense might be equated with returning to pre-Industrial Revolution times, we now have added ecological balancing with environmental protection. The need to maintain and sustain the world as we know it cannot be ignored; indeed, we should aim to improve our world—pollution reduction/prevention being one of the higher priorities. The basis for green lubrication lies in *green chemistry* which needs to be used to support green engineering. Green tribology, therefore, is the science and technology of the tribology of ecological balance in relation to environmental and biological impacts [16,21,22].

Thus green tribology is a subdiscipline of tribology, such as nanotribology and biotribology. In general, green tribology involves tribology for life (human biotribology), biomimetic tribology, renewable energy tribology, and a part of geotribology. Green tribology could be defined as the science and technology of research based on the tribological theories and technologies and the practices related to a sustainable society and nature, and might also be termed tribology for sustainability or sustainable tribology [23]. Green tribology is governed by green philosophy.

10.2.7 THE TWELVE PRINCIPLES OF GREEN TRIBOLOGY

Nosonovsky and Bhushan offer a modified list of 12 principles [17], which are summarized as follows:

1. *Minimization of heat and energy dissipation* resulting from friction which is the primary source of energy dissipation. They use the term "heat pollution," and maintain it is a critical primary task of tribology to reduce it in all appropriate situations.
2. *Minimization of wear which has relevance to green tribology.* In this case, they discuss equipment lifetime, equipment failure, the issue of recycling, and pollution resulting from defunct equipment and particulate waste.
3. *Reduction or complete elimination of lubrication and self-lubrication.* Any liquid or solid lubricant will itself create issues such as continuous sourcing, lifetime and disposal of the lubricant itself, or heat/wear generated byproducts. Only unreactive gaseous lubrication could perhaps be seen as ideal.
4. *Natural lubrication (e.g., vegetable-oil-based) should be used when possible, since it is usually environmentally friendly.* They did not include water and animal fat lubricants under this heading, despite the fact that these were used hundreds of years ago.
5. *Biodegradable lubrication should be used.* The focus here is pollution prevention.
6. *Sustainable chemistry and green engineering principles should be used* in all areas of tribology.
7. *Biomimetic approaches should be used whenever possible.* Any lubricant that is natural or close to natural in all aspects should be used. This means the focus must be particularly on the removal from the wear site as fast as possible and recycling or rapid degradability must be advantageous whenever possible.
8. *Surface texturing should be applied to control surface properties.* The goal here is to reduce wear and friction.

9. *Environmental implications of coatings and surface modification.* These processes generate a lot of pollution and use of energy.
10. *Design for degradation of surfaces, coatings, and tribological components.* Biodegradable materials contribute to reductions in environmental pollution.
11. *Real-time monitoring, analysis, and control of tribological systems* will assist in reducing waste, waste disposal issues, and pollution.
12. *Sustainable energy applications should become the priority of tribological design as well as engineering design in general.* In other words, the longer the lifetime of equipment, the smaller the wear issues and hence a reduction in pollution and disposal issues.

Green tribology principles are extended to biometrics and can be seen to affect blood flow control [24]. Overall, the more we apply these 12 rules, the more likely we will be able to make the world's limited supply of resources last longer.

10.2.8 FUNDAMENTALS OF GREEN TRIBOLOGY

Green tribology has been described as having one high and three low objectives [24]:

- low energy consumption,
- low discharge of the pollutants—for example, carbon dioxide,
- low environmental cost, and
- high quality of life.

The main goal of green tribology is to achieve the following important goals:

- biomimetic and self-lubricating materials/surfaces;
- biodegradable and environmentally friendly lubricants, coatings, and materials; and
- renewable, sustainable sources of energy.

In a review entitled "Green Tribology: Fundamentals and Future Development," Zhang presented four broad areas of tribological concern [25]:

1. Conservation of energy and materials while aiming for increased lifetime for all components in a tribological system.
   a. Engine fuel efficiency studies including new biolubricants, nanoparticle polyetheretherketone (PEEK), and a low-friction tapered roller bearing for cars [26].
   b. Super low friction and hence increased wear resistance can be achieved with various films and coatings: for example, hydrogenated carbon films, $AlMgB_{14}$-TiB_2, and diamond-coated ultra-high molecular weight polyethylene (UHMWPE).

2. Reduction, preferably elimination, of toxic effects of tribological systems on the environment.

   a. Biolubricants that are economical and degrade readily are of paramount importance. Considerations in this area spread from sulfur/phosphorus-free lubricants from amino acids to marine carbohydrates, such as chitin/chitosan, and modifications of them.

   b. Ozone-modified, sulfur-modified and water-vegetable oil emulsions have also been studied (Table 10.2) [27]. Under advantages one might expect low pollution leading to low environmental toxicity, and in principle, low production costs from processes such as pressing/filtering of stalks/seeds to give oils. Under disadvantages, low oxidative stability of some vegetable oils has to be addressed.

   c. Biometric studies aimed at elucidating natural erosion reduction technologies have included, for example, sand erosion resistance capabilities of sandfish skin, dragonfly wings, and desert scorpions.

   d. Tribological studies on noise reduction include the sound of car brake squealing.

   e. Lifecycle assessment or sustainability of a hydraulic fluid can be described in a pyramidal schematic as described by Muller-Zermini and G. Goule [28].

3. Studies on the possible environmental and natural disaster effects on friction-based tribology. In other words, studies on earth movements during earthquakes and slips may help widen our knowledge base of large-scale tribology. Consideration of the chemical nature of the slipping/sliding surfaces, water, and heat are all important in the evaluation of this large-scale natural green tribology [29].

4. Renewable and clean energy sources for tribological applications. A study with a REWITEC showed the positive effects of protecting metal surfaces. In this case, the REWITEC specializes in gearboxes in wind turbines [29].

TABLE 10.2

Advantages and Disadvantages of Vegetable Oils as Lubricants

Advantages	Disadvantages
High biodegradability	Low thermal stability
Low pollution	Low oxidative stability
Compatibility with additives	High freezing points
Low production cost	Poor corrosion protection
Wide production possibilities	
High flash points	
Low volatility	
High viscosity indices	

Source: Shashidhara, Y.M. and Jayaram, S.R., *Tribol. Int.*, 43, 1073–1081, 2010.

Zhang [25] then proceeds to list five directions green tribology might aspire to follow:

1. Large-scale deployment of existing knowledge, methods, and technologies of green tribology. He refers to Tzanakis et al. [30] (Figure 10.3). Although focused on a scroll expander, the schematic covers all aspects of any tribological program of study.
2. Research and development of novel green tribological technologies.
    a. Low-carbon bio- and eco-lubricants, halogen-free and biodegradable oils, carbon-neutral vegetable oils.
    b. Environmentally friendly tribological materials and coatings, tribological applications of ecomaterials, biological coatings applied on the surfaces of implants or medical devices.
    c. New tribo-techniques based on bionics.

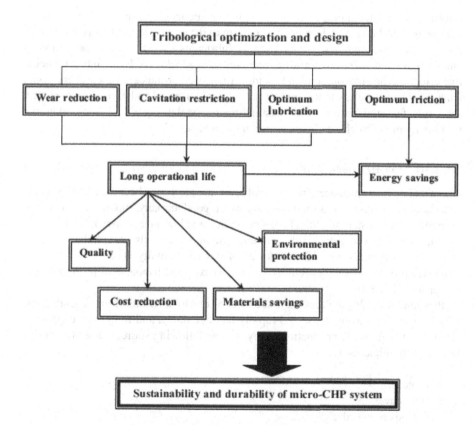

FIGURE 10.3 Scroll expander optimization. CHP: combined heat and power. (From Tzanakis, I. et al., *Renew. Sust. Energ. Rev.*, 16, 4126–4140, 2012.)

3. Making the traditional tribo-materials and lubricating materials "green" in the course of a lifecycle, namely, realizing cleaner production or ecodesign of the above materials.
4. Research and development of tribo-techniques to support diversification and hybridization of renewable and clean energy.
5. Building up the theory and methodology of green tribology.
   a. Setting up the theories and methods of analyses and evaluation of sustainability (including values of environmental and ecological impacts, value of saving energy, etc.) for tribological parts, tribosystems, and tribo-techniques.
   b. Research on the theories and methods of integration of different green tribological techniques, and on the effects of coupling and coordinating among various areas of green tribology.

10.2.9 GREEN LUBRICANT TECHNOLOGIES

Green lubricant technology [18,31] is one that reduces abrasion and wear, saves energy, is recyclable, and does not cause harm to the environment. Efforts have been made to develop environmentally friendly lubricants [32,33] such as phosphorous and/or heavy metal-free materials to prevent the production of lubricants with negative effect on the environment and biologically toxic substances. Researchers began studying lubricating oils that were non-petrochemical-based and derived from nature; such oils featured natural and outstanding bioavailability and biodegradability, making them an excellent choice as green lubricants.

10.2.10 GREEN LUBRICANTS

Green biolubricants produced from vegetable oils are promising because of their specific functional attributes such as high viscosity index, good lubricity, superior anticorrosion properties, high flash point, high biodegradability, and low aquatic toxicity [34,35].

The structure of vegetable oil is amphiphilic in nature. Oils are suitable as a base stock for easily blending different kinds of environmentally acceptable additives. Additives prepared from vegetable oils showed excellent tribological properties and perform well at extreme pressures.

Panchal et al. [36] address the issue of eco- or biolubricants from vegetable oils which should be eco-nontoxic and rapidly biodegradable, and moreover, capable of sustainable large-scale production. They list the following sources of green biomaterials for multiple uses:

* protein,
* tree leaves,
* various seaweeds,
* vegetable oils,
* coffee pulp,
* paper mill sludge, and
* lignocelluloses and other agro-residues

TABLE 10.3

Vegetable Oil Oxidation and Lubrication Impact

Stage	Product	Lubrication
Primary	Hydroperoxide	Pro-wear effect
Secondary	Volatile	Likely negligible
	Nonvolatile	Epoxides—high viscosity, high oxidation stability
		Similar boundary lubrication
		Majority of compounds uncertain
	High molecular weight	Increased viscosity
		Uncertain impact on lubrication
	Free fatty acid	Improve boundary lubrication properties
		Lower oxidation stability

Source: Fox, N.J. and Stachowiak, G.W., *Tribol. Int.*, 40, 1035–1046, 2007.

Table 10.3 shows the oxidation properties of vegetable oils used as lubricants [37].

As noted earlier, we might say that we have come full circle. i.e. we are returning to natural bioavailable renewable materials for tribological applications. Of course, we have now benefited from centuries of advances in chemistry. Hence, we can source natural green materials and modify them for tribological uses. This in turn will enable the preparation of effective long-lasting lubricants, which in their natural green state do not achieve the performance levels of the refined petroleum oil–based materials (fluids, greases, films) [38,39]. Petroleum products have survived to date for several reasons, not least of which is their lifetime, especially in heavy machinery [40]. Nevertheless, they present toxicity and environmental problems as well as having limited sources. In other words, one day there will no longer be any subterranean sources of petroleum [41]. In addition, if suitable green and bio-friendly tribological replacements can be developed, the risks associated with dumping, evaporation, and overall lack of breakdown capability will be considerably reduced [38].

As shown in Table 10.4, a number of vegetable-based oils are currently in use [27].

The entry for soybean oil, for example, illustrates the range of applications that can be achieved from this single oil [42]. Nevertheless, much more needs to be achieved with vegetable oils to increase their resistance to oxidation and to enhance their high temperature stability; as well, vegetable oils have excellent viscosity indices [39,43,44].

Modification of some vegetable oils has been addressed. Transesterification or direct esterification, either acid or base catalyzed, has been employed in effort to improve the tribological characteristics of plant-based oils. Epoxidation of vegetable oils has also been employed to modify them with a view to improving their performance as biolubricants.

TABLE 10.4

Potential Applications for Some Vegetable Oils

Vegetable Oil	Applications
Canola oil	Hydraulic oils, tractor transmission fluids, metalworking fluids, food grade lubes, penetrating oils, chain bar lubes
Castor oil	Gear lubricants, greases
Coconut oil	Gas engine oils
Olive oil	Automotive lubricants
Palm oil	Rolling lubricant, steel industry, grease
Rapeseed oil	Chain saw lubricants, air compressor farm equipment, biodegradable greases
Safflower oil	Light-colored paints, diesel fuel, resins, enamels
Linseed oil	Coatings, paints, lacquers, varnishes, stains
Soybean oil	Lubricants, biodiesel fuel, metal casting/working, printing inks, paints, coatings, soaps, shampoos, detergents, pesticides, disinfectants, plasticizers, hydraulic oil
Jojoba oil	Grease, cosmetic industry, lubricant applications
Crambe oil	Grease, intermediate chemicals, surfactants
Sunflower oil	Grease, diesel fuel substitutes
Cuphea oil	Cosmetics and motor oil
Tallow oil	Steam cylinder oils, soaps, cosmetics, lubricants, plastics

Source: Shashidhara, Y.M. and Jayaram, S.R., *Tribol. Int.*, 43, 1073–1081, 2010.

In their review of biolubricants resourced from vegetable oils, Panchal et al. assembled a number of sources on two main methods of modifying biolubricants [36]. They listed the following sources of esters and oils for transesterification or direct esterification as follows:

Jathropha oil	Palm oil methyl ester	Rapeseed oil methyl ester
Sunflower oil		Karanja oil methyl ester
Oleic acid	Linoleic acid	Stearic acid

The oils were transesterifies or directly esterified with alcohols varying in length and or shape from butanol to n-octanol to trimethylol propane. Various ratios of oil/acid to alcohol were reported and the catalyst varied from basic to acidic to acidic zirconia. In one instance an immobilized lipase was used as the catalyst. Temperatures used for these reactions ranged from 40°C to 170°C. The 40°C was the immobilized enzyme example. Reaction times varied from 1 to 8 hours. The reaction yields varied from 72% to 99%.

Panchal et al. then proceeded to present collated data on epoxidation of vegetable oils as means to improve their performance as biolubricants [36]. The range of oils listed was as follows:

Cotton seed	Soya bean	Jathropha oil
Hemp	Mahua	Canola
Rubber seed	Karaja	Neem oil
Tobacco oil	Sunflower oil	Rice oil

The acids used for the epoxidation were confined to formic, acetic, and glacial acetic acids. The hydrogen peroxide used was at either 30% or 50% concentration. Various ranges of unsaturation (ethylene):acid:hydrogen peroxide were presented and were quite varied. The samples were largely catalyzed with 2% sulfuric acid, although two used Amberlite IR-120 and one used the enzyme Novozyme 435 Lipase b. 20.8%. Reaction conditions varied from 40°C to 75°C for 3.5–24 hours, with and without stirring. Yields varied from 81% to 96.3%.

They also presented the NLGI grease grading code that relates the fluidity of various greases to food consistency. In other words, vegetable-based lubricants can also be developed in a more solid state—grease. Table 10.5 lists the National Lubricating Grease Institute grades for vegetable bio-greases [36].

TABLE 10.5

NLGI Grading of Grease with Penetration Depths in Tenths of a Millimeter

NLGI Consistency Numbers			
NLGI Number	**ASTM Worked (60 strokes) Penetration at 25°C**	**Appearance**	**Consistency Food Analog**
000	445–475	Fluid	Cooking oil
00	440–430	Semi-fluid	Apple sauce
0	355–385	Very soft	Brown mustard
1	310–340	Soft	Tomato paste
2	265–295	"Normal grease"	Peanut butter
3	220–250	Firm	Vegetable shortening
4	175–205	Very firm	Frozen yogurt
5	130–160	Hard	Smooth pate
6	85–115	Very hard	Cheddar cheese

Source: Panchal, T.M., et al., *Renew. Sust. Energ. Rev.*, 70, 65–70, 2017.

TABLE 10.6
Oil Content of Some Nonedible and Edible Oil Seeds

S1. No.	Nonedible Species	Oil Content (% of volume)	Edible Species	Oil Content (% of volume)
1	Jathropha	40–60	Rapeseed	38–46
2	Neem	30–50	Palm	30–60
3	Karanja	30–50	Peanut	45–55
4	Castor	45–60	Olive	45–70
5	Mahua	35–50	Corn	48
6	Linseed	34–45	Coconut	63–65
7	Moringa	20–36	—	—

Source: Karmarkar, G. et al., *Lubricants*, 5, 44, 2017.

10.2.11 BIOLUBRICANTS

Many biolubricants are from vegetable sources [45–55]. Over 300 natural plants sources exist that can release lubricating oils, some of which are nonedible, and hence would not compromise food supply (Table 10.6). Nonedible oil seeds that can grow on marginal land therefore offer lubricant sources of considerable appeal.

10.2.12 GREEN LUBRICANTS

As indicated above, a green lubricant's basic starting materials are sourced from plants or animals.

Green vegetable biolubricants possess or can be modified to produce as needed the following:

- High viscosity indices
- High pressure tolerances
- Long lasting, antiwear lubrication
- Corrosion resistance
- High flash point
- Rapid biodegradability
- Low toxicity especially if entering the soil or water systems
- Resistant to oxidation [56,57]

Many plant materials are amphiphilic—polar at one end and nonpolar at the other. Hence the basic extracts lend themselves to chemical modification that can be directed at specific applications. As an example, eicosanoic and octadecanoic acids present in castor and jojoba oil, respectively, enhance stability. For automobile use, derivatives of poly(hydroxy thioether) vegetable oils were used as antiwear/antifriction additives [58].

10.2.13 CHEMICAL AND PHYSICAL PROPERTIES OF VEGETABLE OILS

The sourcing of oil-bearing crops (real or genetically modified, area grown and seasonal) will define the chemical and physical properties of plant oils. Many oils contain unsaturated fatty acids. Fatty acids present in vegetable oils are mostly long and straight-chained with unconjugated double bonds, and most of these unsaturated fatty acids possess a cis configuration (Figure 10.4) [46].

However, some fatty acid chains, such as ricinoleic (Figure 10.5) and vernolic acids (Figure 10.6) contain hydroxyl and epoxy functional groups, respectively [59,60].

The triglycerides of many vegetable oils show a predominance of linoleic and linolenic acids with unconjugated cis double bonds. However in castor oil, ricinoleic acid (Figure 10.5) and in vernonia oil, vernolic acid (Figure 10.6) predominate, respectively.

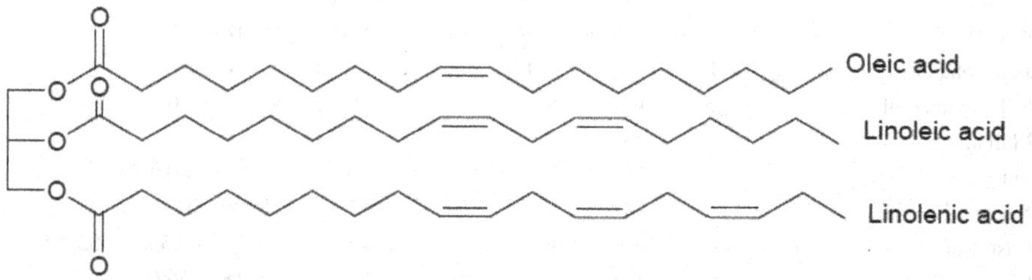

FIGURE 10.4 General structure of triglyceride esters of edible vegetable oils. (From Karmarkar, G. et al., *Lubricants*, 5, 44, 2017.)

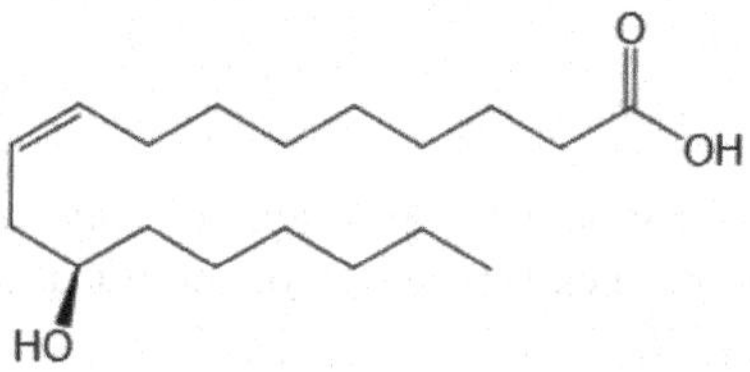

FIGURE 10.5 Ricinoleic acid, the major content of castor oil. (From Borugadda, V.B. and Goud, V.V., *Energy Procedia.*, 54, 75–84, 2014.)

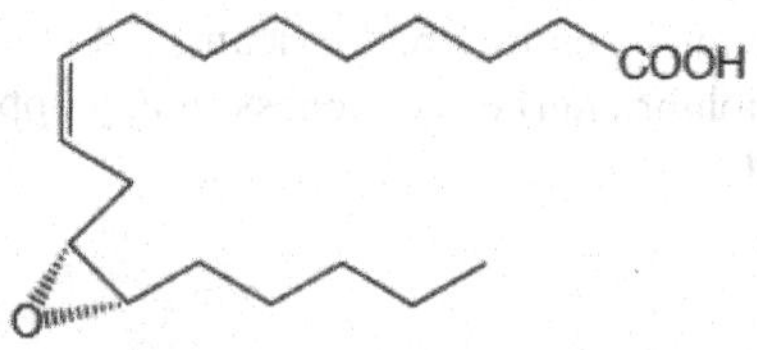

FIGURE 10.6 Vernolic acid, the major content of vernonia oil. (From Baye, T. and Becker, H.C., *Genet. Resour. Crop Evol.*, 52, 805–811, 2005.)

TABLE 10.7
Fatty Acids in Vegetable Oils

Vegetable Oils	C12:0	C14:0	C16:0	C18:0	C16:1	C18:1	C18:2	C18.3	Others
Soybean oil	—	—	11–12	3	0.2	24	53–55	6–7	—
Sunflower oil	—	—	7	5	0.3	20–25	63–68	0.2	—
Rapeseed oil	—	—	4–5	1–2	0.21	56–64	20–26	8–10	9.1(20:1)
Palm oil	—	1	37–41	3–6	0.4	40–45	8–10	—	—
Rice bran oil	—	—	20–22	2–3	0.19	42	31	1.1	—
Cotton seed oil	—	1	22–26	2–5	1.4	15–20	49–58	—	—
Coconut oil	44–52	13–19	8–11	1–3	—	5–8	0–1	—	—
Corn (maize) oil	—	—	11–13	2–3	0.3	25–31	54–60	1	—
Peanut/Ground nut	—	—	10–11	2–3	0	48–50	39–40	—	—
Sesame oil	—	—	7–11	4–6	0.08	40–50	35–45	—	—
Safflower oil	—	—	5–7	1–4	0.8	13–21	73–79	—	—
Karanja oil	—	—	11–12	7–9	—	52	16–18	—	—
Jathropha oil	—	1.4	13–16	6–8	—	38–45	32–38	—	—
Rubber seed oil	—	2–3	10	9	—	25	40	16	—
Mahua oil	—	—	28	23	—	41–51	10–14	—	—
Tung oil	—	—	2.67	24	—	7.88	6.6	80.46[a]	—
Neem oil	—	—	18	18	—	45	18–20	0.5	—
Castor oil	—	—	0.5–1	0.5–1	—	4–5	2–4	0.5–1	83–85[a]
Linseed oil	—	—	4–5	2–4	0–0.5	19.1	12–18	56.6	—
Olive oil	—	—	13.7	2.5	1.8	71	10	0–1.5	—

Source: Karmarkar, G. et al., *Lubricants*, 5, 44, 2017.

[a] Alpha-eleostearic acid. # Ricinoloic acid.

Table 10.7 shows the fatty acid compositions of some common vegetable oils. The level of unsaturation defines the degree of modification that can be achieved with any of these oils [61–73].

Vegetable oils can be classified as edible or nonedible. Coconut, olive, soybean, sunflower, palm, peanut, canola, and corn belong to the class of edible oils. Since they are also used to formulate biolubricants, their demand is always very high. The nonedible vegetable oils, such as neem, castor, mahua, rice bran, karanja, jathropha, and linseed, are comparatively less expensive and therefore, have a price advantage over edible oils for the production of biofuel/biolubricants [74–76].

Green lubricant (or biolubricant) effectiveness in any application will depend on properties such as [77–80]:

• Viscosity index
• Pour point
• Flash point
• Cloud point

- Thermal stability
- Oxidation stability
- Shear stability
- Iodine value
- Density

Biolubricants need to meet and preferably exceed the performance of petroleum-based lubricants in their viscosity index, low carbonation, overall stability, oxidation stability, volatility, and compatibility with additives [81]. The need for improvement of classical lubricants and development of new lubricant applications produces a constant tribological demand [82]. The demands of modern automotive engine oils are now being met with biolubricants.

10.2.14 CHEMICAL MODIFICATION/DERIVATIZATION OF VEGETABLE OILS (VOs)

Vegetable oils have been used as biofuel/biolubricants but these must be chemically modified. This can be performed in two different ways: reactions at the carboxyl groups of fatty acids/esters/triglycerides of vegetable oils or reactions at the olefinic functionalities of the fatty acid chain [83–87].

10.3 GREEN NANOTRIBOLOGY

10.3.1 NANOTECHNOLOGY

Nanoscience is the study of phenomena and manipulation of materials at the atomic, molecular, and macromolecular level, where properties differ significantly from those at bulk scales. Therefore, nanotechnology is the design, characterization, production, and application of material structures, devices, and systems with target objectives developed by controlling shapes and sizes at the nanoscale.

In the natural world there are many examples of structures that exist with nanoscale dimensions, including essential molecules within the human body and components of foods [88]. Many technologies have included nanoscale structures for a long time, but only in the last quarter century has it been possible to actively and intentionally modify molecules and structures within this size range. An ability to control size at the nanometer scale distinguishes nanotechnology. Two basic approaches for the preparation of nanostructures exist:

1. Bottom-up—preparations starting from atomic and molecular building units
2. Top-down—generation of nanoparticles from the breakdown of larger (micro-macro) units

Both methods have their advantages. Bottom-up, in many cases, can start with the basic units in a clean, ready-to-go form. However, depending on basic building unit cost and objectives, the top-down approach is often cheaper. Top-down methods are currently superior, for example, in the electronic circuitry assembly.

Nanotechnology is an interdisciplinary science, defined by the National Nanotechnology Initiative as "the manipulation of matter with at least one dimension sized from 1 to 100 nanometers" [88]. Nanotechnology involves surface science, organic chemistry, molecular biology, semiconductor physics, and microfabrication [88].

Nanotechnology has given the ability to consider the small to large (bottom-up) approach as opposed to large to small (top-down) modeling approach to produce smaller nanoparticles.

The range of current applications for nanotechnologies is vast. Analytical nanotechniques used include atomic force microscopy (AFM), scanning tunneling microscopy (STM), scanning confocal microscopy, and scanning acoustic microscopy (SAM)—all are scanning probe techniques [89].

10.3.2 GREEN NANOTECHNOLOGY

Green nanotechnology as discussed above (Section 10.2) is founded on the 12 principles of green chemistry [11] and aimed at the design of new nanotechnologies for combined economic, social, and health/environmental benefits [90]. The negative impacts of nanotechnology must, however, be acknowledged as this science develops [91]. Green technology, and hence green nanotechnology, still needs global acceptance from both the scientific and business aspects. The overall acceptance of green still awaits widespread acceptance in the real world, which at the very least involves scientists, engineers, regulatory bodies, and entrepreneurs.

As for any new endeavor, nanoscale inventions have the challenges of management, production, funding, and deployment, as well as scale-up. One basic tenet for green nanotechnology is the design of new materials that will promote health and environmental safety [91,92].

Examples of commercial nanotechnological applications at the micro/nanoscale level include magnetic hard disks, magnetic tapes, micro/nanoelectromechanical systems (MEMS/NEMS), micromirror components of commercial digital light processing (DLP) equipment, and the drive mechanism of green laser in micro projectors [93–99]. For example, adhesion is the major cause of failure of accelerometers used in automobile air bag triggering mechanisms and the micromirrors in the DLP [100–102]. Wear has been found to compromise the performance of NEMS-based AFM data storage systems [103,104].

10.3.3 NANOTRIBOLOGY

Krim et al. (1991) introduced the term "nanotribology" with their study on atomic scale friction of a krypton monolayer. Nanotribology deals with the study of tribologically interesting materials, structures, and processes using methods of nanotechnology [105]. Nanotribology is very important to tribology especially when the area of contact between two surfaces can be very small. Rigid surfaces, rugged or not, are usually uneven and rough even at the atomic sale. The term used to describe such

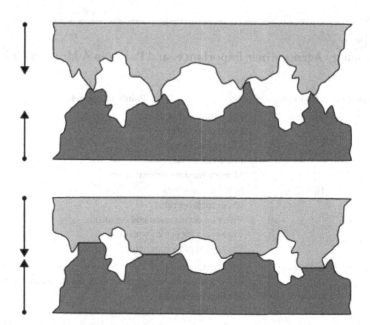

FIGURE 10.7 The top image shows asperities under no load; the bottom image depicts the same surface after a load is applied. (From https://en.wikipedia.org/wiki/Asperity_(materials_science.)

a situation is a collection of asperities (rough contact points) on the surfaces [105]. Figure 10.7 [106] illustrates the interaction of asperities before and after a load is applied.

Nanotribology is a branch of tribology that studies friction, adhesion, thin film lubrication, and wear between sliding surfaces at the molecular and atomic scale [107]. In order to visualize such surfaces one needs to use scanning probe microscopy and nano-indentation techniques and evaluate, for example, sub-nanometer distances, one atom thick separations, and monomolecular lubricant layers. Atomic force Kelvin probe microscopy can be used for the measurement of surface charges. Friction force microscopy can measure stick-slip interactions of single atoms on a crystalline surface.

Any tribological interaction can involve two or more surfaces at any or all possible sizes from macro- to micro- to nanoscale [108]. For example, areas of interest such as MEMS and NEMS are getting considerable attention [97]. Important application fields of nanotribology include molecular dynamics studies, MEMS, hard disks, and diamond-like carbon [109].

Nanotribology is the science that studies friction, wear, adhesion, and lubrication phenomena in the 1–100 nm range. Nanotribology deals with nanostructured surfaces, nanoagents (ingredients, additives), and nanoprocesses (Table 10.8).

TABLE 10.8

Nanotribology Agents, Their Importance, and Points to Address in Going Green

Nanotribology	Importance	Points to Address
Nanosurfaces	Medium	Nanostructured surfaces
		Hierarchical surfaces
		Material selection
		Coated materials
		Monomolecular lubricant layers
Nanoagents	High	Physical properties
		Chemical properties
		Effect on environment and organisms
		Changes in properties with time
		Changes in properties in the triboprocess
Nanoprocesses	Medium to low	Energy efficiency
		Share between process relevant energy, destructive energy and waste and reusable energy
		Effectiveness of reusing process energy

Source: Gebeshuber, I.C., *Nanotribology: Green Nanotribology and Related Sustainability Aspect*, Taylor & Francis Group, LLC, 2016.

10.3.4 GREEN NANOTRIBOLOGY

Gebeshuber [110] reported that green nanotribology is a sustainable green technology dealing with friction, adhesion, wear, and lubrication of interacting surfaces in relative motion at the nanometer scale. Green nanotribology includes the following:

1. Biomimetic tribological nanotechnology
2. Sustainable control of friction, wear and lubrication on the nanoscale
3. Environmental aspects of nanoscale lubrication layers
4. Environmental aspects of nanotechnological surface modification techniques
5. Nanotribological aspects of green applications such as artificial photosynthesis

Green nanotribology should provide technical support for the preservation of resources and energy as well as long-term sustainability [111,112].

The components of green nanotribology are nanostructured surfaces, nanoagents ingredients, additives, products of the additives, and by-products that appear in the system after the technological application, and nanoprocesses as detailed in Table 10.9 [113].

For green nanosurfaces, points such as nanostructured surfaces, hierarchical surfaces, material selection, coated materials and monomolecular lubricant layers need to be addressed.

TABLE 10.9

Contributions of Green Nanotribology to Address Issues Arising from the Nine Planetary Boundaries

Planetary Boundary	Green Nanotribology Solution
Climate change	Energy materials with optimized tribological performance, reduced information
Rate of biodiversity loss	and communication technology (ICT), global greenhouse gas emission by optimized micro- and nanomechanics, less CO_2 waste of machines due to
Interference with the nitrogen and phosphorus cycles	optimized tribology (less fuel consumption in production and use). Less consumption of resources via optimized nanotribology in production and use allows for less deforestation and more places for wildlife to thrive.
Stratospheric ozone depletion	Optimized nanotribosystems with reduced NO_2 emissions and phosphate release.
Ocean acidification	Usage of non-amiable, non-ozone-depleting solvents.
Global freshwater use	Less CO_2 waste from machines due to optimized tribology (less fuel consumption in production and use).
Change in land use	Optimized nanotechnological desalination processes, wastewater treatments,
Chemical pollution	and sewage treatments.
Atmospheric aerosol loading	Optimized nanotribological systems for optimized land use (more efficient production, storage, novel batteries that take less space, etc.)
	Reduced lubricant spillage by usage of molecularly thin lubricants instead of bulk lubrication.
	Optimized machines giving off less wear particles and less polluting exhaust fumes.

Source: Gnanasekaran, D. and Chavidi, V.P., Nanomaterials as an Additive in Biodegradable Lubricants, in: *Vegetable Oil Based Bio-lubricants and Transformer Fluids, in the series Materials Forming, Machining and Tribology*, Springer Nature, Singapore, 2018.

Reaction products (which can be harmful) have to be either chemically inert after use or are fed back to the system for further usage (waste-to-wealth concept). Unused nanoagents need to be either inert or fed back to the system. Potentially harmful by-products that have nothing to do with the initial nanoagent need to be either neutralized or recycled. Biomimetic tribological nanotechnology might help to turn nanotribology green, but it cannot be stated often enough that biomimetics does not automatically yield sustainable, or even simply green, products [111].

Of very high importance for green nanotribology are nanoagents. The points to address here are physical and chemical properties, the effect on the environment and biology, and the changes in the properties during the triboprocess. Regarding green nanoprocesses, the importance of points to address is in the medium to low range. Points to address comprise energy efficiency, and the share between process relevant energy, destructive energy, and waste and reusable energy, as well as the effectiveness of reusing process energy [114].

Nanotribology faces major challenges; for example, in miniaturized devices with moving parts, such as MEMS/NEMS, the small length scale and high surface-high area to volume ratio make interfacial phenomena dominant [115]. Therefore, a good understanding is required of how surface interactions, such as van der Waals and capillary forces, affect friction and adhesion at the molecular and atomic scale [95,107,116]. Bowden and Tabor [117] demonstrated that the real contact between two solids in contact is only a fraction of the apparent contact area due to the surface roughness. At the nanometre/micrometer scale, all surfaces are rough and they contact at some microscopic points (asperities). Therefore, the study of the surface interactions between these asperities at the molecular and atomic scale would provide an enhanced understanding of nanotribology. Since the development of AFM by Binnig et al. (1986) [118], much effort has been devoted to the study of the interactions that govern nanotribology. Figure 10.8 shows the principle of the atomic force microscope [119].

AFM allows the study of the interactions that occur between the tip and the surface, which is used to mimic the interaction between the asperities [120]. Therefore, this technique has provided new insights at length scales not previously accessible, which are essential in order to understand nanotribology.

Goals for effective green nanotribology are categorized under three main headings in three main areas:

1. Production (agents)
2. Reaction (agents; object to nanoproduct; waste agents direct effects)
3. Nanoproduct life cycle (effects on the environment during the service period and during degradation)

Major green goals include

- minimum pollution from the reaction,
- tight control of the reaction—no leakage or toxic byproducts affecting the environment, and
- readily degradable (as demanded by green policies) materials.

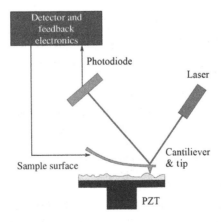

FIGURE 10.8 Atomic force microscope using beam deflection detection. As the cantilever is displaced via its interaction with the surface, so too will the reflection of the laser beam be displaced on the surface of the photodiode. (From https://en.wikipedia.org/wiki/Atomic_force_microscopy.)

10.3.5 GREEN NANOTRIBOLOGY AND SUITABILITY ASPECTS

Tribologists are already used to the inherent interconnectivity of various aspects of their profession, and it is easier for them to adopt new holistic concepts such as green and sustainable as opposed to most other people working in technology fields, and to relate environmentally acceptable tribological practices to saving of resources of energy and reducing the impact on the environment [121]. In the ecobalance tree, Bartz [121] mentions important parameters that need to be considered in the development of ecotribology: the raw materials of their transport, the production of these materials, the transport of these materials, the application, and the disposal. Important aspects are the exploitation of resources, waste management, emissions, recycling, combustion, and reuse.

Tzanakis et al. 2012 [122] reported the future perspectives of sustainable tribology and presented three interesting case studies from diverse areas of interest in tribology micro-CHP (combined heat and power) systems, slipways for lifeboats, recycled plastics for skateboard wheels. They performed tribological analyses and sustainability considerations.

Also, sustainable nanotribological systems based on the sustainability concept of design from living nature, the principles of life as introduced by the Biomimicry Guild [113]. The six basic categories are:

1. "Survival because of evolution" (in the context of sustainable nanotribology, this means continuous incorporation and anchoring of information)
2. Resource efficiency regarding material and energy (closed tribosystems, recycling and reuse of substances and waste energy)
3. Adaptation to changing conditions (reactive nanosurfaces, nanoagents, and nanoprocesses that change depending on the environment and that are used in as small amounts as necessary)
4. Integration of development with growth (in the context of sustainable nanotribology, this means achieving a balance between development and growth)
5. Responsiveness and being locally attuned (tribological systems not as good as possible but as good as necessary, with additional benefits regarding energy savings and environmental compatibility)
6. Usage of life-friendly chemistry (water-based chemistry, green chemistry— one of the basic requirements of green nanotribological systems) (Table 10.9)

The improvement in the tribological behavior of the major deficiency of a vegetable oil as a biolubricant is caused by poor oxidation stability and cold flow behavior due to the presence of unsaturated carbon bond in its atomic structure. The improvement in the thermo-oxidative stability of vegetable oil–based lubricants can be achieved by a chemical modification [124–127]. It was reported that the additives in lubricating oil play an important role in the reduction of the friction and wear of two mating surfaces [128].

Recently, there are few interesting reports confirming that the effect of nanoparticles dispersed in lubricating oil showed higher friction reduction and wear resistance between two mating surfaces [129–136]. It is evident from the literature that the tribological characteristics of various nanoparticles dispersed in an base oil showed reduced friction and reduced wear between the rubbing surfaces [137,138] and the addition of nanoparticles increases the viscosity of the base oil [141–145].

10.3.5.1 Evaluation of the Tribological Behavior and Viscosity of Nano Biolubricant Using a Four-Ball Tribometer and Viscometer, Respectively

In this study, chemically modified rapeseed oil (CMRO) was used as the base oil. Rapeseed oil was modified via an epoxidation, hydroxylation, and esterification processes in order to improve its thermo-oxidative stability and cold flow behavior. The detailed procedure for chemical modification process is reported from the study of Baskara et al. [137]. The various nanoparticles CuO, WS_2, and TiO_2 of 0.5 wt% were dispersed in CMRO with sizes of approximately CuO, 40–70 nm, WS_2 40–80 nm, and TiO_2 30–50 nm, respectively, using an ultrasonic sonicator. They used a four-ball tribometer that consisted of a fixed three-ball pot with the fourth ball rotating on a vertical shaft. Colors, morphologies, and bulk densities were recorded [137]. In addition, SAE20W40 mixtures with CMRO + the three metal oxides are tabled. The properties of these mixtures such as viscosity at 100°C, pour point °C, flash point, viscosity index, specific gravity at 15°C and wear scar index (mm) are comprehensively recorded [137].

The results of the study are detailed in the following sections.

10.3.5.2 Frictional Behavior

The coefficients of friction of three lubricating oils, CMRO containing nano CuO, nano WS_2, and nano TiO_2, are 0.0814, 0.0841, and 0.0825, respectively, whereas for the synthetic lubricant (SAE20W40) it is 0.1009. The coefficient of friction is lowered by about 19%, 16%, and 18% for CMRO containing nano CuO, nano WS_2, and nano TiO_2, respectively, as compared to the synthetic lubricant [137]. From these data it can be concluded that the CMRO containing nano CuO exhibits minimum friction torque and coefficient of friction as compared to other lubricating oils [137]. The nanoparticles that are added to the CMRO increase the viscosity, thereby increasing the oil film thickness which reduces the contact between the ball surfaces. Although all the nanoparticles (CuO, WS_2, and TiO_2) used in this study are of spherical morphology that is responsible for reduced coefficient of friction, the nanoparticles start to diffuse and shrink in volume when the lubricant temperature increases from room temperature to operating temperature. When the oil temperature increases, the nanoparticles are assembled further and diffuse to become comparatively tight clusters. Since the density of bulk nano CuO is higher than the other two nanoparticles in this investigation as reported by Baskara et al. [137], the CuO nanoparticles can maintain their spherical profile even after diffusion and particles were considered assembled is the other important reason for lower friction coefficient.

The above findings are also consistent with the previous study by Wu et al. [125]. They reported that the addition of nano CuO increases the tribological behavior of API-SF engine oil and base oil than with lubricant containing nano TiO_2 and nano diamond particles.

10.3.5.3 Viscosity of Various Lubricating Oils

Baskara et al. report the mean wear scar diameter of tested balls with the lubricants [137]. The mean wear scar diameter is lowered by about 39%, 36%, and 34% for CMRO containing nano CuO, nano WS_2, and nano TiO_2, respectively, as compared to that of the synthetic lubricant. The CMRO containing nano CuO exhibited minimum wear scar diameter compared to other lubricating oils. It is evident the figures shown that wear scar image of CMRO containing nano CuO is smoother, clearer, and circular in shape, whereas for the synthetic lubricant the wear scar image is unclear, rough, and oval in shape [137].

Comparison of the viscosity values of various lubricating oils shows that the viscosity of CMRO containing nanoparticles was higher than the synthetic lubricating oil. Hence viscosity was seen to be independent of the type of nanoparticle. Wear scar images of lubricating oils (50X) and the viscosity of lubricating of the lubrication oils is measured in centistokes (cSt) at the various temperatures is clearly shown [137].

From the tests performed on a four-ball tribometer and a viscometer with synthetic lubricant and CMRO containing various nanoparticles, the conclusions are as follows:

1. The coefficient of friction of CMRO containing nano CuO is lower than the synthetic lubricant and CMRO containing nano WS_2 and nano TiO_2.
2. The wear scar diameter of CMRO containing nano CuO is lower than synthetic lubricant and CMRO containing nano WS_2 and nano TiO_2 [137].

10.3.6 RECENT STUDIES OF GREEN NANOTRIBOLOGY

10.3.6.1 Tribological Performance of Nanoparticles as Lubricating Oil Additives

Gulzar et al. [146] reported in their review that nanoparticles are emerging lubricant additives with friction and wear reduction potentials, and a variety of nanoparticles (NPs) have been used as lubricant additives with potentially interesting friction and wear properties. To date, there has been a great deal of experimental research on NPs as lubricating oil additives. NPs can exist in spherical, rod, platelike, and other shapes. They have a larger surface area to volume ratio than their bulk (larger-sized) material, leading to the surface properties dominating the bulk properties.

When NPs are mixed with liquids, colloids are formed. Colloids should be stable during their use in machinery for lubrication purposes. It has been observed that the mechanical properties of NPs can affect the tribological behavior of lubricants with added NPs. For example, the hardness of NPs, compared to that of the lubricated sliding contact interfaces, would influence the behavior of NPs. When the contact pressure is high, the NPs may deform or indent into the surface. The hardness and elastic modulus of NPs are different from those of their bulk materials, depending on size. The spherical NPs of gold (Au), silver (Ag), and silicon (Si) are reported to exhibit such size-dependent behaviors. In general, the tribological performance of nanocomposites like SiO_2, is enhanced by the change of their surface properties from hydrophilic to lipophilic. Three lubricating mechanisms of NPs in a lubricant are proposed as follows:

1. Metallic and metallic oxides NPs can form adsorption films on tribo-pair surfaces.
2. NPs can roll between two sliding surfaces.
3. NPs can fill wear track and undergo sintering by frictional heat and pressure.

Figure 10.9 lists a main classification of nanoparticles of engineered nanoparticles relevant to tribological studies [146].

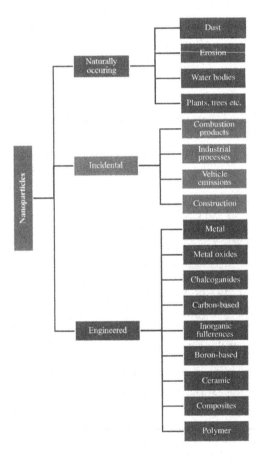

FIGURE 10.9 A classification of nanoparticles with subcategories of engineered nanoparticles relevant to tribological studies. (From Gulzar, M. et al., *J. Nanoparticles. Res.*, 18, 223, 2016.)

10.3.6.2 Novel Strategy for the Design and Preparation of High-Performance Water-Based Green Nanolubricants

The dispersion of two-dimensional (2D) nanosheets in water is conducive to their practical applications in fundamental science communities due to their abundance, low cost, and ecofriendliness. However, it is difficult to achieve stable suspension of aqueous 2D material because of the intrinsic hydrophobic properties of the layered materials. Thus Zang et al. [147] reported an effective and economic way for the synthesis of aqueous dispersions with 2D nanosheets decorated with carbon quantum dots (CQDs) (Figure 10.10) [145].

The prepared 2D nanosheets were characterized by UV–Vis, HRTEM, AFM, Raman spectroscopy, XPS, and XRD. Figure 10.11 shows friction coefficient curves of self-mated AISI 52100 bearing steels in (a) MoS_2/CQDs- and (b) $MoSe_2$/CQDs-based aqueous dispersions under a load of 5 N based on a ball-on-disk apparatus (ball diameter, 10 mm). MoS_2 is molybdenum disulfide and $MoSe_2$ is molybdenum diselenide. Their excellent lubricating performances could be attributed to their dispersion stability and the synergistic effect of CQDs and 2D nanosheets owing to their easy-sliding interlayer properties. Moreover, the small size of these lubricating additives allows them to enter the surface of the contact area and form a lubricating film, resulting in favorable wear resistance. The liquid-phase exfoliation method provides a facile and economic way to achieve large-scale, water-based lubricant additives. Figure 10.11 shows proposed lubrication models for CQDs and 2D nanosheets decorated by CQD-based aqueous dispersions [147].

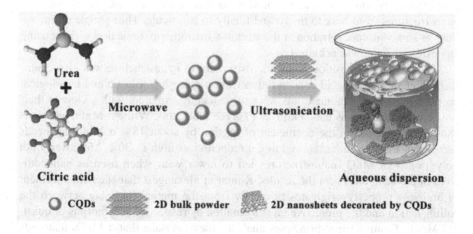

FIGURE 10.10 Schematic of the generation of 2D nanosheets decorated by carbon quantum dots (CQDs). (From Zhang, W. et al., *ACS Appl. Mater. Interfaces*, 8, 32440–32449, 2016.)

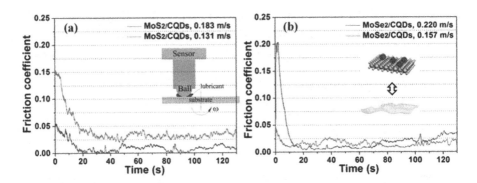

FIGURE 10.11 Friction coefficient curves of self-mated AISI 52100 bearing steels in (a) MoS$_2$/CQDs- and (b) MoSe$_2$/CQDs-based aqueous dispersions under a load of 5 N based on a ball-on-disk apparatus (ball diameter 10 mm). MoS$_2$ is molybdenum disulfide and MoSe$_2$ is molybdenum diselenide. (From Zhang, W. et al., *ACS Appl. Mater. Interfaces*, 8, 32440–32449, 2016.)

10.3.6.3 One-Pot Synthesis and First-Principles Elasticity Analysis of Polymorphic MnO$_2$ Nanorods for Tribological Assessment as Friction Modifiers

Kumar et al. [148] reported the one-pot synthesis of single-crystalline α and β-MnO$_2$ nanorods by selectively varying the acidic concentration. They used the use of MnO$_2$ nanorods as additive in the base oil (palm oil) for evaluating its tribological properties. They developed a common procedure requiring the common time, temperature, and precursors. This enabled checking of rich polymorphism by creating two different polymorphs, α and β, of MnO$_2$ and structure flexibility by drawing 1D nanostructures from bulk to micro and finally to nanoscale. Thus simple optimizations at an acidic concentration at the start of a hydrothermal reaction without using any external agent was achieved.

Morphological transition from microstructure to nanostructure was also examined by changing the acid concentration from high to low. Elastic and tribological properties of these nanomaterials were subsequently explored, with a view to their possible applications as nanoadditives in green lubricants. While β-MnO$_2$ nanorods showed a reduction in the coefficient of friction by about 15%, α-MnO$_2$ nanorods turned out to be even better, yielding a reduction as high as 30%. Moreover, both polymorphs of MnO$_2$ nanostructures led to lower wear when used as nanoadditives in the base oil. From the results, Kumar et al. suggest that such enhancement of antiwear property originates primarily from the mutual interplay between the rolling action and the protective layer formation by respective polymorphs of quasi-1D MnO$_2$. From the first principles analysis, they envisage that α-MnO$_2$ nanorods may potentially serve as an efficient nanoadditive compared to β-MnO$_2$ nanorods due to superior elastic properties of the former. Figures 10.12 and 10.13 display the results of this study [148].

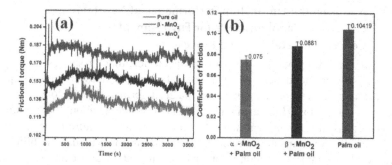

FIGURE 10.12 (a) Frictional torque with respect to time observed using four-ball test technique; (b) coefficient of friction of pure palm oil β-MnO$_2$ added palm oil and α-MnO$_2$ added palm oil. (From Kumar, N. et al., *RSC Adv.*, 7, 34138–34148, 2017.)

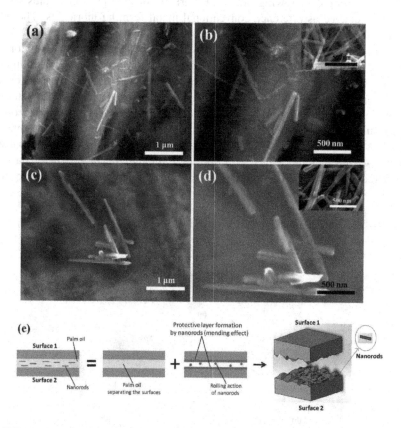

FIGURE 10.13 FESEM images of the deposited nanorods on the scar surfaces of balls tested using (a, b) α-MnO$_2$ added palm oil with FESEM image of α-MnO$_2$ nanorods as inset in (b); (c, d) β-MnO$_2$ added palm oil with FESEM image of β-MnO$_2$ nanorods as inset in (d); and (e) schematic of the mechanisms responsible for enhancement of antiwear property of the base oil by nanorods. (From Kumar, N. et al., *RSC Adv.*, 7, 34138–34148, 2017.)

10.3.6.4 Assessing the Lubrication Performance of Vegetable Oil–Based Nanolubricants for Environmentally Conscious Metal Forming Processes

Zareh-Desari and Davoodi [149] reviewed metal forming processes employing lubricants. They provided desirable tribological circumstances at tool-workpiece interface, which are essential to increase material formability, prolong tool life, and suppress energy loss. Increasing awareness and concerns about the substantial hazardous effects of conventional metal forming lubricants require replacing them with benign and environmentally compatible lubricants. The biodegradability and nontoxicity of vegetable oils, along with friction reduction and extreme pressure capabilities of nanoparticle additives, gave a high performance lubricant that was environmentally friendly. For example, CuO and SiO_2 nanoparticles were dispersed in soybean and rapeseed oils in various concentrations and their properties and their lubrication properties were investigated relative to conventional metal forming lubricants. CM202A press drawing oil, solid lubricant of zinc phosphate plus sodium soap were selected as reference conventional metal forming lubricants. The lubricants were tested on a standard ring compression test and evaluated for their friction coefficient and deformation load. The necessary friction calibration curves for estimating the friction coefficient were provided by employing finite element simulation. The study concluded that the lubricity of vegetable oil was dependent on its chemical structure. As the nanoparticles' concentration increased, the friction factor and deformation load reached a minimum value for certain concentrations. According to the results, the most efficient concentration of nanoparticles was in the range of 0.5–0.7 wt%, which provides 21%–31% enhancement. It was demonstrated that vegetable oil–based nanolubricants with optimum amounts of nanoparticles have the friction reduction capability comparable to conventional lubricants. Figure 10.14

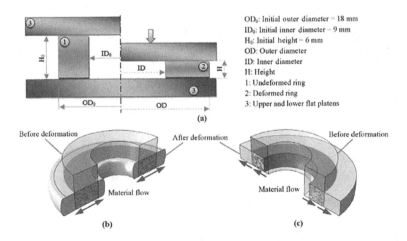

FIGURE 10.14 (a) A schematic display of ring compression test and the effect of frictional circumstances on dimensional modifications of the ring specimen: (b) high friction (poor lubrication) and (c) low friction (appropriate lubrication). (From Zareh-Desari, B. and Davoodi, B., *J. Cleaner Prod.*, 135, 1198–1209, 2016.)

illustrates the compression ring test and the effects of high and low friction resulting from the use of inefficient and efficient lubrication [149].

10.3.7 Recent Studies of Green Nanotribology

10.3.7.1 Life Cycle Assessment of Lubricants

A robust quantification of the overall environmental impact of lubricants would entail a detailed life cycle assessment (LCA) [149] with a scope covering manufacturing, use, and fate at end of life, and with system boundaries encompassing petroleum, petrochemical, oleochemical, and engineering industry activities. This would be a dauntingly complex process, due to the very broad scope required, and also to some particular issues which are characteristic of the industry and the applications.

The first complication is that lubricants are typically manufactured as coproducts in integrated product networks, based on petroleum refining, oleochemical refining, or chemical processing. Consequently, allocation of resource requirements and environmental impacts to the lubricant elements of these networks is necessarily somewhat arbitrary.

Second, since lubricants differ greatly in their performance and the amount required for a particular purpose, a detailed knowledge of application performance is required in order to define an appropriate functional unit for comparison. LCA comparisons are made on the basis of equivalent outputs so a simple comparison of different lubricant types, based only on their resource requirements per kg or per liter, would therefore give misleading results.

Third, because lubricants are used so widely in many different products and applications, tracing the fate at the end of life is very difficult. End-of-life treatment of industrial lubricants used at a single manufacturing site can be controlled and monitored, but following the fate of used engine oils is a more challenging problem. The eventual fate of a large proportion of the overall lubricant production is not accurately known. Finally, even where data are available, they may considered confidential to the manufacturers or users, and not made available. Publishable LCAs generally require coordination with an independent body to facilitate pooling of commercially sensitive information.

Practical illustrations of the magnitude of the task are provided by studies from related sectors in the chemicals industry. These include the 1995 European Life Cycle Inventory for detergent surfactants production [150] and the Eco-profiles of the European plastics industry [151]. The scope of the former study was an inventory of energy and resource requirements for production of seven major surfactant types, with no consideration of impact assessment. Even with these restrictions, the study required two years to prepare with the help of technical professionals from many companies [152,153]. A key conclusion was that no technical basis existed to support a general environmental superiority claim either for an individual product type or for the various sources of raw materials from petrochemical, agricultural, or oleochemical feedstocks.

For these reasons, no such comprehensive lubricant LCA has yet been attempted. Lubricating oils and related fluids have been considered as elements of LCA studies focusing on particular application areas, such as hydraulic equipment for forestry applications and municipal cleaning and domestic refrigerators [154–159].

Furthermore, most companies in the industry carry out more or less detailed analyses of environmental impact in order to guide their strategic and internal decision making. However, the results of these analyses are not generally made public, although some limited studies have been published [160–162]. Single company studies are limited by access to the data of the company involved, and may also raise concerns as to the independence of any conclusions from the commercial interests of that company.

10.4 THREE RECENT GREEN TRIBOLOGY STUDIES

A short selection of recent green tribology and green nanotribology studies shows the breadth of current thinking.

While it would be ideal to have all lubricants green and renewable (organic), for the foreseeable future the need remains to conserve the less renewable (inorganic) components in some tribological cases. Hence green tribology or nanotribology not only seeks to achieve better, safer, more economical, and renewable (organic) lubricants, it also will need to conserve inorganic additives that are not renewable (inorganic).

10.4.1 GREEN LUBRICANTS FROM VEGETABLE OILS

In 2016, Raghunanan and Narine published a series of papers based on vegetable oils for engineering green lubricants. In their first report [163] they investigated the properties of a series of linear diesters separated by a varying number of methylene (–CH$_2$–) groups as shown in Figure 10.15 and represented by (^)$_n$ or –n– in Figures 10.15 through 10.19.

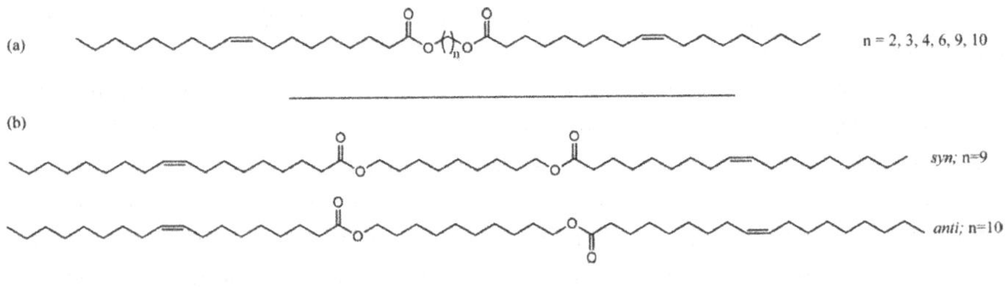

FIGURE 10.15 Linear diesters synthesized from oleic acid—a vegetable-sourced material. (a) Linear aliphatic diol-derived diesters, generalized as 18-n-18 diesters, investigated in this work; n = 2–10. and (b) Odd–even influence of diol chain segment on molecular conformation; odd alkyl moieties give syn-conformation, while even alkyl moieties result in anti-conformation. (From Raghunanan, L. and Narine, S. S., *ACS Sustainable Chem. Eng.*, 4, 686–692, 2016.)

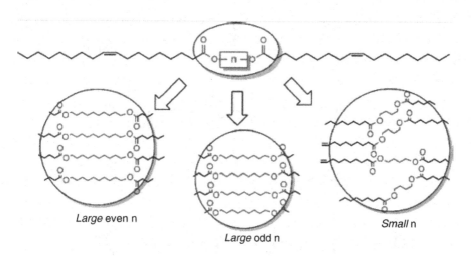

FIGURE 10.16 Variations of the size and number of the CH_2 linker n. (From Raghunanan, L. and Narine, S.S., *ACS Sustain. Chem. Eng.*, 4, 686–692, 2016.)

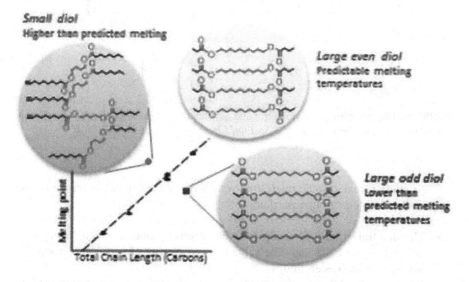

FIGURE 10.17 . Effect of diol number on diester melting point. (From Raghunanan, L. and Narine, S.S., *ACS Sustain. Chem. Eng.*, 4, 3, 693–700, 2016.)

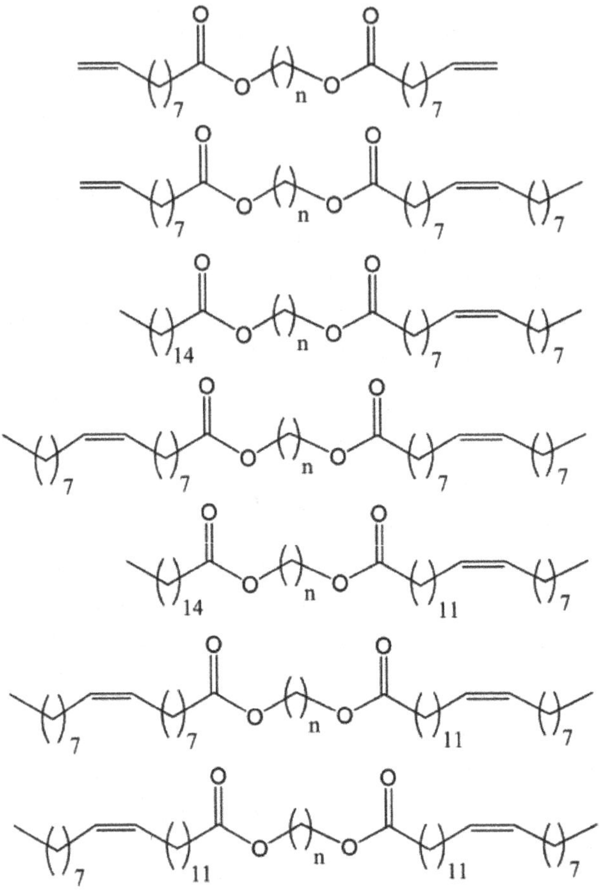

FIGURE 10.18 Structures of symmetric and asymmetric, linear aliphatic diesters. (From Raghunanan, L. and Narine, S.S., *ACS Sustain. Chem. Eng.*, 4, 3, 693–700, 2016.)

Their study revealed some interesting results that were dependent on the seemingly small methylene group insertion:

- Thermal transition—steric repulsion of various n values affected the crystallization of the diesters, The value n is the number of methylene groups between the two ester groups.
- Newtonian fluid flow for the 18-n-18 diesters. Newtonian flow is maintained when the viscosity of a fluid is not altered by stress and or strain.
- Viscosity—scaled linear for n = even, complex for n = odd number.

Overall their study showed that in principle their diesters could be tailored for various green applications: lubrication, heat transfer, oil and/or wax synthesis. The overall

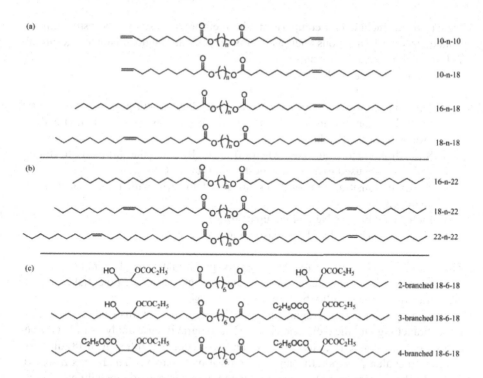

FIGURE 10.19 General structures of model diesters investigated: (a) low molecular weight linear aliphatic diesters investigated for evaporative behavior; (b) high molecular weight linear aliphatic diesters investigated for decomposition behavior; and (c) propionic-branched polyesters of 18-*n*-18 investigated for decomposition behavior. (From Raghunanan, L. and Narine, S.S., *ACS Sustain. Chem. Eng.*, 4, 4868–4874, 2016.)

scope of their study can be summarized as follows: the value of "n" whether large even, large odd, and small governed the properties of their oils.

A second report [164] dealt with novel diol, diacid and di-isocyanate derivatives from oleic acid.

In a continuation of their study [165], they prepared diesters C10–C22 by reacting C2–C10 diols:

- 9-Decenoic acid (C10—unsaturated terminal)
- Palmitic acid (C16—saturated)
- Oleic acid (C18—unsaturated internal)
- Erucic acid (C22—unsaturated internal)
- 1,2-ethanediol (C2)
- 1,3-propanediol (C3)
- 1,4-butanediol (C4)
- 1,6-hexanediol (C6)
- 1,9-nonanediol (C9)
- 1,10-decanediol (C10)

Oleic acid was included for comparison with their previous study. Their study investigated structural variations such as asymmetry, chain length mismatch, terminal and internal double bonds, and saturation.

The results were as follows:

- Thermal transition properties were influenced by structure.
- Crystallization and melting temperatures of the diesters with methyl end groups were predictable.
- Steric hindrance is an issue when (n) is small. The value (n) refers to the diols that are used to make the diesters in Figure 10.18.
- Thermal transition temperatures increased linearly with total molecular chain length.
- Flow behavior was Newtonian up to 100°C.
- Viscosity increased with increasing molecular mass.

This study showed promise for the design of green materials. The authors noted, however, that tribological studies need be carried out to evaluate potential lubricant properties including stability both in use and storage.

In a fourth report [166], Raghunanan and Narine extended this study by incorporating branching into the oleic acid chains. Thermogravimetric analyses (TGA) were carried out on symmetrical (ex. 10-n-10) and nonsymmetrical (ex. 10-n-18) diesters in order to evaluate evaporative and decomposition behaviors. The results revealed that a number of factors affect evaporation and/or decomposition including:

- Chain length
- Presence or absence of double bonds
- Presence or absence of polar group branching
- Side chain derivatives chemical structure (hydroxyl and/or ester)

Even though, no tribology applications were included in this study these green oils hold promise for high-temperature applications. In other words, we should be able to dial up green lubricants to fit any particular application as implied by their evaporation/decomposition graph (Figure 10.20).

10.4.2 Fatty Acid Green Lubricants

In 2017, Fan et al. [167] reported the use of fatty acid ionic liquids (FAILs) as *good and universal* lubricants that had the capability to adsorb onto metallic surfaces, as shown in Figure 10.21. They combined quaternary ammonium cations with a range of biobased fatty acid anions.

Their ionic liquids (ILs) were found to be useful lubricants for the following metal interacting combinations: steel-steel, steel-copper, and steel-aluminum. The ionic liquids showed resistance to corrosion and their viscosity index and thermal

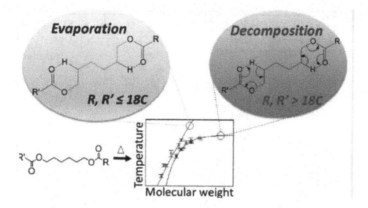

FIGURE 10.20 Evaporation versus decomposition in relation to temperature and molecular weight. (From Raghunanan, L. and Narine, S.S., *ACS Sustain. Chem. Eng.*, 4, 4868–4874, 2016.)

Fatty acid ionic liquids

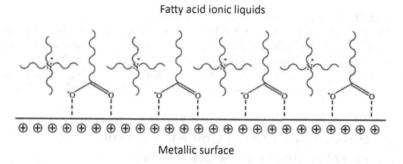

Metallic surface

FIGURE 10.21 Physical adsorption of FAILs on a metallic surface. (From Fan, M. et al., *Tribol. Trans.*, 61, 195–206, 2018.)

decomposition were relatively moderate compared to PAO 10 (a synthetic polyalphao-lefin) and the known ionic liquid coded L-F104 (1-butyl-3-methylimidazolium bis (tri-fluoromethyl sulfonyl)imide). Their ILs coded N_{1444}Oct (methyltributyl ammonium octanoate) and N_{4444}Oct (tetrabutyl ammonium octanoate) had much lower toxicity levels than L-F104. They maintain that the sliding capability of the metals was aided by ionic adsorption of the ILs forming an aliphatic slippery film and hence keeping the metals apart as indicated in the Figure 10.22 [167]. The anions used were octa-noic acid, 2-ethylhexanoic acid, decanoic acid, and hexadecenoic acid. Hence this hydrophobic film is seen as a means of reducing friction and wear, especially in the hard-soft metal (Fe-Al and Fe-Cu) combinations.

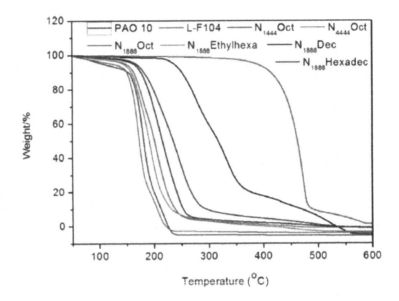

FIGURE 10.22 Weight loss versus temperature from thermogravimetric analysis curves for ionic liquids [166]. PAO 10 (poly-α-olefin), L-F104 (1-butyl-3-methylimidazolium bis (trifluoromethyl sulfonyl)imide), N_{1444}Oct (methyltributyl ammonium octanoate), N_{4444}Oct (tetrabutyl ammonium octanoate), N_{1888}Oct (methyltrioctyl ammonium octanoate), N_{1888}Ethylhexa (methyltrioctyl ammonium 2-ethylhexanoate), N_{1888}Dec (methyltrioctyl ammonium decanoate), N_{1888}Hexadec (methyltrioctyl ammonium hexadecanoate). (From Fan, M. et al., *Tribol. Trans.*, 61, 195–206, 2018.)

10.4.3 HYDROXYPROPYLMETHYLCELLULOSE (HMPC) FILMS WITH MoS$_2$ FOR ENHANCING LUBRICATION

Shi et al. in 2016 reported a series of the papers dealing with hydroxypropylmethyl-cellulose (HMPC) for producing green lubricant biofilms and molybdenum disulfide (MoS$_2$) as an inorganic lubricant modifier. Their first paper [168] reports the preparation of HMPC and detailed analytical data showing controllability of film thickness levels and uniformity, good wear resistance capability and low toxicity.

They then proceeded to add MoS$_2$ to their HMPC to ultimately give an inorganic fullerene molybdenum disulfide, IF-MoS$_2$ [169]. They also included in their study a molybdenum trioxide-HMPC (MoO$_3$)-HMPC) complex. A fullerene is a carbon only molecule that can be hollow and spherical, elliptical or tubal. Other shapes are also possible. SEM from images from this study can be seen in Figure 10.23. Here the clear layer of tribofilm has reduced friction and thus has increased wear resistance, which is the best of all the four examples presented (Figure 10.24).

Overall, Shi states that "the study showed that IF-MoS$_2$ NPs have multiple lubricating mechanisms, and superior and more stable lubricating properties than 2H-MoS$_2$ MPs and MoO$_3$ NPs" [169].

Two further papers by Shi et al. add further support for their HMPC-MoS$_2$ lubricant coatings claims [170,171].

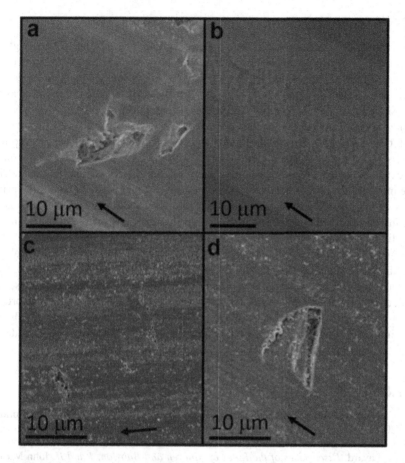

FIGURE 10.23 SEM images of worn surfaces: (a) pure HMPC; (b) IF-MoS$_2$-; (c) 2H-MoS$_2$-; and (d) MoO$_3$-added HMPC. The arrows show the sliding direction. The effect of the IF-MoS$_2$ is seen in the tribotest results given in Figure 24.

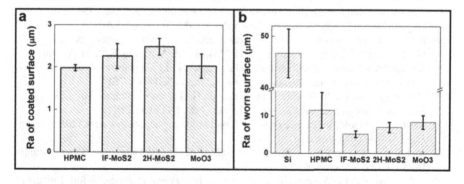

FIGURE 10.24 Surface roughness of (a) coated surface and (b) worn surface after a tribotest. (From Shi, S.-C., *Materials*, 9, 856, 12, 2016.)

10.5 SUMMARY

Green tribology is about overall improvement in energy sourcing to negate global warming. Green tribology and green nanotribology have become key subjects to improve energy efficiency and conserve nonrenewable resources in a wide spectrum of industries. The basic goals of green tribological efforts include reduction of material loss, reduction of energy loss, and life-cycle extension.

The only constant, renewable sources of power that we can totally rely on are solar, wind, biofuels, and hydro (tidal, flowing rivers, and waterfalls). These green power sources will not run out like mineral-sourced oils and fuels will. Of course, biodiesel will extend the story of fuel-driven machinery of all kinds for some time yet. Indeed, there may remain certain needs for biodiesel even if solar energy becomes the prime source of energy. Even though the need for liquid fuels will decrease, as long as we have moving parts in any apparatus, we will need lubricants whether they be in gaseous, liquid, or solid form. As we have detailed in this chapter, biosources for lubricants are many and now, unlike in ancient times, we have a large chemical capability to modify green natural-based materials such as animal fats and vegetable oils. Such modifications can be engineered to maintain rapid biodegradability. Of course, the most accessible renewable lubricant is water. However, water is not always appropriate for the friction problem at hand. Of concern, then, is that we have metals and other inorganic elements, such as halogens, sulfur, and phosphorus, to consider as ultimately limited sourced materials for green tribological applications.

REFERENCES

1. The Sustainable Development Goals (SDGs) Are 17 Goals to Transform Our World by 2030. (September 2016), https://blog.aiesec.org/sustainable-development-what/.
2. D. Dowson, *History of Tribology*, Second edition, Wiley, New York (1998).
3. G. Layard, *Discoveries of the Rules of Nineveh and Babylon, I and II*, John Murray, Albemarle Street, London, UK (1853).
4. J. Li, H. Chen and H. A. Stone, "Ice lubrication for moving heavy stones to the Forbidden City in 15th- and 16th-century China," *Proc. National Acad. Sci. USA*, 110, 20023–20027 (2013).
5. W. J. Bartz, History of tribology—The bridge between classical antiquity and the 21st century. Wilfried J. Bartz, Technische Akademie Esslingen, Ostfildern, Germany. www.oetg.at/fileadmin/Dokumente/oetg/Proceedings/.../M-00-01-001-BARTZ.
6. B. Bhushan, *Introduction to Tribology*, Second Edition, John Wiley & Sons, New York (2013).
7. H. M. Mobarak, E. Niza Mohamad, H. H. Masjuki, M. A. Kalam, K. A. H. Al Mahmud, M. Habibullah and A. M. Ashraful, "The prospects of biolubricants as alternatives in automotive applications," *Renew. Sust. Energ. Rev.*, 33, 34–43 (2014).
8. N. S. Ahmed and A. M. Nassar, "Lubrication and Lubricants," Chapter 2 in: *Tribology—Fundamentals and Advancements*, J. Gegner (Ed.), INTECH (2013).
9. M. Iqbal, "Tribology: Science of lubrication to reduce friction and wear," *Int. J. Mech. Eng.*, 3, 648–668 (2014).
10. R. Scruton, *Green Philosophy, How to Think Seriously About the Planet*. Atlantic Books, Great Britain, UK (2012).
11. P. T. Anastas and J. Warner, *Green Chemistry: Theory and Practice*, Oxford University Press, Oxford, UK (1998).

12. P. Robbins (Ed.), *Green Technology: An A-to-Z Guide. The SAGE Reference Series on Green Society: Toward a Sustainable Future, Book 10*. Sage Publications, Los Angeles, CA (2011).

13. Brundtland report: World Commission on Environment and Development 1987. file:///I:/TB6%2008%202018/refs%20notes%20re.%20R!,%20R2%20fix/our-common-future.pdf.

14. N. A. Ashford and R. P. Hall, *Technology, Globalization, and Sustainable Development. Transforming the Industrial State*, Yale University Press, New Haven, CT (2010).

15. H. P. Jost, *Green Tribology—A Footprint where Economics and Environment Meet, presented at 4th World Tribology Congress*, Kyoto, 6–11 September (2009).

16. S. W Zhang, *Tribological Application in China and Green tribology*. Institution of Engineering and Technology, London, UK (2009).

17. N. M. Nosonovsky and B. Bhushan, "Green tribology: Principles, research areas and challenges," *Philos. Trans. Royal Soc.*, 368, 4677–4694 (2010).

18. M. Nosonovsky and B. Bhushan (Eds.), *Green Tribology, Green Energy and Technology*. Springer-Verlag, Berlin, Germany (2012).

19. Anonymous, Green Engineering (2010) http://www.epa.gov/oppt/green-engineering/

20. J. A. Linthorst, "An overview: Origins and development of green chemistry," *Found. Chem.*, 12, 55–68 (2010).

21. D.T. Allen and D. R. Shonnard, *Green Engineering: Environmentally Conscious Design of Chemical Processes*, Prentice Hall, Upper Saddle River, NJ (2001).

22. 30th Anniversary and "Green Tribology—Report of a successful Chinese Mission to the United Kingdom (7th to 14th June 2009), Tribology Network of Institution of Engineering and Technology (2009).

23. S. W. Zhang, Green tribology—The way forward to a sustainable society, in: *Proceedings of the International Tribology Congress—ASIATRIB 2010*, Perth, Australia, p. 6 (2010).

24. N. Maani, V. S. Rayz and M. Nosonovsky, "Biomimetic approaches for green tribology: From the lotus effect to blood flow control," *Surf. Topog. Metrol Prop.*, 3, 034001 (2015).

25. S.-W. Zhang, "Green tribology: Fundamentals and future development," *Friction*, 1, 186–194 (2013).

26. H. Matsuyama, K. Kawaguchi, A. Uemura and N. Masuda, Development of super-low friction torque tapered roller bearing for high efficiency axle differential, in: *Proceedings of the 4th World Tribology Congress*, Kyoto, 591 (2009).

27. Y. M. Shashidhara and S. R. Jayaram, "Vegetable oil as a potential cutting fluid—An evolution," *Tribol. Int.*, 43, 1073–1081 (2010).

28. B. Muller-Zermini and G. Goule, Environmental approach to hydraulic fluids, in: *Proceeding of the 18th International Colloquium Tribology*, Stuttgart/Ostfildern, Germany (2012).

29. S. Bill, BEWITEC surface technology-Reconditioning and durable wear protection for high loaded gearboxes and bearings in wind turbines, in: *Proceedings of the 18th International Colloquium on Tribology*, Stuttgart/Ostfildern, Germany (2012).

30. I. Tzanakis, M. Hadfield, B. Thomas, S. M. Noya, I. Henshaw and S. Austen, "Future perspectives on sustainable tribology," *Renew. Sust. Energ. Rev.*, 16, 4126–4140 (2012).

31. M. Palacio and B. Bhushan, "A review of ionic liquids for green molecular lubrication in nanotechnology," *Tribol. Lett.*, 40, 247–268 (2010).

32. M. Lovell, C. Higgs, P. Deshmukh and A. Mobley, "Increasing formability in sheet metal stamping operations using environmentally friendly lubricants," *J. Mater. Process. Technol.*, 177, 87–90 (2006).

33. M. R. Lovell, M. Kabir, P. L. Menezes and C. F. Higgs, "Influence of boric acid additive size on green lubricant performance," *Philos. Trans. R. Soc. Lond. A*, 368, 4851–4868 (2010).

34. J. Salimon, N. Salih and E. Yousif, "Biolubricants: Raw materials, chemical modifications and environmental benefits," *Eur. J. Lipid Sci. Technol.*, 112, 519–530 (2010).

35. J. Salimon, B. M. Abdullah, R. M. Yusop and N. S. Salih, "Synthesis, reactivity and application studies for different biolubricants," *Chem. Cent. J.*, 8, article 16 (2014).

36. T. M. Panchal, A. Patel, D. D. Chauhan, M. Thomas and J. V. Patel, "A methodological review on bio-lubricants from vegetable oil based resources," *Renew. Sust. Energ. Rev.*, 70, 65–70 (2017).

37. N. J. Fox and G. W. Stachowiak, "Vegetable oil-based lubricants—A review of oxidation," *Tribol. Int.*, 40, 1035–1046 (2007).

38. T. Mang and W. Dresel, *Lubricants and Lubrication*, second edition, Wiley Interscience, Weinheim, Germany (2007).

39. M. Kučera, M. Hnilicová, J. Turis and P. Semanova, "The experimental research of physicochemical and tribological characteristics of hydraulic oils with a low environmental impact," *Acta Facultatis Technicae Zvolen.*, 2, 25–31 (2013).

40. V. Stepina, "Lubricants and special fluids," *Tribol. Series*, 23, 125–254 1992. https://www.elsevier.com/books/lubricants-and-special-fluids/stepina/978-0-444-98674-0.

41. R. Luther, Lubricants, 3. environmental aspects, in: *Ullmann's Encyclopedia of Industrial Chemistry*. Wiley—VCH, Weinheim, Germany. pp. 455–460 (2012).

42. J. Salimon and N. Smith, "Chemical modification of oleic acid oil for biolubricant," *Aust. J. Basic Appl. Sci.*, 4, 1999–2003 (2010).

43. T. Panchal, D. Chauhan, M. Thomas and J. Patel, "Synthesis and characterization of bio lubricants from tobacco seed oil," *Res. J. Agric. Environ. Manag.*, 3(2), 97–105 (2013).

44. P. Nagendramma and S. Kaul, "Development of ecofriendly/biodegradable lubricants: An overview," *Renew. Sust. Energ. Rev.*, 16, 764–774 (2012).

45. H. M. Mobarak, E. N. Mohamad, H. H. Masjuki, M. A. Kalam, K. A. H. Al Mahmud, M. Habiibullah and A. M. Ashraful, "The prospects of biolubricants as alternatives in automotive applications," *Renew. Sust. Energ. Rev.*, 33, 34–43 (2014).

46. G. Karmarkar, P. Ghosh and B. J. K. Sharma, "Chemically modifying vegetable oils to prepare green lubricants," *Lubricants*, 5, 44 (2017).

47. A. Kumar and S. Sharma, "Potential non-edible oil resources as biodiesel feedstock: An Indian perspective," *Renew. Sust. Energ. Rev.*, 15, 1791–800 (2011).

48. Y. Sharma, and B. Singh, "An ideal feedstock, kusum (*Schleichera triguga*) for preparation of biodiesel: Optimization of parameters," *Fuel*, 89, 1470–1474 (2010).

49. F. Karaosmanoglu, M. Tuter, E. Gollu, S. Yanmaz and E. Altintig, "Fuel properties of cottonseed oil," *Energy Sources*, 21, 821–828 (1999).

50. N. Usta, B. Aydogan, E. Uguzdogan and S. Ozkal, "Properties and quality verification of biodiesel produced from tobacco seed oil," *Energy Convers. Manag.*, 52, 2031–2039 (2011).

51. D. Singh and S. Singh, "Low cost production of ester from non edible oil of *Argemone mexicana*," *Biomass Bioenergy*, 34, 545–549 (2010).

52. R. Wang, M. Hanna, W. Zhou, P. S. Bhadury, Q. Chen, B. A. Song and S. Yang, "Production and selected fuel properties of biodiesel from promising non-edible oils: *Euphorbia lathyris L.*, *Sapium sebiferum L.* and *Jatropha curcas L.*," *Bioresource Technol.*, 102, 1194–1199 (2011).

53. X. Li, X. Y. He, Z. L. Li, Y. D. Wang, H. Shi and F. Wang, "Enzymatic production of biodiesel from *Pistacia chinensis bge* seed oil using immobilized lipase," *Fuel*, 92, 89–93 (2012). R1 *bge* is short for Bunge a taxonomist's name. It follows many Latin Chinese plant names he categorised.

54. R. Wang, W. Zhou, M. Hanna, Y.-P. Zhang, P. S. Bhadury, Y. Wang, B-A. Song and S. Yang, "Biodiesel preparation, optimization, and fuel properties from non-edible feedstock, *Datura stramonium L.*," *Fuel*, 91, 182–186 (2012).

55. M. Mofijur, H. H. Masjuki, M. A. Kalam, M. A. Hazrat, A. M. Liaquat, M. Shahabuddin and M. Varman, "Prospects of biodiesel from Jatropha in Malaysia," *Renew. Sustain. Energy Rev.*, 16, 5007–5020 (2012).

56. W. Li, C. Jiang, M. Chao and X. Wang, "Natural garlic oil as a high-performance environmentally friendly, extreme pressure additive in lubricating oils," ACS Sustain. *Chem. Eng.*, 2, 798–803 (2014).

57. V. Ossia, H. G. Han and H. Kong, "Additive properties of saturated very long chain fatty acids in castor and jojoba oils," *J. Mech. Sci. Technol.*, 22, 1527–1536 (2008)

58. S. Z. Erhan, A. Adhvaryu and B. K. Sharma, "Poly(hydroxy thioether) Vegetable Oil Derivatives Useful as Lubricant Additives," U.S. Patent 7,279,448 B2, Lubricants 5, 44 12 of 17 (2017).

59. V. B. Borugadda and V. V. Goud, "Epoxidation of castor oil fatty acid methyl esters (COFAME) as a lubricant base stock using heterogeneous ion-exchange resin (IR-120) as a catalyst," *Energy Procedia.*, 54, 75–84 (2014).

60. T. Baye and H. C. Becker, "Exploration of *vernonia galamensis* in Ethiopia, and variation in fatty acid composition of seed oil," *Genet. Resour. Crop Evol.*, 52, 805–811 (2005).

61. V. Corma, S. Iborra and A. Velty, "Chemical routes for the transformation of biomass into chemicals," *Chem. Rev.*, 107, 2411–2502 (2007).

62. Y. Xia and R. C. Larock, "Vegetable oil-based polymeric materials: Synthesis, properties, and applications," *Green Chem.*, 12, 1893–1909 (2010).

63. Y. H. Hui, *Bailey's Industrial Oil and Fats Products, Edible Oil and Fat Products: General Application*, Fifth Edition, Wiley, Blackwell, UK (1996).

64. M. N. Belgacem and A. Gandini, *Monomers, Polymers and Composites from Renewable Resources*, pp. 39–66, Elsevier, Amsterdam, the Netherlands (2008).

65. Y. N. Wang, M. H. Chen, C. H. Ko, P. J. Lu, J. M. Chern, C. H. Wu and F. C. Chang, "Lipase catalyzed transesterification of tung and palm oil for biodiesel," in: *Proceedings of the World Renewable Energy Congress*, Linköping, Sweden, pp. 8–13 (2011).

66. S. Güner, Y. Yağcı and A. T. Erciyes, "Polymers from triglyceride oils," *Prog. Polym. Sci.*, 31, 633–670 (2006).

67. U. Rashid, F. Anwar and G. Knothe, "Evaluation of biodiesel obtained from cottonseed oil," *Fuel Process. Technol.*, 90, 1157–1163 (2009).

68. S. Rani, M. L. Joy and K. P. Nair, "Evaluation of physiochemical and tribological properties of rice bran oil—Biodegradable and potential base stoke for industrial lubricants," *Ind. Crops Prod.*, 65, 328–333 (2015).

69. P. S. Chauhan and V. K. Chhibber, "Epoxidation in karanja oil for biolubricant applications," *Int. J. Pharm. Biol. Sci. Arch.*, 1, 61–70 (2013).

70. L. C. Meher, S. N. Naik and L. M. Das, "Methanolysis of *Pongamiapinnata* (Karanja) oil for production of biodiesel," *J. Sci. Ind. Res.*, 63, 913–918 (2014).

71. Z. Akbar, Z. Yaakob, S. K. Kamarudin, M. Ismail and J. Salimon, "Characteristic and composition of Jatropha Curcas oil seed from Malaysia and its potential as biodiesel feedstock," *Eur. J. Sci. Res.*, 29, 396–403 (2009).

72. Aigbodion and C. K. S. Pillai, "Preparation, analysis and applications of rubber seed oil and its derivatives in surface coatings," *Prog. Org. Coat.*, 38, 187–192 (2000).

73. E. F. Aransiola, E. Betiku, D. I. O. Ikhuomoregbe and T. V. Ojumu, "Production of biodiesel from crude neem oil feedstock and its emissions from internal combustion engines," *Afr. J. Biotechnol.*, 11, 6178–6186 (2012).

74. R. K. Singh and S. K. Padhi, "Characterization of jatropha oil for the preparation of bio-diesel," *Natural Prod. Radiance*, 8, 127–132 (2009).

75. A. B. Chhetri, M. S. Tango, S. M. Budge, K. C. Watts and M. R. Islam, "Non-edible plant oils as new sources for biodiesel production," *Int. J. Mol. Sci.*, 9, 169–180 (2008).

76. M. Murugesan, C. Umarani, T. R. Chinnusamy, M. Krishnan, R. Subramanian and N. Neduzchezhain, "Production and analysis of bio-diesel from non-edible oils— A review," *Renew. Sustain. Energ. Rev.*, 13, 825–834 (2009).

77. S. Tanveer and R. Prasad, "Enhancement of viscosity index of mineral base oils," *Indian J. Chem. Technol.*, 13, 398–403 (2006).

78. P. Ghosh, M. Hoque, G. Karmakar and M. K. Das, "Dodecyl methacrylate and vinyl acetate copolymers as viscosity modifier and pour point depressant for lubricating oil," *Int. J. Ind. Chem.*, 8, 197–205 (2017).

79. B. K. Sharma and A. J. Stipanovic, "Development of a new oxidation stability test method for lubricating oils using high-pressure differential scanning calorimetry," *Thermochim. Acta*, 402, 1–18 (2003).

80. P. Ghosh, A. V. Pantar, U. S. Rao and A. S. Sarma, "Shear stability of polymers used as viscosity modifiers in lubricating oils," *Indian J. Chem. Technol.*, 5, 309–314 (1998).

81. T. Mang and G, Lingg, "Base oils, lubricants and lubrication," in: *Lubricants and Lubrication*, Second Edition, Chapter 4, pp. 34–35, I. T. Mang and W. Dresel (Eds.), Wiley-VCH, Weinheim, Germany (2007).

82. A. Bouffet, P. Duteurte and P. Airard, "A challenge for the lubricant industries," *Petrol. Technol.*, 362, 65–79 (1991).

83. S. Z. Erhan, K. M. Doll and B. K. Sharma, Method of Making Fatty Acid Ester Derivatives, U.S. Patent 8,173,825B2 (2012).

84. S. Z. Erhan, K. M. Doll and B. K. Sharma, Method of Making Fatty Acid Ester Derivatives, U.S. Patent 20,080,154,053 A1 (2008).

85. K. M. Doll, B. K. Sharma and P. A. Suarez, Process to Prepare a Phosphorous Containing Vegetable Oil Based Lubricant Additive, U.S. Patent 8,822,712 B1 (2014).

86. A. Biswas, K. M. Doll, H. N. Cheng and B. K. Sharma, Process for Preparation of Nitrogen-Containing Vegetable Oil-Based Lubricant Additive, U.S. Patent 8,841,470 B1 (2014).

87. G. L. Heise, B. K. Sharma and S. Z. Erhan, Boron Containing Vegetable Oil Based Antiwear/Antifriction Additive and their Preparation, U.S. Patent 9,156,859 B2 (2015).

88. http://en-wikipedia.org/wiki/National_Nanotechnology_Initiative

89. R.V. Lapshin, "Feature—Oriented scanning methodology for probe microscopy and nanotechnology," *Nanotechnology*, 15, 1135–1151 (2004).

90. J. E. Hutchison, "Greener nanoscience: A proactive approach to advancing applications and reducing implications of nanotechnology," *ACS Nano.*, 2, 395–402 (2008).

91. L. C. McKenzie and J. E. Hutchison, "Green nanoscience: An integrated approach to greener products, processes, and applications," *Chimica Oggi Chem. Today*, 22, 30–33 (2004).

92. J. A. Dahl, B. L. S. Maddux and J.E. Hutchison, "Toward greener nanosynthesis," *Chem. Rev.*, 107, 2228–2269 (2007).

93. B. Bhushan, *Tribology and Mechanics of Magnetic Storage Devices*, Second Edition. Springer, New York (1996).

94. B. Bhushan, (Ed.), *Tribology Issues and Opportunities in MEMS*, Kluwer Academic, Dordrecht, the Netherlands (1998).

95. B. Bhushan, *Handbook of Micro/Nanotribology*, Section Edition. CRC Press, Boca Raton, FL (1999).

96. B. Bhushan, *Introduction to Tribology*, Wiley, New York (2002).

97. B. Bhushan, Nanotribology and Nanomechanics—An Introduction, Second Edition. Springer-Verlag, Heidelberg, Germany (2008).

98. B. Bhushan, *Springer Handbook of Nanotechnology*, Third Edition. Springer-Verlag, Heidelberg, Germany (2010).

99. B. Bhushan, H. Lee, S. C. Chaparala and V. Bhatia, "Nanolubrication of sliding components in adaptive optics used in microprojectors," *Appl. Surf. Sci.*, 256, 7545–7558 (2010).

100. S. A. Henck, "Lubrication of digital micromirror devices," *Tribol. Lett.* 3, 239–247 (1997).

101. M. R. Douglass, Lifetime estimates and unique failure mechanisms of the digital micromirror device (DMD), in: *Proceedings of the 36th Annual International Reliability Physics Symposium*. IEEE Press (1998).

102. R. E. Sulouff, MEMS opportunities in accelerometers and gyros and the microtribology problems limiting commercialization, in: Bhushan, B. (ed.) Tribology Issues and Opportunities in MEMS, KluwerAcademic, Dordrecht, the Netherlands (1998).

103. P. Vettiger, J. Brugger, M. Despont, U. Dreschler, U. Durig, W. Haberle, M. Lutwyche, H. Rothuizen, R. Stuz, R. Widmer and G. Binnig, "Ultrahigh density, high data rate NEMS-based AFM data storage system," *Microelectron. Eng.*, 46, 11–17 (1999).

104. B. Bhushan, K. J. Kwak and M. Palacio, "Nanotribology and nanomechanics of AFM probe-based data recording technology," *J. Phys. Condens. Matter*, 20, 365207 (2008).

105. J. Krim, D. H. Solina and R. Chiarello, "Nanotribology of a Kr monolayer: A quartz-crystal microbalance study of atomic scale friction," *Phys. Rev. Lett.*, 66, 181–184 (1991).

106. https://en.wikipedia.org/wiki/Asperity_(materials_science).

107. C. M. Mate, *Tribology on the Small Scale, A Bottom Up Approach to Friction, Lubrication, and Wear*, Oxford: Oxford University Press, 333 (2008).

108. A. Tomala, H. Goecerler, and I. C. Gebeshuber, Bridging nano and microtribology in mechanical and biomolecular layers, in: *Scanning Probe Microscopy in Nanoscience and Nanotechnology III*, B. Bhushan (Ed.), Springer, Berlin, Germany (2013).

109. B. Elango, P. Rajendran and L. Bornmann, "Global nanotribology research output (1996–2010): A scientometric analysis," *PLoS ONE*, 8, e81094 (2013).

110. I. C. Gebeshuber, Nanotribology: Green nanotribology and related sustainability aspect, in: *CRC Concise Encyclopedia of Nanotechnology*, B. I. Kharisov, O. V. Kharissova, U. Ortiz-Mendez (Ed.), pp. 871–875. Taylor & Francis Group, Boca Raton, FL (2016).

111. I.C. Gebeshuber; P. Gruber and M. Drack, "A gaze into the crystal ball—biomimetics in the year 2059," *IMechE. Part C, J. Mech. Eng. Sci.*, 223(12), 2899–2918 (2009).

112. I. C. Gebeshuber, M. Drack and M. Scherge, "Tribology in biology," *Tribol. Mater. Surf. Interfaces*, 2, 200–212 (2008).

113. I. C. Gebeshuber, Green nanotribology and sustainable nanotribology in the frame of the global challenges for humankind, in: *Green Tribology—Biomimetics, Energy Conservation and Sustainability*, B. Bhushan (Ed.), pp. 105–125, Springer, Heidelberg, Germany (2012).

114. I. C. Gebeshuber, Green nanotribology, *Proceedings of the Institution of Mechanical Engineers Part C: Journal of Mechanical Engineering Science*, Special Issue, Guest Editors: Ille C. Gebeshuber, Manish Roy (2012).

115. S. H. Kim, M. T. Dugger and K. L. Mittal (Eds.) *Adhesion Aspects in MEMS and NEMS*, CRS Press, Boca Raton, FL (2018).

116. S. Izabela, C. Michael and W. C. Robert, "Recent advances in single asperity nanotribology," *J. Phys D Appl. Phys.*, 41, 123001 (2008).

117. F. P. Bowden and D. Tabor, "Mechanism of metallic friction," *Nature*, 150, 197–199 (1942).

118. G. Binnig, C. F. Quate and C. Gerber, "Atomic force microscope," *Phys. Rev. Lett*, 56, 930–933 (1986).

119. https://en.wikipedia.org/wiki/Atomic_force_microscopy.

120. J. Ralston, I. Larson, M. W. Rutland, A. A. Feiler and M. Kleijn, "Atomic force microscopy and direct surface force measurements—(IUPAC technical report)," *Pure App. Chem.*, 77, 2149–2170 (2005).

121. W. J. Bartz, "Ecotribology: Environmentally acceptable tribological Practices," *Tribol. Int.*, 39, 728–733 (2006).

122. I. Tzanakis, M. Hadfield, B. Thomas, S. M. Noya, I. Henshaw and S. Austen, "Future perspectives on sustainable tribology," *Renew. Sustain. Energy Rev.*, 16, 4126–4140 (2012).

123. D. Gnanasekaran and V. P. Chavidi, Nanomaterials as an Additive in Biodegradable Lubricants, in: *Vegetable Oil Based Bio-lubricants and Transformer Fluids, in the series Materials Forming, Machining and Tribology*, J. P. Davim (Ed.), Springer Nature, Singapore. (2018).

124. S. Z. Erhan and S. Asadauskas, "Lubricant basestocks from vegetable oils," *Indust. Crops Prod.*, 11, 277–282 (2000).

125. X. Wu, X. Zhang, S. Yang, H. Chen and D. Wang, "The study of epoxidized rapeseed oil used as a potential biodegradable lubricant," *J. Amer. Oil Chem. Soc.*, 77, 561–563 (2000).

126. S. Arumugam and G. Sriram, "Synthesis and characterisation of rapeseed oil bio-lubricant—Its effect on wear and frictional behaviour of piston ring-cylinder liner combination," *J. Eng. Tribol.*, 227, 3–15 (2013).

127. S. Arumugam and G. Sriram, "Effect of biolubricant and biodiesel-contaminated lubricant on tribological behavior of cylinder liner-piston ring combination," *Tribol. Trans.*, 55, 438–445 (2012).

128. G. Liu, X. Li, B. Qin, Y. Xing, Y. Guo and R. Fan, "Investigation of the mending effect and mechanism of copper nano particles on a tribologically stressed surface," *Tribol. Letts.*, 17(4), 961–966 (2004).

129. Y. Y. Wu, W. C. Tsui and T. C. Liu, "Experimental analysis of tribological properties of lubricating oils with nanoparticle additives," *Wear*, 262, 819–825 (2007).

130. Y. Choi, C. Lee, Y. Hwang, M. Park, J. Lee, C. Choi and M. Jung, "Tribological behavior of copper nanoparticles as additives in oil," *Current Appl. Phys.*, 9, 124–127 (2009).

131. S. Arumugam and G. Sriram, "Synthesis and characterization of rapeseed oil bio-lubricant dispersed with nano copper oxide; its effect on wear and frictional behavior of piston ring cylinder liner combination," *J. Eng. Tribology*, 228, 1308–1318 (2014).

132. I. Sudeep, C. Archana, K. Amol, S. S. Umare, D. V. Bhatt and M. Jyoti, "Tribological behavior of nano TiO_2 as an additive in base oil," *Wear*, 301, 776–785 (2013).

133. S. Arumugam and G. Sriram, "Preliminary study of nano and microscale TiO_2 additives on tribological behavior of chemically modified rapeseed oil," *Tribology Trans.*, 56, 797–805 (2013).

134. L. Rapoport, V. Leshchinsky, I. Lapsker, Y. Volovik, O. Nepomnyashchy, M. Lvovsky, R. Popovitz, Y. Biro, Y. Feldman and R. Tenne, "Tribological properties of WS_2 nanoparticles under mixed lubrication," *Wear*, 255, 785–793 (2003).

135. R. Greenberg, G. Halperin, I. Etsion and R. Tenne, "The effect of WS_2 nanoparticles on friction reduction in various lubrication regimes," *Tribol. Lett.*, 17, 179–186 (2004).

136. I. Lahouij, F. Dassenoy, L. De Knoop, J. M. Martin and B. Vacher, "In situ TEM observation of the behaviour of an individual fullerene-like MoS_2 nanoparticle in a dynamic contact," *Tribol. Lett.*, 42, 133–140 (2011).

137. S. Baskara, G. Sriram and S. Arumugam. "Experimental analysis on tribological behavior of nano based bio-lubricants using four ball tribometer," *Tribol. Ind.*, 37, 449–454 (2015).

138. T. C. Ing, A. K. M. Rafiq, Y. Azli and S. Syahrullail, "Tribological behaviour of refined bleached and deodorized palm olein in different loads using a four-ball tribotester," *Sci. Iran. B*, 19, 1487–1492 (2012).

139. H. Unal, M. Sincik and N. Izli, "Comparison of some engineering properties of rapeseed cultivars," *Ind. Crops Prod.*, 30, 131–136 (2009).

140. L. R. Rudnick, *Synthetics, Mineral Oils, and Bio-Based Lubricants: Chemistry and Technology*, Second Edition, CRC Press/Taylor & Francis Group, Boca Raton, FL (2006).

141. P. K. Namburu, D. P. Kulkarni, D. Misra and D. K. Das, "Viscosity of copper oxide nanoparticles dispersed in ethylene glycol and water mixture," *Experi. Thermal Fluid Sci.*, 32, 397–402 (2007).

142. M. N. Rashin and J. Hemalatha, "Viscosity studies on novel copper oxide-coconut oil nanofluid," *Expl. Thermal Fluid Sci.*, 48, 67–72 (2013).

143. M. Kole and T.K. Dey, "Viscosity of alumina nanoparticles dispersed in car engine coolant," *Expl. Thermal Fluid Sci.*, 34, 677–683 (2010).

144. T. Yiamsawas, O. Mahian, A. S. Dalkilic, S. Kaewnai and S. Wongwises, "Experimental studies on the viscosity of TiO_2 and Al_2O_3 nanoparticles suspended in a mixture of ethylene glycol and water for high temperature applications," *Appl. Energy*, 111, 40–45 (2013).

145. B. Aladag, S. Halelfadl, N. Doner, T. Maré, S. Duret and P. Estellé, "Experimental investigations of the viscosity of nanofluids at low temperatures," *Appl. Energy*, 97, 876–880, (2012).

146. M. Gulzar, H. H. Masjuki, M. A. Kalam, M. Varman, N. W. M. Zulkifli, R. A. Mufti and R. Zahid, "Tribological performance of nanoparticles as lubricating oil additives," *J. Nanoparticles. Res.*, 18, 223 (2016).

147. W. Zhang, Y. Cao, P. Tian, F. Guo, Y. Tian, W, Zheng, X. Ji and J. Liu, "Soluble, exfoliated two-dimensional nanosheets as excellent aqueous lubricants," *ACS Appl. Mater. Interfaces*, 8, 32440–32449 (2016).

148. N. Kumar, S. Bhaumik, A. Sen, A. P. Shukla and S. D. Pathak, "One-pot synthesis and first-principles elasticity analysis of polymorphic MnO_2 nanorods for tribological assessment as friction modifiers," *RSC Adv.*, 7, 34138–34148 (2017).

149. B. Zareh-Desari and B. Davoodi, "Assessing the lubrication performance of vegetable oil-based nano-lubricants for environmentally conscious metal forming processes," *J. Clean. Prod.*, 135, 1198–1209 (2016).

150. Environmental Management—Life Cycle Assessment—Principles and Framework, International Organisation for Standardisation, ISO 14040 (1997).

151. M. Stalmans, H. Berenbold, J. L. Berna, L. Cavalli, A. Dillarstone, M. Franke, F. Hirsinger et al., "European Life Cycle Inventory for detergent surfactants production," *Tenside Surf. Det.*, 32, 84 (1996).

152. I. Boustead, Eco profiles of the European plastics industry, APME Brussels (1999).

153. L. R. Rudnick and R. L. Shubkin (Eds.), *Synthetic Lubricants and High Performance Functional Fluids*, Marcel Dekker, New York (1999).

154. C. Kajdas, Tribochemistry, in: *Tribology 2001*, F. Franek et al. (Eds.), p. 29, OTG, Vienna, Austria (2001).

155. C. R. Burrows, G. P. Hammond and M. C. McManus, *The Sixth Scandinavian International Conference on Fluid Power*, p. 1163, SICFP 99, Tampere; Finland (1999).

156. M. C. McManus, G. P. Hammond and C. R. Burrows, *Environmental Impact of Fluid Power Systems*, IMechE, London, UK (1999).

157. M. C. McManus, G. P. Hammond and C. R. Burrows, "Life cycle assessment of alternative hydraulic fluids for municipal cleaning equipment." Paper presented at *Proceedings of the National Fluid Power Association International Exposition for Power Transmission*, NFPA, Chicago, p.21 (2000).

158. C. R. Burrows, G. P. Hammond, and M. C. McManus, Life-cycle assessment of oil "Hydraulic" systems for environmentally-sensitive applications, in S. S. Nair and S. I. Misstry (Eds.), *Fluid Power Systems and Technology*, FPST-Vol. 5, ASME, New York (1998).

159. C. Ciantar and M. Hadfield, "Investigating the sustainable development of domestic refrigerating systems working in an HFC-134a environment, in Recent Developments in Refrigeration and Heat Pump Technologies," *IMechE Semin. Publ.*, 10, 47, 1992.

160. C. Vag, A. Marby, M. Kopp, L. Furberg and T. Norrby, A comparative life cycle assessment (LCA) of the manufacturing of base fluids for lubricants, in: *Tribology 200—Plus. Proceedings of 12th International Conference*, p. 2241, Technische Akademie Esslingen, ed. W. J. Bartz (2000).

161. A. Levizzari, M. Voglino and P. Volpi, Environmental and Economic Impact of re-refined products: A Life Cycle Analysis, in Lubricants for the Future and Environment, *Proceedings of the 6th International LFE congress*, Brussels (1999).

162. M. Voltz, Ökobilanz fur Schmierstoffe—Grundlagen und Vorgehensweise, in: Lubricants, Materials and Lubrication Engineering. *Proceedings of 13th International Conference*, Technische Akademie Esslingen, ed. W. J. Bartz, p. 1041 (2002).

163. L. Raghunanan and S. S. Narine, "Engineering green lubricants I: Optimizing thermal and flow properties of linear diesters derived from vegetable oils," *ACS Sustain Chem. Eng.*, 4, 686–692 (2016).

164. L. Raghunanan, J. Yue and S. S. Narine, "Synthesis and characterization of novel diol, diacid and di-isocyanate from oleic acid," *J. Am. Oil Chem. Soc.*, 91, 349–356 (2014).

165. L. Raghunanan and S. S. Narine, "Engineering green lubricants II: Thermal transition and flow properties of vegetable oil-derived diesters," *ACS Sustain Chem. Eng.*, 4, 3, 693–700 (2016).

166. L. Raghunanan and S. S. Narine, "Engineering Green Lubricants IV: Influence of structure on the thermal behaviour of linear and branched aliphatic fatty acid-derived diesters," *ACS Sustain. Chem. Eng.*, 4, 4868–4874 (2016).

167. M. Fan, L. Ma, C. Zhang, Z. Wang, J. Ruan, M. Han, Y. Ren, C. Zhang, D. Yang, F. Zhou and W. Liu. "Biobased green lubricants: Physicochemical, tribological and toxicological properties of fatty acid ionic liquids," *Tribol. Trans.*, 61, 195–206 (2018).

168. S.-C. Shi and F.-I. Lu, "Biopolymer green lubricant for sustainable manufacturing," *Materials*, 9, 338, 11 (2016).

169. S.-C. Shi, "Tribological performance of green lubricant enhanced by sulfidation IF-MoS$_2$," *Materials*, 9, 856, 12 (2016).

170. S.-C. Shi, J.-Y. Wu, T.-F. Huang and Y.-Q. Peng. "Improving the tribological performance of biopolymer coating with MoS$_2$ additive," *Surf. Coat. Technol.*, 303, 250–255 (2016).

171. S.-C. Shi, J.-Y. Wu and T.-F. Huang, "Raman, FTIR, and XRD study of MoS$_2$ enhanced hydroxypropyl methylcellulose green lubricant," *Opt. Quant. Electron*, 48, 474, (2016).

Index

Note: Page numbers in italic and bold refer to figures and tables, respectively.

Printed in the United States
by Baker & Taylor Publisher Services